云计算技术应用专业校企合作系列教材

云计算基础架构平台构建与应用

（第2版）

主 编 何 淼 史 律 孙仁鹏
副主编 乔 洁 董志勇 王 晖
　　　　朱 敏

高等教育出版社·北京

图书在版编目（CIP）数据

云计算基础架构平台构建与应用 / 何淼，史律，孙仁鹏主编 . --2 版 . -- 北京：高等教育出版社，2021.11（2023.2 重印）
ISBN 978-7-04-054291-2

Ⅰ. ①云… Ⅱ. ①何… ②史… ③孙… Ⅲ. ①云计算 - 高等职业教育 - 教材 Ⅳ. ① TP393.027

中国版本图书馆 CIP 数据核字 (2020) 第 102361 号

YUNJISUAN JICHU JIAGOU PINGTAI GOUJIAN YU YINGYONG

策划编辑	吴鸣飞	责任编辑	吴鸣飞	封面设计	赵 阳	版式设计	王艳红
插图绘制	于 博	责任校对	吕红颖	责任印制	刘思涵		

出版发行	高等教育出版社	网 址	http://www.hep.edu.cn	
社 址	北京市西城区德外大街 4 号		http://www.hep.com.cn	
邮政编码	100120	网上订购	http://www.hepmall.com.cn	
印 刷	北京汇林印务有限公司		http://www.hepmall.com	
开 本	787mm×1092mm 1/16		http://www.hepmall.cn	
印 张	20.75	版 次	2017 年 3 月第 1 版	
字 数	520 千字		2021 年 11 月第 2 版	
购书热线	010-58581118	印 次	2023 年 2 月第 2 次印刷	
咨询电话	400-810-0598	定 价	55.00 元	

本书如有缺页、倒页、脱页等质量问题，请到所购图书销售部门联系调换
版权所有 侵权必究
物 料 号 54291-00

内容简介

本书以目前 IT 领域最热门技术之一的云计算技术为背景，面向云计算行业运维人员、技术开发人员以及高校云计算专业广大学生的实践教程。书中所用平台以及讲解内容为基于 OpenStack 的云计算产品的基本参数、部署及运维，OpenStack 是当今云计算行业最为热门的开源基础架构云平台之一。

本书以培养云计算行业中云计算系统运维以及云计算系统开发工程师为主要目标，注重实际动手操作。全书分为三大部分、共 14 章，第一部分（第 1~3 章）介绍云计算的定义、层次分类、产业现状及主要云计算产品的开发厂商，第二部分（第 4~12 章）以基于 OpenStack 开源云平台的部署为主线，全面介绍 OpenStack 基础架构平台中每个组件的主要功能及其实现，每章配备详细的实训项目，所有实训项目承上启下，一步一步通过基于 PC 的实训环境构建完整的 OpenStack 云平台；第三部分（第 13、14 章）详细介绍每个模块的运维命令行及其在界面中的相应操作，同时还介绍云平台镜像的制作方法。

本书配套建设微课视频、授课计划、电子教案、授课用 PPT、程序源代码等数字化学习资源。与本书配套的数字课程在"智慧职教"（www.icve.com.cn/）平台上线，学习者可以登录平台进行数字课程的学习，授课教师可以调用本课程构建符合自身教学特色的 SPOC 课程，详见"智慧职教"服务指南。读者可登录平台进行资源的学习及获取，也可发邮件至编辑邮箱 1548103297@qq.com 获取相关教学资源。

本书既可作为高等职业院校计算机、通信、云计算等相关专业课程的教材，也可供广大云计算行业技术人员使用。

Ⅲ 智慧职教服务指南

"智慧职教"是由高等教育出版社建设和运营的职业教育数字教学资源共建共享平台和在线课程教学服务平台,包括职业教育数字化学习中心平台(www.icve.com.cn)、职教云平台(zjy2.icve.com.cn)和云课堂智慧职教 App。用户在以下任一平台注册账号,均可登录并使用各个平台。

● 职业教育数字化学习中心平台(www.icve.com.cn):为学习者提供本教材配套课程及资源的浏览服务。

登录中心平台,在首页搜索框中搜索"云计算基础架构平台构建与应用",找到对应作者主持的课程,加入课程参加学习,即可浏览课程资源。

● 职教云(zjy2.icve.com.cn):帮助任课教师对本教材配套课程进行引用、修改,再发布为个性化课程(SPOC)。

1. 登录职教云,在首页单击"申请教材配套课程服务"按钮,在弹出的申请页面填写相关真实信息,申请开通教材配套课程的调用权限。

2. 开通权限后,单击"新增课程"按钮,根据提示设置要构建的个性化课程的基本信息。

3. 进入个性化课程编辑页面,在"课程设计"中"导入"教材配套课程,并根据教学需要进行修改,再发布为个性化课程。

● 云课堂智慧职教 App:帮助任课教师和学生基于新构建的个性化课程开展线上线下混合式、智能化教与学。

1. 在安卓或苹果应用市场,搜索"云课堂智慧职教"App,下载安装。

2. 登录 App,任课教师指导学生加入个性化课程,并利用 App 提供的各类功能,开展课前、课中、课后的教学互动,构建智慧课堂。

"智慧职教"使用帮助及常见问题解答请访问 help.icve.com.cn。

前言

云计算作为一种弹性IT资源的提供方式，通过近十年的技术发展和经验积累已在各行各业中广泛应用，成为IT信息处理能力的源动力。同时，新一代信息技术包括大数据和人工智能的应用也促进了云计算基础设施软硬件系统不断推陈出新。

OpenStack是当今云计算基础平台架构的行业主流软件，其版本更新速度快，社区活跃度高，从2011年至今共迭代了19个主要版本。华为、腾讯以及国内众多云计算厂商都成为OpenStack基金会的黄金会员。目前，国内大多的私有云系统以及部分公有云系统底层架构都是基于OpenStack开源框架。从岗位需求方面来讲，国内云计算行业市场在未来5年需要大量掌握OpenStack基本原理及技术架构、从事云计算系统部署与运维的从业人员。国内众多高职院校以及应用型本科院校已经开设云计算专业，多数院校已将云计算或大数据相关的课程纳入传统的计算机网络、通信、计算机软件、计算机应用等IT专业的人才培养方案。

本书作为国内高职院校云计算专业课实践教材，经过3年的使用已得到了广大读者，特别是高职院校教师和学生的支持与青睐。同时读者也对教材提出了大量宝贵的意见，这些意见对本次改版工作提供了帮助和依据。为了进一步培养新一代信息技术高技能型人才，提升学生云计算工程、产品部署、实施和应用的专业技能，并使其具备更高层次的产品对接调试、排错、运维等综合能力，编写团队在本书第1版的基础上重新规划和设计了本书的全部内容，并开发了与之配套的软件环境及教学资源。

从整体内容上看，本次改版后的内容紧密围绕教育部高等职业学校"云计算技术应用"专业教学标准中的核心课程"云计算基础架构平台应用"而设计，完整介绍了OpenStack核心组件的内部参数设置、安装实施脚本的编写和配置，以及使用运维的全部过程。同时本书也可以嵌入到大数据应用、人工智能技术服务等新一代IT技术相关专业课程使用。编写团队针对实验实训环节使用了OpenStack基金会截至2019年年底最新Train版本软件包，又将各组件的安装过程加工成脚本模板，在实训环节重点介绍脚本模板中关键参数的配置过程及使用过程，其实施方式完全和生产环境中基于OpenStack的私有云系统部署方式一致，进一步贴近了教育部"云计算技术应用"专业教学标准中云计算系统部署与运维的典型岗位技能的要求。同时，教材还扩展讲解了云系统中非常关键的虚拟机qcow2镜像的制作过程，编写团队对教材内容的选择与把控最大程度保障了教材的先进性与实用性。此外，教材中实训不依赖于硬件服务器，通过普通的高性能PC即可完成。实训中用到的所有软件包无须互联网环境，完全通过配套的本地源安装，最大程度地方便了教师教学及学生实践。

本书由南京信息职业技术学院何淼、史律、孙仁鹏担任主编，乔洁、董志勇、王晖、朱敏担任本书副主编，南京技师学院姜技负责对本书中所有实训进行测试。编写团队成员均具有多年从事云计算产品开发的工程经验和云计算课程的教学经验，连续五年带领学生获得全国职业院校技能大赛"云计算技术与应用"赛项的一等奖，同时还获得过全国信息化教学大赛一等奖。

I

前言

南京第五十五所技术开发有限公司为本书的撰写提供了大力支持，在此表示诚挚的谢意。

教师可发邮件至编辑邮箱 1548103297@qq.com 获取相关教学资源。

鉴于编者水平有限，书中难免存在不足和错误之处，恳望广大读者提出宝贵意见和建议。

编 者

2021 年 9 月

目录

第一部分　云计算技术基础

第 1 章　云计算基本概念 ………………… 3
1.1　计算模式的演变 ……………………… 4
1.2　云计算的定义 ………………………… 5
1.3　云计算的层次以及分类 ……………… 9
1.4　国内外云计算产业现状 ……………… 15

第 2 章　云计算知名厂商及其产品 …… 17
2.1　VMware 的云计算技术及其相关产品 … 18
2.2　Citrix 的云计算技术 ………………… 23
2.3　微软私有云虚拟化产品 Hyper-V …… 25
2.4　国内私有云相关产品 ………………… 26
2.5　知名公有云平台简介 ………………… 30

第 3 章　原生 OpenStack 云平台 ……… 37
3.1　OpenStack 技术简介 ………………… 38
3.2　体验原生 OpenStack 云平台 ………… 40

第二部分　原生 OpenStack 云平台基础环境的构建

第 4 章　原生 OpenStack 云平台的环境准备 ……………………… 51
4.1　原生 OpenStack 云平台的逻辑架构及其实现 ………………………… 52
4.2　终端仿真软件的使用 ………………… 68
4.3　实训项目 1　原生 OpenStack 云平台基本环境配置 ………………………… 86

第 5 章　MySQL 数据库的安装及其配置 ……………………………… 95
5.1　MySQL 数据库功能简介 …………… 96
5.2　实训项目 2　MySQL 数据库的手动安装与配置 ……………………… 101
5.3　MySQL 数据库安装脚本及其解读 … 104

第 6 章　Keystone 的安装及其配置 …… 107
6.1　Keystone 功能详解 ………………… 108
6.2　实训项目 3　Keystone 的手动安装与配置 …………………………… 110
6.3　Keystone 安装脚本及其解读 ……… 116

第 7 章　Glance 的安装及其配置 ……… 119
7.1　Glance 功能简介 …………………… 120
7.2　实训项目 4　Glance 的手动安装与配置 … 121
7.3　Glance 安装脚本及其解读 ………… 131

第 8 章　Placement 的安装及其配置 … 135
8.1　Placement 功能简介 ………………… 136
8.2　实训项目 5　Placement 的手动安装与配置 …………………………… 137
8.3　Placement 安装脚本及其解读 ……… 142

第 9 章　Nova 的安装及其配置 ……… 145
9.1　Nova 功能简介 ……………………… 146
9.2　实训项目 6　Nova 的手动安装与配置 … 151
9.3　Nova 安装脚本及其解读 …………… 162

第 10 章　Neutron 的安装及其配置 …… 167
10.1　Neutron 功能简介 ………………… 168
10.2　实训项目 7　Neutron 的手动安装与外部环境配置 ……………………… 172
10.3　实训项目 8　Neutron 的主要服务组件配置与网络创建 ……………… 180

I

10.4　Neutron 安装脚本及其解读 …………191

第 11 章　Cinder 的安装及其配置 ………199
11.1　Cinder 功能简介 ………………………200
11.2　实训项目 9　Cinder 的手动安装与配置…204
11.3　Cinder 安装脚本及其解读 ……………211

第 12 章　Dashboard 的安装及其配置…215
12.1　Dashboard 功能简介 …………………216
12.2　实训项目 10　Dashboard 的安装与配置…216
12.3　启动虚拟机实例及其排错案例…………220

第三部分　原生 OpenStack 云平台运维详解

第 13 章　原生 OpenStack 云平台各组件运维 ……………………………227
13.1　实训项目 11　Keystone 基本运维命令及其应用 ………………………228
13.2　实训项目 12　Glance 基本运维命令及其应用 …………………………246
13.3　实训项目 13　Nova 基本运维命令及其应用 ……………………………256
13.4　实训项目 14　Neutron 基本运维命令及其应用 …………………………285
13.5　实训项目 15　Cinder 基本运维命令及其应用 ……………………………296

第 14 章　虚拟机镜像文件的制作 ………309
14.1　实训项目 16　准备虚拟机镜像环境……310
14.2　实训项目 17　云平台 qcow2 格式 Windows 镜像制作 …………………312

参考文献 ……………………………………320

第一部分　云计算技术基础

第 1 章 云计算基本概念

 本章导读：

本章介绍云计算的发展以及演变模式，并从不同的角度分析引出云计算的定义及其基本特点；在此基础上，介绍云计算的 3 个层次：基础设施服务层（Infrastructure as a Service，IaaS）、平台服务层（Platform as a Service，PaaS）、软件服务层（Software as a Service，SaaS）及其功能；此后进一步阐述云计算根据其服务范围的分类，即公有云、私有云、混合云；最后介绍国内外云计算产业的现状。

电子资源：

电子教案　云计算基本概念
PPT　云计算基本概念
习题　云计算基本概念

1.1 计算模式的演变

电子教案 计算模式的演变

PPT 计算模式的演变

拓展阅读 我国云计算行业市场现状及前景趋势

IT 技术的发展日新月异，传统的计算模式已经难以适应当今大数据的处理以及各类工程或科学计算任务。事实上，随着计算机的逐步普及和半导体技术的不断进步，计算模式已经经历了几次大的变革，这些变革主要包括 4 个阶段，即"字符哑终端—主机""客户—服务器""集群计算"和"云计算"。

1. 字符哑终端—主机

随着 1964 年第一台基于集成电路的通用电子计算机 IBM 360 问世，20 世纪 60—70 年代，计算环境主要是主机（大型机）环境，字符哑终端—主机成为主要的计算模式。这种计算环境主要由一台功能强大、允许多用户连接的主机（大型机）组成，它不具备客户端。多个哑终端通过网络连接到主机，并可以与主机进行通信。哑终端一般只是主机的扩展，用户从终端键盘输入的信息被传到主机，然后由主机将执行的结果以字符方式返回到终端上。哑终端上没有任何程序和数据，所有的程序和数据都集中在主机上，并在主机上运行。主机处理多个用户发出的指令时，处理的方案一般为分时，即计算机把它的运行时间分为多个时间段，并且将这些时间段平均分配给用户指定的任务，轮流为每一个任务运行一定的时间，如此循环，直至完成所有任务。

"字符哑终端—主机"是一种集中式的计算模式，可以实现集中管理，安全性也较好，但是很多任务如字处理软件的使用等就无法与主机进行交互。

2. 客户—服务器

集成电路的快速发展极大地降低了计算机生产成本，从 20 世纪 70 年代末开始，计算机逐步进入家用市场。到了 20 世纪 90 年代，个人计算机开始普及，并且形成了相对统一的计算机操作系统，有了方便的计算机软件编程语言和工具。但是，由于个人计算机的计算和存储能力有限，仍有一些计算任务无法在单台个人计算机上完成。为此，"客户—服务器"的计算模式逐渐兴起，它允许应用程序分别在客户工作站和服务器上执行。客户工作站向服务器发送处理请求，而服务器处理结束后返回处理结果给客户工作站。

在分布式系统的发展历程中，"客户—服务器"模式扮演了重要角色。20 世纪 90 年代随着个人计算机的兴起，客户端处理能力不断增强，促进了这一计算模式的快速发展。在这一模式中，客户端负责应用的呈现，服务器处理应用的逻辑并承担资源管理的任务。这种计算模式的好处是可以利用客户机的处理能力，降低服务器的运算负担，同时也使得针对不同个性的用户呈现不同的界面内容成为可能。然而，这种计算模式往往会造成客户端和服务器之间耦合紧密、可伸缩性差，服务器往往成为处理瓶颈。此外，一旦应用环境发生变化，需要改变业务逻辑，一般每个客户端的程序都要进行更新，给系统的维护和管理造成一定的困难。

3. 集群计算

"客户—服务器"计算模式可以将在单台客户计算机上无法完成的计算任务交给服务器协同来完成。但是，很多计算任务并不是单台普通的服务器能够完成的。这时，除了采用更高性能的计算机作为服务器之外，性价比更高的办法是采用计算机集群。尤其是近年来随着硬件能力激增、成本大幅下降，使得通过在电力、能源等较为便宜的地方将硬件设备集中起来实现规模效益成为可能。一些有研发实力的机构或组织开始使用大量廉价的个人计算机或普通服务器来建立集群，从而实现大规模数据中心的功能。

计算机集群通过将一组松散集成的计算机软件和硬件连接起来，高度紧密地协作完成计算工作。集群系统中的单个计算机通常被称为节点，一般通过局域网连接。在某种意义上，它们可以被看作一台计算机。然而，由个人计算机或普通服务器构成的大规模集群面临很多具有挑战性的问题，如可用性和可靠性保障等。目前，一台个人计算机或普通服务器平均无故障运行时间通常是几年，而用几千台个人计算机或普通服务器构成的集群平均几个小时就会有一个节点出现故障。这些问题对集群体系结构、硬件和系统软件设计等方面都提出了新的挑战。

4. 云计算

集群计算将计算资源整合在一起，本世纪初，人们便开始研究如何更加合理、高效地利用这样的计算资源，并以服务形式对外共享这些资源。"云计算"便在这样的思想中诞生，它是近十年来在 IT 领域出现并飞速发展的新技术之一。对于云计算中的"计算"一词读者并不陌生，而对于云计算中的"云"可以理解为一种提供资源的方式，或者说提供资源的硬件和软件系统被统称为"云"。"云"中的资源在用户看来是可以无限扩展的，并且可以随时获取，按需使用，随时扩展，按使用付费。"云计算"模式的出现是对计算资源使用方式的一种巨大变革，有个比方，从"传统计算"转向"云计算"就好比是从古老的单台发电机模式转向电厂集中供电的模式。它意味着计算能力也可以作为一种商品进行流通，就像煤气、水电一样，取用方便，费用低廉。最大的不同在于，它是通过互联网进行传输的。所以对于云计算，可以初步理解为通过网络随时随地获取到特定的计算资源。

1.2 云计算的定义

1. 定义

通过上一节的分析，不难看出云计算是一个新技术，同时也是一个新概念，一种新模式，而不是单纯地指某项具体的应用或标准。与此同时许多人对云计算的理解就如同盲人摸象，不同的人从不同的角度出发就会有不同的理解，如图 1.2.1 所示。为了尽量准确而全面地理解云计算，了解云计算产业相关的方方面面，需要进一步了解来自业界的对于云计算的各种说法。

首先是几个产业分析师的看法，他们对于业界的众多厂商有着全面的了解，因而他们的说法有一定的中立性。

Gartner 研究分析师认为，云计算正在成为一个大众化的词语。作为一个对互联网的比喻，"云"是很容易理解的。一旦同"计算"联系起来，它的意义就扩展了，而且开始变得模糊起来。

美林证券认为，云计算是透过互联网从集中的服务器交付个人应用（E-mail、文档

图 1.2.1 对云计算的理解如同盲人摸象

处理和演示文稿）和商业应用（销售管理、客户服务和财务管理）。这些服务器共享资源，如存储、处理能力和带宽。通过共享，资源能得到更有效的利用，而成本也可以降低 80%～90%。

而 InformationWeek 的定义则更加宽泛，云计算是一个环境，其中任何的 IT 资源都可以以服务的形式提供。

就连财经媒体也对云计算很感兴趣，它认为云计算使得企业可以通过互联网从超大数据中心获得计算能力、存储空间、软件应用和数据。客户只需要在必要时为他使用的资源付费，从而可以避免建立自己的数据中心并采购服务器和存储设备。

以下是各个 IT 厂商的看法。

IBM 公司认为，云计算是一种计算风格，其基础是用公共或私有网络实现服务、软件及处理能力的交付。云计算的重点是用户体验，而核心是将计算服务的交付与底层技术相分离。在用户界面之外，云背后的技术对于用户来讲是不可见的，这使得云计算对于用户来说十分友好。云计算的推动力来自接入互联网设备的急剧增长、实时数据流、SOA 及 Web 2.0 应用的广泛出现，如 Mashup、开放式协作、社会网络和移动商务。

Google 公司认为，云计算与传统以 PC 为中心的计算不同，它把计算和数据分布在大量的分布式计算机上，这使计算能力和存储获得了很强的可扩展能力，并方便了用户通过多种接入方式（如计算机、手机等）方便地接入网络获得应用和服务。其重要特征是开放式的，不会有一个企业能控制和垄断它。整个互联网就是一朵云，网民们需要在"云"中方便地连接任何设备，访问任何信息，自由创建内容，与朋友分享。当然，这一切都要在一个安全、快速和便捷的前提下完成。所谓"云计算"，就是要以公开的标准和服务为基础，以互联网为中心，提供安全、快速和便捷的数据存储和网络计算服务，让互联网这片"云"成为每一个网民的数据中心和计算中心。

微软公司认为，如果未来计算能力和软件全集中在云上，那么客户端就不需要很强的处理能力了，Windows 也就失去了大部分的作用。因此，微软公司的提法一直是"云 + 端"。未来的计算模式是云端计算，而不是单纯的云计算。一字之差，带来的含义却大不相同。这里的端是指客户端，也就是说云计算一定要有客户端来配合。

而在学术界，网格计算之父 Ian Foster 认为，云计算是一种大规模分布式计算的模式，其推动力来自规模化所带来的经济性。在这种模式下，一些抽象的、虚拟化的、可动态扩展和被管理的计算能力、存储、平台和服务汇聚成资源池，通过互联网按需交付给外部用户。他认为云计算的几个关键点是：大规模可扩展性；可以被封装成一个抽象的实体，并提供不同的服务水平给外部用户使用；由规模化带来的经济性；服务可以被动态配置（通过虚拟化或者其他途径），按需交付。

来自著名的伯克利（Berkeley）大学的一篇技术报告则指出，云计算既指透过互联网交付的应用，也指在数据中心中提供这些服务的硬件和系统软件。前半部分即是 SaaS，而后半部分则被称为 Cloud。简单地说，云计算就是"SaaS+效用计算（Utility Computing）"。如果这个基础架构可以按使用付费的方式提供给外部用户，那么这就是公共云，否则便是私有云。公共云即是效用计算，SaaS 的提供者同时也是公共云的用户。

根据以上这些来自产业不同的说法，不难发现，人们对于云计算基本上还是有一致的看法，只是在某些范围的划定上有所区别。现阶段广为接受的是美国国家标准与技术研究院（NIST）的定义，即云计算是一种按使用量付费的模式，这种模式提供可用的、便捷的、按需的网络访问，进入可配置的计算资源共享池（资源包括网络、服务器、存储、应用软件、服务），只需投入很少的管理工作，或与服务供应商进行很少的交互，就可以让这些资源能够被快速提供。从以上的分析可以给出一个更加技术性的定义：云计算是一种模式，它实现了对共享可配置计算资源（网络、服务器、存储、应用和服务等）的方便、按需访问；这些资源可以通过极小的管理代价或者与服务提供者的交互被快速地准备和释放。

2. 云计算的特点

云计算具有如下的特点。

① 超大规模。大多数云计算数据中心都具有相当的规模，如图 1.2.2 所示。云计算中心能通过整合和管理这些数目庞大的计算机集群，来赋予用户前所未有的计算和存储能力。

图 1.2.2　超大规模的数据中心

② 虚拟化。云计算支持用户在任意位置使用各种终端获取应用服务。所请求的资源来自云，而不是固定的、有形的实体。资源以共享资源池的方式统一管理，利用虚拟化技术，将资源分享给不同用户，资源的放置、管理与分配策略对用户透明。

云计算是基于网络提供的一种服务，只要有网络，使用任何终端（笔记本电脑或手机等),都可以实时连接到云计算服务器,享受云的服务。在享受服务的时候,用户不知道也没必要知道,这个服务是由哪台服务器提供的。

③ 高可靠性。云计算中心在软硬件层面采用了诸如数据多副本容错、心跳检测和计算节点同构可互换等措施来保障服务的高可靠性,使用云计算比使用本地计算机可靠。此外,它还在设施层面上的能源、制冷和网络连接等方面采用了冗余设计来进一步确保服务的可靠性。由于云计算系统由大量商用计算机组成集群向用户提供数据处理服务,随着计算机数量的增加,系统出现错误的概率大大增加,因而云计算系统在硬件部署上均有冗余设计,软件上也通过数据冗余和分布式存储来保证数据的可靠性。

④ 通用性与高可用性。云计算不针对特定的应用,云计算中心很少为特定的应用存在,但它有效支持业界的大多数主流应用,并且一个云可以支撑多个不同类型的应用同时运行,在云的支撑下可以构造出千变万化的应用,并保证这些服务的运行质量。并且,通过集成海量存储和高性能的计算能力,云能提供较高的服务质量。云计算能容忍节点的错误,因为它可自动检测失效节点,并将失效节点排除,而不影响系统整体的正常运行。

⑤ 高可扩展性。云计算系统可以随着用户的规模进行扩张,可以保证支持客户业务的发展。因为用户所使用的云资源可以根据其应用的需要进行调整和动态伸缩,并且再加上前面所提到的云计算数据中心本身的超大规模,云能够有效地满足应用和用户大规模增长的需要。云计算能够无缝地扩展到大规模的集群之上,甚至包含数千个节点同时处理。

⑥ 按需服务。云是一个庞大的资源池,用户可以支付不同的费用,以获得不同级别的服务。并且,服务的实现机制对用户透明,用户无须了解云计算的具体机制,就可以获得需要的服务。

⑦ 极其经济廉价。由于云的特殊容错措施可以采用极其廉价的节点来构成云,云的自动化集中式管理使大量企业无须负担日益高昂的数据中心管理成本,云的通用性使资源的利用率较传统系统大幅提升,因此用户可以充分享受云的低成本优势。通常只要花费几百美元、几天时间就能完成以前需要数万美元、数月时间才能完成的任务。显然,组建一个采用大量的商业机组成的集群,相对于组建同样性能的超级计算机花费的资金要少很多。

⑧ 自动化。在云中,不论是应用、服务和资源的部署,还是软硬件的管理,主要通过自动化的方式来执行和管理,从而也极大地降低了整个云计算中心的人力成本。

⑨ 节能环保。云计算技术能将许许多多分散在低利用率服务器上的工作负载整合到云中,来提升资源的使用效率,而且云由专业管理团队运维,所以其电源使用效率(Power Usage Effectiveness,PUE)值比普通企业的数据中心出色很多。

⑩ 高层次的编程模型。云计算系统提供高层次的编程模型。用户通过简单学习,就可以编写自己的云计算程序,在云系统上执行,满足自己的需求。现在云计算系统主要采用 MapReduce 模型。

⑪ 完善的运维机制。在云的另一端,有全世界专业的团队来帮用户管理信息,有全世界先进的数据中心来帮用户保存数据。同时,严格的权限管理策略可以保证这些数据的安全。这样,用户无须花费重金就可以享受到最专业的服务。

此外,云计算还以其部署迅速、资源利用率高、易管理、几乎可以提供无限的廉价存

1.3 云计算的层次以及分类

云计算可以按需提供弹性资源，它的表现形式是一系列服务的集合。因此，大多数学者以及工程技术人员将云计算的 3 层体系架构多分为基础设施服务层（Infrastructure as a Service，IaaS）、平台服务层（Platform as a Service，PaaS）、软件服务层（Software as a Service，SaaS），即 3 层 SPI（SaaS、PaaS、IaaS 的首字母缩写）架构，如图 1.3.1 所示。

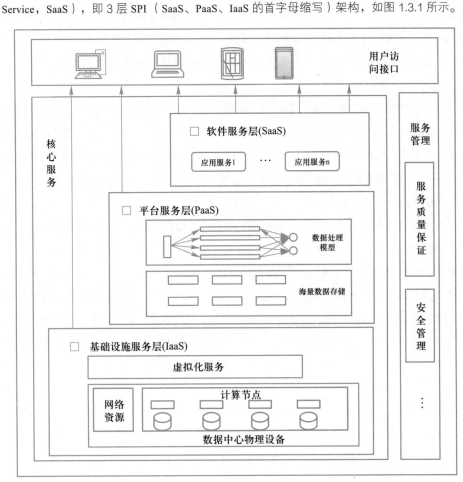

图 1.3.1 云计算的 3 层体 SPI 系架构

1. 云计算的层次架构

（1）基础架构服务层（IaaS）

该层位于云计算 3 层服务的最底端，也是云计算狭义定义所覆盖的范围，就是把 IT 基础设施像水、电一样以服务的形式提供给用户，以服务形式提供基于服务器和存储等硬件资源的可高度扩展和按需变化的 IT 能力。通常按照所消耗资源的成本进行收费。

该层提供的是基本的计算和存储能力，以计算能力的提供为例，其提供的基本单元就是服务器，包含 CPU、内存、存储、操作系统及一些软件。为了让用户能够定制自己的服务器，需要借助服务器模板技术，即将一定的服务器配置与操作系统和软件进行绑定，并提供定制的功能。服务的供应是一个关键点，它的好坏直接影响到用户的使用效率及 IaaS 系统运行和维护的成本。自动化是一个核心技术，它使得用户对资源使用的请求可以以自行服务的方式完成，无须服务提供者的介入。一个稳定而强大的自动化管理方案可以将服务的边际成本降低为 0，从而保证云计算的规模化效应得以体现。在自动化的基础上，资源的动态调度得以成为现实。资源动态调度的目的是满足服务水平的要求。例如，根据服务器的 CPU 利用率，IaaS 平台自动决定为用户增加新的服务器或存储空间，从而满足事先跟用户订立的服务水平条款。在这里，资源动态调度技术的智能性和可靠性十分关键。此外，虚拟化技术是另外一个关键的技术，它通过物理资源共享来极大提高资源利用率，降低 IaaS 平台成本与用户使用成本。而且，虚拟化技术的动态迁移功能能够带来服务可用性的大幅度提高，这一点对许多用户极具吸引力，如 IBM 为无锡软件园建立的云计算中心。

（2）平台服务层（PaaS）

该层位于云计算 3 层服务的最中间，通常也称为"云计算操作系统"。它提供给终端用户基于互联网的应用开发环境，包括应用编程接口和运行平台等，并且支持应用从创建到运行整个生命周期所需的各种软硬件资源和工具。通常按照用户或登录情况计费。在 PaaS 层面，服务提供商提供的是经过封装的 IT 能力，或者说是一些逻辑的资源，如数据库、文件系统和应用运行环境等。

通常又可将 PaaS 细分为开发组件即服务和软件平台即服务。前者指的是提供一个开发平台和 API 组件，给开发人员更大的弹性，依不同需求定制化。一般面向的是应用软件开发商（ISV）或独立开发者，这些用户在 PaaS 厂商提供的在线开发平台上进行开发，从而推出自己的 SaaS 产品或应用。后者指的是提供一个基于云计算模式的软件平台运行环境，让应用软件开发商（ISV）或独立开发者能够根据负载情况动态提供运行资源，并提供一些支撑应用程序运行的中间件支持。目前有能力提供 PaaS 平台的厂商并不多，本部分关于云的产品示例包括 IBM 的 Rational 开发者云、Saleforce 公司的 Force.com 和 Google 的 Google App Engine 等。

这个层面涉及两个核心技术。第一个核心技术是基于云的软件开发、测试及运行技术。PaaS 服务主要面向软件开发者，如何让开发者通过网络在云计算环境中编写并运行程序，在以前是一个难题。如今，在网络带宽逐步提高的前提下，两种技术的出现解决了这个难题：一个是在线开发工具，开发者可通过浏览器、远程控制台（控制台中运行开发工具）等技术直接在远程开发应用，无须在本地安装开发工具；另一个是本地开发工具和云计算的集成技术，即通过本地开发工具将开发好的应用直接部署到云计算环境中，同时能够进行远程调试。第二个核心技术是大规模分布式应用运行环境。它指的是利用大量服务器构建的可扩展的应用中间件、数据库及文件系统。这种应用运行环境可以充分利用云计算中心的海量计算和存储资源，进行充分扩展，突破单一物理硬件的资源瓶颈，满足互联网上百万级用户量的访问要求，Google 的 App Engine 就采用了这样的技术。

（3）软件服务层（SaaS）

该层位于云计算 3 层服务的顶端。用户通过标准的 Web 浏览器来使用互联网上的软件。

服务供应商负责维护和管理软硬件设施，并以免费（提供商可以从网络广告之类的项目中生成收入）或按需租用方式向最终用户提供服务。尽管这个概念之前就已经存在，但并不影响它成为云计算的组成部分。

这类服务既有面向普通用户的，如 Google Calendar 和 Gmail，也有直接面向企业团体的，用以帮助处理工资单流程、人力资源管理、协作、客户关系管理和业务合作伙伴关系管理等。这些产品的常见示例包括 IBM LotusLive、Salesforce.com 和 Sugar CRM 等。这些 SaaS 提供的应用程序减少了客户安装和维护软件的时间和技能等代价，并且可以通过按使用付费的方式来减少软件许可证费用的支出。

在 SaaS 层面，服务提供商提供的是消费者应用或行业应用，直接面向最终消费者和各种企业用户。这一层面主要涉及如下技术：Web 2.0、多项目和虚拟化。Web 2.0 中的 Ajax 等技术的发展使得 Web 应用的易用性越来越高，它把一些桌面应用中的用户体验带给了 Web 用户，从而让人们容易接受从桌面应用到 Web 应用的转变。多项目是指一种软件架构，在这种架构下，软件的单个实例可以服务于多个客户组织（项目），客户之间共享一套硬件和软件架构，它可以大大降低每个客户的资源消耗（即客户成本）。虚拟化也是 SaaS 层的一项重要技术，与多项目技术不同，它可以支持多个客户共享硬件基础架构，但不共享软件架构，这与 IaaS 中的虚拟化是相同的。

以上 3 层，每层都有相应的技术支持提供该层的服务，具有云计算的特征，如弹性伸缩和自动部署等。每层云服务可以独立成云，也可以基于下面层次的云提供的服务。每种云可以直接提供给最终用户使用，也可以只用来支撑上层的服务。

以上云计算的三层架构统一属于云计算的核心服务模块，除此之外完整运营的云计算系统还需要具备服务管理模块以及用户访问接口模块，如图 1.3.1 所示。其中，服务管理模块为核心服务模块提供支持，以进一步确保核心服务的质量、可用性与安全性。服务管理实际内容应包括很多，但主要是服务质量（Quality of Service，QoS）保证和安全管理等。用户访问接口模块实现了云计算服务的泛在访问，通常包括命令、Web 服务、Web 门户等形式。命令和 Web 服务的访问模式既可为终端设备提供应用程序开发接口，又便于多种服务的组合。Web 门户是访问接口的另一种模式，通过 Web 门户，云计算将用户的桌面应用迁移到互联网，从而使用户随时随地通过浏览器就可以访问数据和程序，提高工作效率。

虽然，用户通过访问接口使用便利的云计算服务，但是由于不同云计算服务商提供接口标准不同，导致用户数据不能在不同服务商之间迁移。为此，在 Intel、Sun 和 Cisco 等公司的倡导下，云计算互操作论坛（Cloud Computing Interoperability Forum，CCIF）宣告成立，并致力于开发统一的云计算接口（Unified Cloud Interface，UCI），以实现"全球环境下不同企业之间可利用云计算服务无缝协同工作"的目标。

2. 云计算的分类

依据云计算的服务范围又可以将云计算系统分类为私有云、公有云以及混合云。

（1）公有云

公有云是云基础设施由一个提供云计算服务的运营商（或称云供应商）所拥有，该运营商再将云计算服务销售给一般大众或广大的中小企业群体所共有,是现在最主流的、也是最受欢迎的一种云计算部署模式。

公有云是一种对公众开放的云服务,能支持数目庞大的请求,而且因为规模的优势,其成本偏低。公有云由云供应商运行,为最终用户提供各种各样的 IT 资源。云供应商负责从应用程序、软件运行环境到物理基础设施等 IT 资源的安全、管理、部署和维护。用户在使用 IT 资源时,只需为其所使用的资源付费,而无须任何前期投入,所以非常经济。而且,在公有云中,用户不清楚与其共享和使用资源的还有哪些其他用户,整个平台是如何实现的,甚至无法控制实际的物理设施,所以云服务提供商能保证其所提供的资源具备安全、可靠等非功能性需求。

目前,许多 IT 巨头都推出了自己的公有云服务,包括 Amazon 的 AWS、微软的 Windows Azure Platform、Google 的 Google Apps 与 Google App Engine 等。

① 公有云在许多方面都有其优越性,以下列出了其中的 4 个方面。

● 规模大。因为公有云的公开性,它能聚集来自整个社会并且规模庞大的工作负载,从而产生巨大的规模效应,如能降低每个负载的运行成本或者为海量的工作负载做更多优化。

● 价格低廉。由于对用户而言,公有云完全是按需使用的,无须任何前期投入,所以与其他模式相比,公有云在初始成本方面有非常大的优势。随着公有云的规模不断增大,它将不仅使云供应商受益,而且也会相应地降低用户的开支。

● 灵活。对用户而言,公有云在容量方面几乎是无限的。就算用户的需求量很高,公有云也能非常快地予以满足。

● 功能全面。公有云在功能方面非常丰富全面,如可支持多种主流的操作系统和成千上万的应用。

② 公有云的不足之处有以下几点。

● 缺乏信任。虽然在安全技术方面,公有云有很好的支持,但由于其存储数据并不是在企业本地,所以企业会不可避免地担忧数据的安全性。

● 不支持遗留环境。由于现在公有云技术基本上都是基于 x86 架构的,操作系统普遍以 Linux 或者 Windows 为主,所以对于大多数遗留环境没有很好地支持,如基于大型机的 Cobol 应用。

由于公有云在规模和功能等方面的优势,它受到绝大多数用户的欢迎。从长期而言,公有云将像公共电厂那样毋庸置疑地成为云计算最主流、甚至是唯一的模式,因为它在规模、价格和功能等方面的潜力实在太大。但是,在短期之内,因为信任和遗留等方面的不足,会降低公有云对企业的吸引力,特别是一些大型企业。

(2)私有云

对许多大中型企业而言,因为有很多限制条款,它们在短时间内很难大规模地采用公有云技术,可是它们也期盼云计算所带来的便利,所以引出了私有云这一云计算的部署模式。私有云是云基础设施被某单一组织拥有或租用,可以坐落在本地(on Premise)或防火墙外的异地,该基础设施只为该组织服务。也就是说,私有云主要是为企业内部提供云服务,不对公众开放,大多在企业的防火墙内工作,并且企业 IT 人员能对其数据、安全性和服务质量进行有效的控制。与传统的企业数据中心相比，私有云可以支持动态灵活的基础设施，从而降低 IT 架构的复杂度，使各种 IT 资源得以整合和标准化。

在私有云界，主要有两大联盟：一是 IBM 与其合作伙伴，主要推广的解决方案有 IBM Blue Cloud 和 IBM CloudBurst；二是由 VMware、Cisco 和 EMC 组成的 VCE 联盟，主推的是 Cisco UCS 和 vBlock。在实际的例子方面，已经建设成功的私有云有采用 IBM Blue Cloud 技术的中化云计算中心和采用 Cisco UCS 技术的 Tutor Perini 云计算中心。

① 创建私有云的方式主要有以下两种。

- 独自构建方式。即通过使用诸如 Enomaly 和 Eucalyptus 等软件将现有硬件整合成一个云，这比较适合预算少或者希望重用现有硬件的一些企业。
- 购买商业解决方案。它通过购买 Cisco 的 UCS 和 IBM 的 Blue Cloud 等方案来一步到位，这比较适合那些有实力的企业和机构。

② 由于私有云主要在企业数据中心内部运行，并且由企业的 IT 团队来进行管理，因此这种模式在以下 5 个方面表现了出色的优势。

- 数据安全。虽然每个公有云的供应商都对外宣称，其服务在各方面都非常安全，特别是在数据管理方面。但是，对企业特别是大型企业而言，和业务相关的数据是其生命线，是不能受到任何形式的威胁和侵犯的，而且需要严格控制和监视这些数据的存储方式和位置。因此，短期而言，大型企业不会将其关键应用部署到公有云上，私有云在这方面非常有优势，因为它一般都构筑在防火墙内，企业会比较放心。
- 服务质量（QoS）。因为私有云一般在企业内部，而不是在某个遥远的数据中心，所以当公司员工访问那些基于私有云的应用时，它的服务质量应该会非常稳定，这样不会受到远程网络偶然发生异常的影响。
- 充分利用现有硬件资源。每个公司，特别是大公司，都会存在很多低利用率的硬件资源。这样，就可以通过一些私有云解决方案或者相关软件，让它们重获"新生"。
- 支持定制和遗留应用。现有公有云所支持应用的范围都偏主流，这对于一些定制化程度较高的应用和遗留应用就很有可能束手无策。但是，这些往往都是一个企业最核心的应用，如大型机、UNIX 等平台的应用。这时，私有云可以说是一个不错的选择。
- 不影响现有 IT 管理的流程。对大型企业而言，流程是其管理的核心，如果没有完善的流程，企业将会成为一盘散沙。实际情况是，不仅企业内部和业务有关的流程非常多，而且 IT 部门的自身流程也不少，而且大多都不可或缺，如那些和 Sarbanes-Oxley 相关的流程。私有云的适应性比公有云好很多，因为 IT 部门能完全控制私有云，就有能力使私有云比公有云更好地与现有流程进行整合。

③ 私有云也有其不足之处，具体表现在以下两方面。

- 成本开支高。因为建立私用云需要很高的初始成本，特别是如果需要购买大厂家的解决方案，更是如此。
- 持续运营成本偏高。由于需要在企业内部维持一支专业的云计算团队，因而其持续运营成本也同样会偏高。

在将来很长一段时间内，私有云将成为大中型企业最认可的云模式，而且将极大地增强企业内部的 IT 能力，并使整个 IT 服务围绕着业务展开，从而更好地为业务服务。

（3）混合云

混合云是云基础设施由两种或以上的云（私有云、公有云或行业云）组成，每种云仍然保持独立实体，但用标准的或专有的技术将它们组合起来，具有数据和应用程序的可移植性，可通过负载均衡技术来应对处理突发负载（Cloudburst）等。

混合云虽然不如公有云和私有云常用，但已经有类似的产品和服务出现。顾名思义，混合云是把公有云和私有云结合到一起的方式，即是让用户在私有云的私密性和公有云的低廉性之间做一定权衡的模式。例如，企业可以将非关键的应用部署到公有云上来降低成本而将安全性要求很高、非常关键的核心应用部署到完全私密的私有云上。

现在混合云的案例非常少，最相关的就是 Amazon VPC（Virtual Private Cloud，虚拟私有云）和 VMware vCloud 了。混合云的构建方式有以下两种：

① 外包企业的数据中心。企业搭建了一个数据中心，但具体维护和管理工作都外包给专业的云供应商，或者邀请专业的云供应商直接在厂区内搭建专供本企业使用的云计算中心，并且在建成之后由专业的云供应商负责今后的维护工作。

② 购买私有云服务。通过购买云供应商的私有云服务，能将一些公有云纳入防火墙内。而且，在这些计算资源和其他公有云资源之间进行隔离，同时获得极大的控制权，这样也免去了维护之苦。

通过使用混合云，企业既可以享受私有云的私密性，又可以享受公有云的成本，能快速接入大量位于公有云的计算能力，以备不时之需。但现在可供选择的混合云产品较少，在私密性方面不如私有云好，在成本方面不如公有云低，操作起来较复杂。混合云比较适合那些想尝鲜云计算的企业，以及面对突发流量但不愿将企业 IT 业务都迁移至公有云的企业。虽然，混合云不是长久之计，但是它也会有一定的市场空间，并且，也将会有一些厂商推出类似的产品。

（4）行业云

除了以上 3 类，行业云（Community Cloud）近年来开始被提及。行业云可译成社区云、行业云或机构云，即云基础设施被一些组织共享，并为一个有共同关注点的社区、行业或大机构服务（如任务、安全要求、政策和准则等），这种云可以被该社区、行业或大机构拥有和租用，也可以坐落在本地、防火墙外的异地或多地，它也可能是一组私有云通过 VPN 连接到一起的 NPC，即混合云的一种。

行业云虽然较少提及，但是有一定的潜力，主要指的是专门为某个行业的业务设计的云，并且开放给多个同属于这个行业的企业。虽然，行业云现在还没有成熟的例子，但盛大（游戏行业国内知名企业）的开放平台颇具行业云的潜质，因为它能将整个云平台共享给多个小型游戏开发团队。这样，这些小型团队只需负责游戏的创意和开发，其他和游戏相关的运维，可转交给盛大的开放平台来负责。

在构建方式方面，行业云主要有以下两种方式。

① 独自构建方式。即由某个行业的领导企业，自主创建一个行业云，并与其他同行业的公司分享。

② 联合构建方式。即由多个同类型的企业，联合建设和共享一个云计算中心，或者邀请外部的供应商来参与其中。

行业云的优势是：能为行业的业务作专门的优化，这和其他云计算部署模式相比，能进一步方便用户，为行业的业务作专门的优化，进一步降低成本。

行业云的不足之处是：支持的范围较小，只支持某个行业，建设成本较高，行业云非常适合那些业务需求比较相似，而且对成本非常关注的行业。虽然，现在还没有非常好的示例，但是对部分行业存在一定的吸引力，如游戏业。

1.4 国内外云计算产业现状

目前，全球云计算市场迅速增长，云计算如今已经被确立为一种核心技术，它标志着企业传统运营方式的转变，并且仍在不断发展。将一部分业务迁移到云端的一个令人信服的原因是，企业能够根据业务需求更好地分配资源，以实现灵活性、敏捷性、成本效率方面的收益。Gartner 公司预测，全球公共云服务市场将从 2018 年的 1824 亿美元增长到 2022 年的 3312 亿美元，复合年增长率为 12.6%。尽管公共云的市场价值已经高于私有云，但由于在向公共云模式转换过程中存在组织上的困难，私有云和虚拟私有云继续占有重要的市场份额。尽管如此，公共云将会持续增长。

电子教案　国内外云计算产业现状

世界信息产业强国和地区也对云计算给予了高度关注，已把云计算作为未来战略产业的重点，纷纷研究制定并出台云计算发展战略规划，加快部署国家级云计算基础设施，并加快推动云计算的应用，抢占云计算产业制高点。IBM、Microsoft 等知名公司相继推出云计算产品和服务，Intel、Cisco 等传统硬件厂商也纷纷向云计算服务商转型。云计算受到了国际资本市场的高度关注，VMware 因在云基础架构领域的领先优势成为继 Google 上市后美国融资额排名第二的科技公司，Salesforce 等多家新兴云计算技术和服务企业也凭借先发优势成功在欧美证券市场上市，发展势头劲猛。

PPT　国内外云计算产业现状

云计算产业在我国国内尚处于蓬勃发展阶段，总体规模每年迅速递增，我国云计算产业生态链的构建正在进行中。在政府的监管下，云计算服务提供商与软硬件、网络基础设施服务商以及云计算咨询规划、交付、运维、集成服务商、终端设备厂商等一同构成了云计算的产业生态链，为政府、企业和个人用户提供服务。但云计算的产业化快速发展尚存在诸多障碍，如用户认知不足、标准缺失、数据主权争议、可用性、稳定性担忧、用户锁定、服务质量难以规范等。其中，标准和安全以及相关法律法规的完善是最核心、也是最迫切需要解决的核心问题。

现阶段在政府的大力扶持下，我国已有 20 多个城市开展云计算相关研究和项目建设。北京市发布的"祥云工程"行动计划，建成了亚洲最大的超云服务器生产基地；上海市发布的"云海计划"三年方案，致力打造"亚太云计算中心"，带动信息服务业新增经营收入 1000 亿元；广州市部署的"天云计划"，打造世界级云计算产业基地，达到国内云计算应用领先水平。陕西、福建、天津、黑龙江、重庆、宁波、深圳、武汉、杭州、无锡、廊坊、南京等省市均加强了对云计算产业的研究与部署，并联合大型信息技术企业积极推动云计算产业发展，加强云计算基础设施建设，重点搭建商务云平台、开发云平台和政务云平台三大云计算服务平台。预计未来几年,我国云计算市场规模年均复合增长率将超过80%,主要以政府、电信、教育、医疗、金融、石油石化、电力等行业为应用重点。相信在未来的数年乃至数十年的时间,云计算产业将成为 IT 行业中增长速度最快、人才需求量最大的一个发展方向。

云计算产业在国外尤其是美国为代表的市场已经非常成熟了。在我国云计算产业保持了较好的发展态势,创新能力显著增强、服务能力大幅提升、应用范畴不断拓展,已成为提升信息化发展水平、打造数字经济新动能的重要支撑。

IDC 预计政务、教育、银行、电信等行业的云计算至少在从 2020 年起的未来 5~10 年还将保持双位数增长。其中,在政策驱动下,中国的政务云近年来实现高增长,政务云规模目前占据了中国私有云市场的 53%,是私有云增长最快的子行业。而随着 5G 和物联网技术的发展,大量 IoT 设备所产生的庞大数据量也将推动包括金融、教育、电信、公共事业、安全在内的传统行业采纳物联网云计算,提升数据采集、存储、分析的能力,以提升经营效率。

以目前国内云计算市场占比最大的 IaaS 收入来看,在 2017 年中国公有云 IaaS 增速中,AWS、腾讯云的增速较快,达到了 152.33% 和 136.76%,阿里云由于基数较大,虽然增速低于 100%,但份额占有量依然很可观。而电信运营商却正在由 2015 年的公有云 IaaS 市场份额 20.7%,降到 2019 年底的不足 10%。伴随着 IaaS 市场的逐步饱和,产业结构持续优化,PaaS 和 SaaS 市场需求的不断增加,考虑到运营商在上层云服务领域的劣势,在今后的一段时间内将会面临更为严酷的挑战。但同时由于 5G 驱动云计算产业升级,电信运营商还是面临新的机遇。

习题 云计算基本概念

面向 5G 时代,电信运营商的云化之路需要从网络架构、基础设施、业务服务和运营模式等 4 个方面全面提升,在满足未来融合应用场景的网络需求的同时,以网络能力开放为基础,在能力平台和云服务领域必须加大创新投入,加快推出面向垂直行业领域的云服务,以适应即将到来的数字经济时代。

5G 时代即将来临,云计算又将迎来新一轮的爆发,对于电信运营商来说,机遇与挑战并存,要想在国内云计算领域占据一席之地,在坚持创新驱动发展的同时,还要准确把握市场需求,寻求多元化合作,打造良性的生态体系,从而实现业务收入的持续增长。

第 2 章 云计算知名厂商及其产品

 本章导读：

　　本章介绍云计算产业内知名的厂商及其相关的云计算产品，其中包括业界最知名的 VMware 以及 Citrix；随着云计算产业的迅速崛起，国内不少大型 IT 企业也开始推出自己的云计算产品，这里介绍其中的典型代表，有华为以及 H3C 的云计算相关产品；最后，还将介绍知名的公有云平台。

电子资源：
　　电子教案　云计算知名厂商及其产品
　　PPT　云计算知名厂商及其产品
　　习题　云计算知名厂商及其产品

第 2 章 云计算知名厂商及其产品

2.1 VMware 的云计算技术及其相关产品

电子教案 VMware 的云计算技术及其相关产品

PPT VMware 的云计算技术及其相关产品

拓展阅读 信创云及其分类和特征

1. VMware 公司及其战略理念

VMware 公司是全球领先的虚拟化和云基础架构提供商,为客户提供经过验证的解决方案,显著降低 IT 复杂性,实现更灵活、更敏捷的 IT 服务提供。借助全球客户广泛采用的云计算解决方案——VMware vSphere,VMware 帮助客户在充分利用现有投资并提升安全性和控制力的同时,加速实现向云计算的转变。利用 VMware vCloud 解决方案和服务,VMware 实现了私有云、公有云和混合云的联合部署模式,确保应用的可迁移性和便携性。VMware 从保护现有投资和降低技术风险的角度协助客户制定云计算发展战略,该战略具有以下特点。

(1)渐进的技术实现

VMware 提供了一种务实的途径帮助客户通过渐进的方式来实现云计算愿景,通过封装遗留应用并将它们迁移至现代云计算环境中,确保安全性、可管理性、服务质量和法规遵从。采用 VMware 解决方案,将逐步实现云计算模型所定义的包括可按需提供服务、高可用性和高安全性在内的多种优势;借助自动化的服务级别管理和标准化的访问,VMware 确保在迈向云计算的旅途中实现成本效益和业务敏捷性双方面的成功。

(2)开放的技术平台

VMware 平台是业界领先的平台,已经有众多企业和服务提供商选用了这一平台,采用它,就可以按照业务需求将应用部署在最佳场所(私有云或公共云),并且可以利用混合私有云环境,使应用在跨私有云和公共云的基础架构上迁移。

2. VMware 相关产品

VMware 公司根据云计算的层次和结构也把产品线分成 3 个层次。第一是面向 SaaS 的产品,主要是针对终端应用设备的虚拟化或者说是云时代的终端设备。VMware View,是一个典型的面向用户的桌面虚拟化产品。第二是面向 PaaS 的产品,主要是针对软件开发环境的,无论当前 .NET 还是 Java 开发者,都可以做到快速的部署和升级,应对各种各样的问题。这中间除了耳熟能详的 Spring,还包括像 GemFire 这样用于提供快速、安全、可靠和可扩展的数据访问理想解决方案。第三是面向 IaaS 平台的产品,无论公有云还是私有云,都是云计算和虚拟化的基础,都是基于现有数据中心构造虚拟的数据中心。这是基于云计算架构的基础结构,VMware vSphere 系列产品提供了很好的数据中心云计算基础环境的解决方案,该产品也常用于企业私有云的架设。

VMware 公司产品丰富,产品之间功能交错融合,能够满足不同企业,不同应用场景下的需求,其中目前应用最多的为 VMware View、VMware vSphere 以及 VMware WorkStation,以下分别介绍这些产品。

(1)VMware vSphere

vSphere 是 VMware 推出的基于云计算的新一代数据中心虚拟化套件,提供了虚拟化基础架构、高可用性、集中管理、监控等一整套解决方案。VMware 于 2001 年正式推出了企业级虚拟化产品 ESX(Esx 和 Esxi 都是 vSphere 的组件),该产品到现在历经了 5 代演进。

而整个架构功能经过不断扩展，也越来越完善了。利用 VMware vSphere 的数据中心虚拟化平台，可使所有应用程序和服务具备高级别的可用性和响应速度。它通过将关键业务应用程序与底层硬件分离来实现前所未有的可靠性和灵活性，从而优化 IT 服务的交付，使每种应用程序工作负载均能够以最低的总体成本履行最高级别的应用程序服务协议。

VMware vSphere 构建了整个虚拟基础架构，将数据中心转化为可扩展的聚合计算机基础架构。虚拟基础架构还可以充当云计算的基础。完美的 VMware vSphere 架构是由软件和硬件两方面组成的。

拓展阅读 品高信创云解决方案

各大服务器厂商都针对虚拟化提出了自己的解决方案，并针对虚拟化架构进行了优化，每个厂家都有自己的特点和卖点。VMware vSphere 的物理结构由 x86 虚拟化服务器、存储器网络和阵列、IP 网络、管理服务器和桌面客户端 4 部分组成，如图 2.1.1 所示。

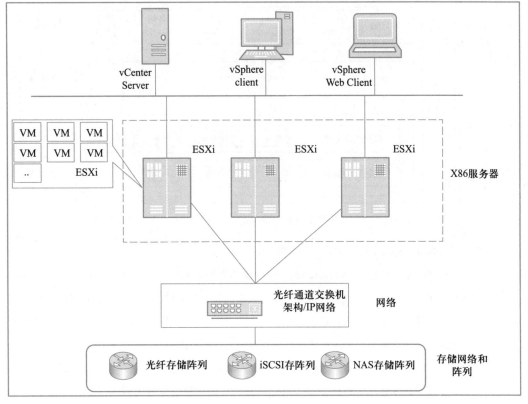

图 2.1.1　VMware vSphere 的物理结构

x86 虚拟化服务器是 VMware vSphere 提供的一种虚拟化资源，可以运行虚拟机。x86 服务器由多个相同的 x86 平台服务器组成，每台服务器相互独立。在硬件上直接安装 ESXi 操作系统，通过网络提供虚拟化的资源。x86 服务器主要为虚拟化提供 CPU 计算能力和内存等资源。

存储是虚拟化的基石，用于存放大量虚拟化数据。存储资源是由 vSphere 来分配的，这些资源在整个数据中心的虚拟机之间共享。存储网络和阵列可以是应用了光纤通道 SAN 阵列、iSCSI SAN 阵列和 NAS 阵列的存储技术。

IP 网络是连接各种资源和对外服务的通道，每台 x86 服务器和存储器都处于不同的网络。IP 网络为整个虚拟化数据中心提供可靠的网络连接。

管理服务器提供了基本的数据中心服务，如访问控制、性能监控和配置功能。它将各个 x86 服务器中的资源统一在一起，使这些资源在整个数据中心中的各虚拟机之间共享。

客户端是用户通过连网设备连接到虚拟机的管理控制服务器，用来进行资源的部署和调配，或向虚拟机发出控制命令等，是人机交互的通道。

VMware vSphere 平台从其自身的系统架构来看，可分为虚拟化层、管理层、接口层 3 个层次。这 3 层构建了 VMware vSphere 平台的整体，如图 2.1.2 所示。VMware vSphere 平台充分利用了虚拟化资源、控制资源和访问资源等各种计算机资源，同时还能使 IT 组织提供灵活可靠的 IT 服务。

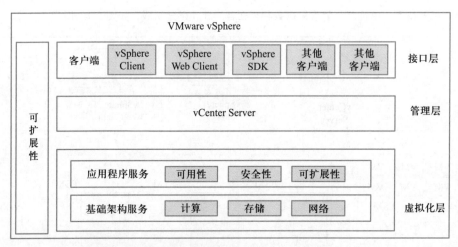

图 2.1.2　VMware vSphere 系统架构的 3 个层次

VMware vSphere 有如下几个核心组件。

· VMware ESXi：是 VMware vSphere 的操作系统，其他组件都建立在它之上。VMware ESXi 服务器是在通用环境下分区和整合系统的虚拟主机软件，是具有高级资源管理功能的、高效灵活的虚拟主机平台。它运行在物理硬件上的虚拟化层，将计算机资源分成若干个逻辑资源，将处理器、内存、存储器和资源虚拟化为多个虚拟机。

· VMvare vCenter Server：一个管理端 VMware vSphere 的操作系统，可以同时管理数据中心中所有的 VMware ESXi 主机，它是配置、置备和管理虚拟化 IT 环境的中心，是调度资源的总控，该软件既可以构建在 Linux 环境下，也可以构建在 Windows Server 环境下。

· VMware vSphere Client：允许用户从任何 Windows PC 上安装的专用客户端远程连接到 vCenter Server 或 ESXi 的界面。

· VMware vSphere Web Client：允许用户通过 Web 浏览的方式访问 vCenter Server 或 ESXi 的界面。

（2）VMware View

VMware View 5.0 是建立在 VMware vSphere 上，提供集中、自动化的桌面管理，通过一台终端实现数万个虚拟桌面的可扩展管理。VMware View 可以简化桌面和应用管理，同时可加强安全性和控制力，可跨会话和设备为终端用户提供个性化的高保真体验，可实现传统 PC 难以企及的更优异的桌面服务可用性和敏捷性。终端用户可以享受到新的工作效率级别以及从更多设备及位置访问桌面的自由，同时为 IT 部门提供更强的策略控制能力。如今，企业正陷入"桌面困境"。一方面，IT 组织面临着成本、合规性、可管理性和安全性方面的压力。以 PC 为中心的现有计算模式加剧了这种态势，这种模式的管理成本高昂，限制了 IT 的敏捷性，使其难以应对不断变化的业务形势。另一方面，终端用户越来越需要从更多设备和位置自由、灵活地访问其应用程序和数据。此"桌面困境"（终端用户自由和 IT 控制之间的两难困境）抬高了管理成本，影响了用户桌面的安全，并令 IT 资源不堪重负。为了摆脱这种困境，各个组织都在寻求既灵活敏捷又适应性强的计算方法，从而使 IT 能够平衡企业的需要和终端用户的需要，实现灵活的高性能计算体验。

使用 VMware View 可实现桌面虚拟化，使各个组织能够达到事半功倍的效果。VMware View 5.0 建立在业界最广泛部署的虚拟平台 VMware vSphere 上，简化了 IT 管理控制，同时为最终用户提供最高保真度的体验，无论用户采用的是 WAN 还是 LAN 连接。VMware View 5.0 改进了 PCoIP 协议，可将 LAN 和 WAN 的连接带宽降低 75%；提供对 3D 图形的高级支持；可扩展的统一通信集成，用于 Avaya 与 Mitel 等业界领先的语音与视频媒体服务；具有集成化角色管理的虚拟桌面个性化。VMware View 5.0 提供集中的、自动化的桌面管理，能够通过一台终端实现对数万个虚拟桌面的可扩展管理。VMware View 在使其可用性、可靠性以及安全级别远超出传统 PC 水平的同时，最高能将运营成本降低 50%。

VMware View 能够简化 IT 并自动执行成千上万个桌面的管理，并且以服务的形式从中央位置安全地向用户交付桌面，可实现传统 PC 无法企及的可用性和可靠性级别。通过根据用户需要随时随地为任何设备提供对应用程序和数据的安全访问，VMware View 可为终端用户提供最高级别的流动性和灵活性。

VMware View 由 VMware vSphere Desktop、VMware vCenter Server for Desktops、Manager、VMware ThinApp、VMware View 角色管理、VMware View Composer 组成，如图 2.1.3 所示。

● VMware vSphere Desktop：此版本的 VMware vSphere 专门为桌面而设计，它为运行拟桌面和应用程序提供了一个高度可扩展、高度可靠的强健平台。它具有内置的业务连续性能，能够保护桌面数据和可用性，而且不像传统解决方案那样昂贵和复杂。

● VMware vCenter Server for Desktops：此版本的 VMware vCenter Server 是 VMware vSphere 的集中管理中心，可让用户完全控制和查看虚拟基础构架中的集群、主机、虚拟机存储、网络连接和其他关键元素。

● VMware View Manager：使 IT 管理员能够通过单一控制台集中管理数千个虚拟界面，简化了虚拟桌面的管理、调配和部署。此外，终端用户通过 View Manager 连接安全而轻松地访问 VMware View 虚拟桌面。

● VMware ThinApp：一种无代理的应用虚拟化解决方案，它可以简化应用程序交付，同时避免冲突。作为 VMware View 的一部分，ThinApp 可保持应用程序独立于底层操作系统，从而简化重复性管理任务，减少虚拟桌面的存储需求。

● VMware View 角色管理：角色管理可动态地将用户角色与无状态流动桌面相关联，管理员可轻松部署低成本、无状态流动桌面池，让用户能够在不同会话之间保持其指定设置。

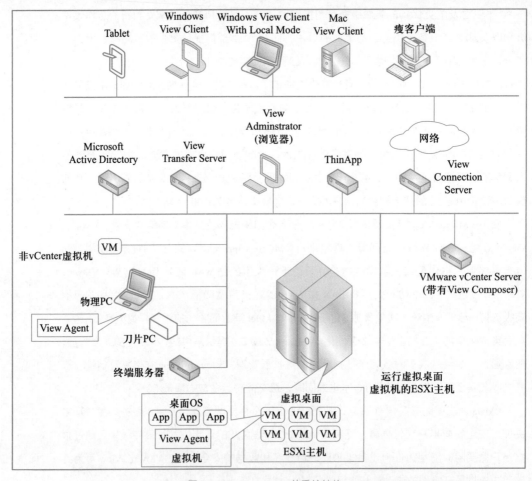

图 2.1.3　VMware View 的系统结构

● VMware View Composer：是通过创建可共享通用虚拟磁盘的黄金主映像，可让用户轻松管理"相似"桌面的池。使用 VMware View Composer，只需通过更新单个主映像就可以对链接到主映像的所有克隆桌面进行修补或更新，而不会影响用户的设置、数据或应用程序。

● VMware View Client：支持从 Windows PC、Mac、瘦客户端、零客户端、iPad 和基于 Android 的客户端访问集中托管的虚拟桌面。使用 View Client With Local Mode，无论网络是否可用，都可以访问在基于 Windows 的本地端点上运行的虚拟桌面。

● VMware View Server：根据部署的性能、可用性和安全性需求的不同，安装时的选项也不同，可以有 4 种角色，分别是标准服务器、副本服务器、安全服务器和传输服务器。各个角色的功能各不相同。

（3）VMware WorkStation

VMware WorkStation 确切地说并不是云计算基础架构的产品，它构建于现有的操作系统（Windows Linux 或是苹果的 OS X）之上，相当于是操作系统上的一个应用软件，但这款应用软件很特殊，它可以帮助开发者和系统管理员进行软件开发、测试以及配置的强大虚拟机，或者说它是一款非常好用的虚拟机管理软件。目前 VMware WorkStation 最新的版本为 15.5.1，内部虚拟化的功能已经非常完善，特别虚拟化网络使得内部的虚拟机和真实网络中物理机一样可以被访问、被远程操作。后面的章节将会介绍使用该产品构建实验环境。

2.2 Citrix 的云计算技术

Citrix 即美国思杰公司，是一家致力于云计算虚拟化、虚拟桌面和远程接入技术领域的高科技企业，致力于帮助企业通过利用虚拟化、网络、协作和云技术来充分适应并利用消费化趋势，从根本上转变企业拓展业务的模式。

Citrix 关于云计算的产品以及解决方案面向不同的用户和不同的应用场景主要有以下 3 个，XenServer、XenDesktop 以及 XenApp。

1. Citrix XenServer

XenServer，是除 VMware vSphere 外的另一种服务器虚拟化平台，其功能强大、丰富，具有卓越的开放性架构、性能、存储集成和总拥有成本。XenServer 基于 Xen 的开源设计，是一种高度可靠、可用而且安全的虚拟化平台。XenServer 利用 64 位架构提供接近本地的应用性能和无与伦比的虚拟机密度。XenServer 可通过一种直观的向导工具（可帮助用户在 10min 内完成 Xen 部署），帮助轻松完成服务器、存储设备和网络设置。磁盘快照和恢复可创建虚拟机和数据的定期快照，在出现故障的情况下轻松恢复到已知的工作状态。磁盘快照还可以克隆，以加快系统置备。XenCenter 还可以管理虚拟服务器、虚拟机（VM）模板、快照、共享存储支持、资源池和 XenMotion 实时迁移。Citrix XenServer 是 Citrix 推出的完整服务器虚拟化平台。XenServer 软件包中包含创建与管理在 Xen（性能接近本机性能的开源半虚拟化虚拟机管理程序）上运行的虚拟 x86 计算机部署所需的全部内容。XenServer 已针对 Windows 和 Linux 虚拟服务器进行了优化。Citrix XenServer 是一种全面的企业级虚拟化平台，用于实现虚拟数据中心的集成、管理和自动化。一整套服务器虚拟化工具可在整个数据中心内实现成本节约，更高的数据中心灵活性和可靠性可为企业提供高性能支持。此外，多种新特性可以有效地管理虚拟网络，将所有虚拟机连接在一起，并为各应用用户分配虚拟机管理接入权限。

电子教案　Citrix 的云计算技术

从管理基础架构到优化长期 IT 运营，从实现关键流程的自动化到交付 IT 服务，XenServer 都能够提供必要的功能来满足任何企业的 IT 要求，帮助企业将数据中心转变为 IT 服务交付中心。

PPT　Citrix 的云计算技术

利用 Citrix XenServer，可以自动完成关键 IT 流程，改进虚拟环境中的服务交付，提高业务连续性，节约时间和成本，同时提供响应更灵敏的 IT 服务。XenServer 的业务连续性包括以下几点。

① 站点恢复。为虚拟环境提供站点间灾难恢复规划和服务。站点恢复易于设置，恢复操作非常快速，而且可以定期测试，以确保灾难恢复计划的有效性。

② 动态工作负载均衡。可在一个资源池中的两台虚拟机之间自动均衡负载，进而提高系统利用率和应用性能。工作负载均衡可对应用要求和可用的硬件资源进行匹配，进而智能地将虚拟机放置到资源池中最适当的主机上。

③ 高可用性。当虚拟机、虚拟机管理系统或服务器发生故障时自动重启虚拟机，这种自动重启功能使用户可以保护所有虚拟化应用，同时为企业带来更高的可用性。

④ 挂机电源管理。利用嵌入式硬件特性，动态地将虚拟机整合到数量更少的系统中，在服务需求波动时关闭未得到充分利用的服务器，进而降低数据中心的功耗。

⑤ 自动 VM 保护和恢复。利用简便易用的设置向导，管理员可以创建快照并对策略进行存档。定期快照可在出现虚拟机故障时帮助防止数据丢失。制定的策略基于快照类型、频率、所保存的历史数据量以及归档位置。只需选择最后一个良好的已知档案就可以删除虚拟机，优化内存。在主机服务器上的虚拟机之间共享未使用的服务器内存，进而降低成本，提高应用性能，并实现更有效的保护。

XenServer 架构与 VMware 完全不同，因为 XenServer 是利用虚拟化感知处理器和操作系统进行开发的。XenServer 的核心是开源 Xen Hypervisor，在基于 Hypervisor 的虚拟化中，有两种实现服务器虚拟化的方法：一种方法是将虚拟机器产生的所有指令都翻译成 CPU 能识别的指令格式，这会给 Hypervisor 带来大量的工作负荷；另一种方法是直接执行大部分子机 CPU 指令，直接在主机物理 CPU 中运行指令，性能负担很小（VMware ESXServer 采用的方法）。XenServer 采用了超虚拟化和硬件辅助虚拟化技术，客户机操作系统清楚地了解它们是基于虚拟硬件运行的。操作系统与虚拟化平台的协作进一步简化了系统管理程序开发，同时改善了性能。Linux 发行版是第一批采用 Xen 进行超虚拟化的操作系统。XenServer 为许多 Linux 发行版提供超虚拟化支持，包括 RedHat Enterprise Linux、NovellSUSE、Debian、Oracle Enterprise Linux 和 CentOS。对于不能完全进行超虚拟化的客户机操作系统（如 Windows），XenServer 将采用硬件辅助虚拟化技术（如 Intel VT 和 AMD-V 处理器）来进行虚拟化。在 Xen 环境中，主要有两个组成部分。一个是虚拟机监控器，又称 Hypervisor。Hypervisor 层在硬件与虚拟机之间，是必须最先载入到硬件的第一层。Hypervisor 载入后，就可以部署虚拟机。在 Xen 中，虚拟机又称 Domain。在这些虚拟机中，其中一个扮演着很重要的角色，就是 Domain0，它具有很高的管理员权限。通常，在任何虚拟机之前安装的操作系统才有这种权限。

Domain0 要负责一些专门的工作。由于 Hypervisor 中不包含任何与硬件对话的驱动，也没有与管理员对话的接口，这些驱动就由 Domain0 来提供。通过 Domain0，管理员可以利用一些 Xen 工具创建其他虚拟机。Xen 术语叫 domainU，这些 domainU 也叫无特权 Domain。这是因为在基于 i386 的 CPU 架构中，它们绝不会享有最高优先级，只有 Domain0 才可以。在 Domain0 中，还会载入一个 xend 进程，这个进程会管理所有其他虚拟机，并提供这些虚拟机控制台的访问。在创建虚拟机时，管理员使用配置程序与 Domain0 直接对话。

XenServer 的设备驱动方式也与 VMware 迥异。采用 XenServer，所有虚拟机与硬件的互操作行为都通过 Domain0 控制域进行管理，而 Domain0 控制域本身就是基于 Hypervisor 运行的、具有特定权限的虚拟机。Domain0 运行的是安全加固型和优化型 Linux 操作系统。对管理员来说，Domain0 是整个 XenServer 系统的一部分，不需要任何安装或管理。因此，XenServer 可采用任意标准的开源 Linux 设备驱动，从而实现对各种硬件的广泛支持。

2. Citrix XenDesktop

Citrix XenDesktop 是一种桌面虚拟化解决方案，类似于 VMware View 的功能，它可以将 Windows 桌面作为一种按需服务随时随地交付给任何用户。XenDesktop 快速而安全地向企业内的所有用户交付单个应用或整个桌面，用户可以通过任何设备灵活地访问他们的桌面，同时获得真正的高清体验。IT 运维部门只需管理操作系统、应用和用户配置文件的单个实例，这大大简化了桌面管理。

利用 XenDesktop 的 FlexCast 交付技术，可以交付所有类型的虚拟桌面，无论是托管桌面还是本地桌面，无论是物理桌面还是虚拟桌面。XenDesktop 支持全部桌面虚拟化技术（例如基于服务器的模型，在该模型中，一台物理服务器上可以托管多达 500 个共享虚拟桌面）以及 VDI（虚拟桌面基础结构，在该结构中，桌面在数据中心内服务器上的虚拟机内运行）。借助 Citrix HDX 技术，XenDesktop 网络和显示优化措施以及性能增强技术可通过任何网络（低带宽和高延迟的 WAN 连接）提供最佳性能。数据中心的 HDX 利用服务器的处理能力和可扩展性来提供高级图形和多媒体性能（无论用户设备的功能如何）。网络上的 HDX 集成了高级优化和加速能力，从而在所有网络（包括高延迟、低带宽环境中的远程桌面访问）上提供绝佳的用户体验，维护数据中心的单一主桌面映像不仅可在用户每次登录时为其提供最新的原始桌面，还可以大量减少修补程序和升级维护工作，最多可节省 90% 的存储成本。

3. Citrix XenApp

Citrix XenApp 是一款按需应用交付解决方案，允许在数据中心对任何 Windows 应用进行虚拟化、集中保存和管理，然后随时随地通过任意设备按需交付给用户。在数据中心集中实现应用管理，将相关成本降低 50%，随时随地向用户快速、安全地交付各种 Windows 应用。无论是采用在线方式还是离线方式，虚拟应用交付均经过特殊优化，Citrix XenApp 虚拟化应用可通过任意设备、任意网络实现高性能、高清用户体验，甚至包括采用大量图形的多媒体内容，保证用户可以获得无缝体验，实现零宕机时间，提高整体工作效率。应用虚拟化将应用程序与操作系统解耦合，为应用程序提供了一个虚拟的运行环境。在这个环境中，不仅包括应用程序的可执行文件，还包括所需要的运行时环境。从本质上说，应用虚拟化是把应用对底层的系统和硬件的依赖抽象出来，可以解决版本不兼容的问题。

在数据中心集中配置、存储、维护单一应用镜像，可使用多用户终端服务器或集中式虚拟机，能够通过任意操作系统实现无缝交付。系统智能和可配置的访问控制功能，能够根据用户所处场景、设备功能、网络性能、连接位置以及安全状况，自动选择交付每个应用的最佳方式。用户能够在统一的企业应用"店面"中预订所需应用，通过 PC 机、Mac 机、笔记本电脑、平板电脑或者智能电话等方便使用的设备快速访问应用。Citrix XenApp 通过高速交付协议交付应用，供在线用户使用，也可通过思杰应用虚拟化解决方案或 MicrosoftApp-V，采用流技术直接将应用交付给设备，方便用户离线使用。

2.3 微软私有云虚拟化产品 Hyper-V

Hyper-V 是微软公司提出的一种系统管理程序虚拟化技术，能够实现桌面虚拟化。Hyper-V 设计的目的是为广泛的用户提供更为熟悉以及成本效益更高的虚拟化基础设施软件，这样可以降低运作成本、提高硬件利用率、优化基础设施并提高服务器的可用性。

Hyper-V 网络虚拟化为虚拟机提供"虚拟网络"（称为 VM 网络），这类似于服务器虚拟化（虚拟机监控程序）为操作系统提供"虚拟机"的方式。网络虚拟化将虚拟网络

电子教案 微软私有云虚拟化产品 Hyper-V

PPT 微软私有云虚拟化产品 Hyper-V

与物理网络基础结构脱耦，并摆脱了 VLAN 以及虚拟机配置的分等级 IP 地址分配的限制。这种灵活性使客户容易移至 IaaS 云中，并且使主机和数据中心管理员更有效地管理其基础结构，同时能保持必要的多项目隔离、安全要求，还能支持重叠的虚拟机 IP 地址。

客户想要将其数据中心无缝地延伸到云中，然而构建这种无缝混合云架构面临一些技术挑战。客户面临的最大障碍之一是在云中重复使用现有网络拓扑（子网、IP 地址、网络服务等），并在本地资源与云资源之间进行桥接。Hyper-V 网络虚拟化提供了一种独立于底层物理网络的 VM 网络的概念。VM 网络由一个或多个虚拟子网组成，在这种概念下，连接到虚拟网络的虚拟机在物理网络中的确切位置与虚拟网络拓扑相互脱耦。因此，客户可轻松地将其虚拟子网移至云中，同时在云中仍保持现有 IP 地址和拓扑，这样现有的服务程序就能继续运作，而不会察觉到子网的物理位置。也就是说，Hyper-V 网络虚拟化可实现无缝混合云的建立。

除了混合云，许多组织正在整合其数据中心，建立私有云，以享受云架构所带来的效率和扩展性。Hyper-V 网络虚拟化通过将业务部门的网络拓扑与实际的物理网络拓扑进行分离，使私有云的灵活性和效率更高。这样的话，业务部门就能轻松共享一个内部私有云，同时又相互独立，并继续保持现有的网络拓扑。数据中心操作小组能灵活部署及实时迁移工作负荷于数据中心的任何地方，而且不会出现服务器中断，能提高操作效率并带来一个整体更加有效的数据中心。

对于工作负荷所有者，主要优势在于，他们现在可以将其工作负荷"拓扑"移到云中，而无须更改其 IP 地址或重写其应用程序。例如，典型的 3 层 LOB 应用程序由前端层、业务逻辑层以及数据库层构成。通过策略，Hyper-V 网络虚拟化能让客户将这 3 层的整体或者部分装载到云中，同时保持服务程序的路由选择拓扑以及 IP 地址（即虚拟机 IP 地址），而不需更改应用程序。

对于基础结构拥有者，虚拟机布置的额外灵活性使其可以将工作负荷移至数据中心的任何地方，而不用改变虚拟机或者重新配置网络。例如，通过 Hyper-V 网络虚拟化，可以实现跨子网实时迁移，这样虚拟机就能在数据中心实时迁移到任何地方，并且不会发生服务中断现象。原本实时迁移只限于在同一子网中，这样就限制了虚拟机可处的位置。跨子网实时迁移可让管理员在动态资源需求以及能源效率的基础上整合工作负荷，也可以在不干扰客户工作负荷正常操作的条件下适应基础结构维护。

2.4 国内私有云相关产品

国内私有云相关产品起步较晚，目前北京易捷思达科技发展有限公司的 EasyStack 与华云数据集团都有相关的商用私有云产品，从整个云计算的产业结构来说，国内私有云产品目前主要定位于中小企业的云平台。面对 VMware，以及 Citrix 几十年的私有云产品的积累，国内云计算产品有着后来居上的趋势，无论从性能上还是稳定性上都有着非常卓越的表现，以下是国内的私有云产品。

1. 易捷行云 EasyStack 的私有云产品线

EasyStack 是中国开源云计算的领导者，由 IBM 中国研发中心 OpenStack 核心研发团队创建,基于 OpenStack、Ceph、Kubernetes、Docker 等一系列开源软件为企业级客户提供开放、稳定可靠、高性能的云计算产品与服务。EasyStack 坚持开源云计算理念,持续参与社区核心功能开发,在多个开源社区代码贡献全球领先,拥有 OpenStack 基金会独立董事 1 名、核心 PTL2 名,以及多个开源社区核心成员。自成立以来,EasyStack 已经为中国移动、中国电信、中国银联、中国邮政储蓄银行、兴业银行、兴业数金、国家电网、清华大学、TCL 等超过 500 家企业级客户提供开源云计算产品和服务,客户覆盖金融、电信、能源、教育等多个关键行业。EasyStack 坚持不懈地推动生态发展,已与百余家企业达成合作伙伴关系。

EasyStack 的主要产品与方案如下。

(1) ECS 易捷行云企业云

相较传统私有云的碎片化、定制化、不可进化,新一代私有云是以企业客户防火墙内的复杂环境和数据需求为设计初衷,构建以客户为中心的、具备多云管理能力的私有云。它拥有更为广泛的硬件和软件的生态兼容性,兼顾企业级新一代应用和传统应用,同时具备应对企业复杂环境下的可进化特性,还能提供公有云似的消费级体验。易捷行云 ECS 正是按照新一代私有云理念和特性交付的私有云。新旧私有云的对比,如图 2.4.1 所示。

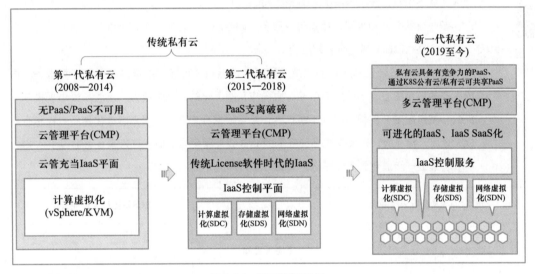

图 2.4.1　新旧私有云对比

新一代私有云 ECS 易捷行云企业云的核心特性是平滑无感可进化,其三要素包括业务无感知、数据不迁移、服务不中断,可进化的 3 个维度包括产品形态可进化、云服务能力可进化和支持场景可进化,如图 2.4.2 所示。

新一代私有云 ECS 产品优势如下。

① 可持续进化能力。具有服务引入、架构迭代、功能升级的特征,能为客户提供融合、分离、边缘部署场景下的新一代云解决方案。

② 分布式微服务架构。全新设计的微服务架构控制平面,实现全组件微服务化,使产品具备可持续升级进化能力。

③ 全栈一体化设计。计算、存储、网络、监控、运维一体化设计、交付与升级,云服务间充分协同,为用户提供一体化全栈云体验。

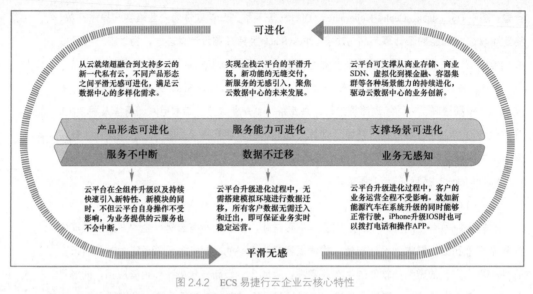

图 2.4.2　ECS 易捷行云企业云核心特性

④ 平滑无感升级。云平台控制平面组件可实现微服务粒度的平滑无感升级，升级过程保障业务不中断，云平台服务平滑无感知。

新一代私有云 ECS 可以提供计算服务、网络服务、存储服务、运维与升级服务等。

（2）ECS Stack 易捷行云超融合

ECS Stack 超融合具备自助式流水线服务、感知式可视化编排、双引擎计算服务、面向生态的应用中心、高级服务引擎等面向应用服务化的云就绪能力，还能从最小规模 3 节点的云就绪超融合部署为起点不断扩容，并可根据客户需求平滑无感进化到新一代私有云 ECS 企业云标准版，以及支撑多种场景的 ECS 企业云场景化版，在未来还可通过引入 EMS 多云管理进一步进化到多云形态，如图 2.4.3 所示。

拓展阅读　智网未来私有云产品

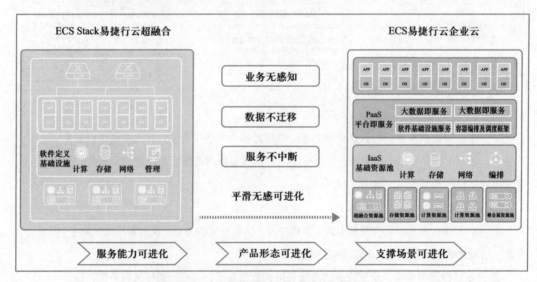

图 2.4.3　云就绪超融合向私有云的进化过程

ECS Stack 超融合采用业界首创的全对称分布式微服务架构，一方面实现了超融合平台的高可靠架构，另一方面在新功能、新组件、新服务持续快速的引入，以及从云就绪超融合向私有云的进化过程中,实现了平滑无感的消费级体验,即业务无感知、数据不迁移、服务不中断,是真正以用户和应用为核心的、高可靠、易部署、轻运维的新一代云就绪超融合。

(3) EKS 易捷行云容器云

EasyStack Kubernetes Service(EKS)易捷行云容器云通过深度融合 Kubernetes 与 OpenStack 两个平台,从编排、调度、安全、运维等多个管理维度实现应用与基础设施资源的融合,打通云数据中心新一代应用交付的"最后一公里",发挥"1+1>2"的效能优势。EKS 产品以面向应用角度进行设计,提供应用管理服务,方便用户在平台中集中管理发布容器化应用。

拓展阅读 智网未来私有云典型部署架构及特点

其产品主要特征如下:

① 统一权限。EKS 可使用 OpenStack Keystone 组件提供统一的用户管理服务,打通两个平台的权限控制体系。

② VM、容器统一 SDN 网络。使用 Neutron 组件为 EKS 和 ECS 平台提供统一的网络服务,实现 SDN 网络统一管理,并可以将 Neutron 的高级特性引入容器网络中,如安全组、LBaaS 等。

③ 多用户隔离。EKS 支持每个项目单独创建并管理 Kubernetes 集群,借助 OpenStack 多项目隔离特性,增强容器的安全隔离性。

④ VM、容器统一 SDS 存储。EKS 可使用 OpenStack Cinder 组件作为 Kubernetes 的持久化存储驱动,后端可对接 OpenStack 中已广泛支持的开源存储平台或主流商业存储设备,充分支持有状态的容器服务。

⑤ 二层网络性能提升。EKS 直接使用 Neutron 二层网络作为网络驱动,避免嵌套 Overlay 网络带来的性能损失,并且可以实现容器和虚拟机之间的二层网络直连。

2. H3C 公司私有云产品

H3C CAS 云计算管理平台是 H3C 公司推出的构建云计算基础架构的管理软件,它为数据中心云计算基础架构提供最优化的虚拟化和云业务运营管理解决方案,用于实现数据中心云计算环境的中央管理控制。通过简洁的管理界面,轻松地统一管理数据中心内所有的物理资源和虚拟资源,不仅能提高管理员的管控能力、简化日常例行工作,更可降低 IT 环境的复杂度和运维成本。

H3C CAS 云计算管理平台由以下 3 个组件构成。

① 虚拟化内核平台(Cloud Virtualization Kernel,CVK),运行在基础设施层和上层操作系统之间的"元"操作系统,用于协调上层操作系统对底层硬件资源的访问,减轻软件对硬件设备以及驱动的依赖性,同时对虚拟化远行环境中的硬件兼容性、高可靠性、高可用性、可扩展性、性能优化等问题进行加固处理。

② 虚拟化管理系统(Cloud Virtualization Manager,CVM),主要实现对数据中心内的计算、网络和存储等硬件资源的软件虚拟化,形成虚拟资源池,对下层应用提供自动化服务。其业务范围包括虚拟计算、虚拟网络、虚拟存储、高可靠性、动态资源调度、虚拟机容灾与备份、虚拟机模板管理、集群文件系统、虚拟交换机策略等。

③ 云业务管理中心(Cloud Intelligence Centre,CIC),由一系列云基础业务模块组成,

通过将基础架构资源（包括计算、存储和网络）及其相关策略整合成虚拟数据中心资源池，并允许用户按需消费这些资源，从而构建安全的多项目混合云。其业务范围包括组织（虚拟数据中心）、多项目数据和业务安全、云业务工作流、自助式服务门户、兼容 OpenStack 的 REST API 接口等。

3. 华云数据私有云产品

华云数据集团成立于 2010 年，以推动中国行业数字化转型为己任，专注于为企业级用户提供应用创新的云计算服务，以帮助用户采用云计算提升 IT 能力，实现业务变革。华云数据主要向用户提供定制化私有云、混合云解决方案，同时还可以提供大数据服务、超融合产品、公有云和 IDC 转云等服务。目前，华云数据凭借定制化私有云、混合云服务，在能源电力、国防军工、教育医疗、交通运输、政府金融等十几个行业打造了行业标杆案例，目前客户总量超过 30 万。

华云数据的主要产品与方案如下。

（1）企业级云平台

华云数据为客户提供一站式全方位混合云 IT 管理解决方案，是集公有云、私有云、灾备、大规模集群、秒级扩容、面向应用优化、高性能虚拟化、多云管理、运维支撑、统计分析、大屏展示、自服务于一体的新一代企业级云平台。

使用场景包括负载弹性扩展、灾难恢复、数据备份、多云管理等。

（2）企业级大数据

助力企业实现大数据业务从 0 到 1，在大数据平台构建、大数据分析和 AI 方面提供持续的技术支撑。

使用场景包括大数据服务部署、大数据服务管理和持续大数据服务技术支持。

（3）企业级超融合

华云企业级超融合产品预安装了计算虚拟化软件、存储虚拟化软件以及云管理平台等，融合计算、存储、网络、安全于一体，硬件采用 Intel 的 Purley 平台，全新的性能表现，秒级创建云计算环境，轻松构建开箱即云的数据中心。

使用场景包括小型私有云、数据保护与容灾、混合云环境搭建、远程分支办公室 IT 基础架构等。

此外，华云数据也拥有高品质公有云、云操作系统等其他产品。

2.5 知名公有云平台简介

公有云目前国内市场已经开始全面普及，用户对象从大型的国际级企业到中小型企业乃至个人用户。通常用户可以在共有云平台上注册账户，根据自身的需要定制计算资源，如自行选择 CPU 核心的数量、内存的大小、硬盘空间以及所用的操作系统等，所定制的资源规格越高费用也相应越高。公有云通常会自动为用户提供可靠性的保护机制，包括定期的数据备份、操作系统的快照等，主要公有云平台分为外资在境内合资运营的公有云平台以及国内企业独立运营的共有云平台。

1. 外资联合国内合资运营的公有云

Windows Azure 是微软公司基于云计算的操作系统，现在更名为 Microsoft Azure，与 Azure Services Platform 一样，是微软"软件和服务"技术的名称。Windows Azure 的主要目标是为开发者提供一个平台，帮助开发可运行在云服务器、数据中心、Web 和 PC 上的应用程序。云计算的开发者能使用微软全球数据中心的存储、计算能力和网络基础服务。Azure 服务平台包括了以下主要组件 Windows Azure、Microsoft SQL 数据库服务、Microsoft .Net 服务、用于分享、存储和同步文件的 Live 服务、针对商业的 Microsoft SharePoint 和 Microsoft Dynamics CRM 服务。

2012 年 11 月 1 日，微软公司宣布与国内互联网基础设施服务提供商世纪互联达成合作，微软公司向世纪互联授权技术，由世纪互联在中国运营 Windows Azure。

Azure 是一种灵活和支持互操作的平台，它可以被用来创建云中运行的应用或者通过基于云的特性来加强现有应用。它开放式的架构给开发者提供了 Web 应用、互联设备的应用、个人计算机、服务器或者提供最优在线复杂解决方案的选择。Windows Azure 以云技术为核心，提供了"软件+服务"的计算方法，它是 Azure 服务平台的基础。Azure 能够将处于云端的开发者个人能力，同微软全球数据中心网络托管的服务，如存储、计算和网络基础设施服务，紧密结合起来。

微软公司会保证 Azure 服务平台自始至终的开放性和互操作性。人们确信企业的经营模式和用户从 Web 获取信息的体验将会因此改变。最重要的是，这些技术将使用户有能力决定，是将应用程序部署在以云计算为基础的互联网服务上，还是将其部署在客户端，或者根据实际需要将二者结合起来。

2013 年 5 月 22 日，微软公司宣布 Windows Azure 公有云服务正式落地中国，成为第一个在中国落地的国际化公有云服务平台。2014 年 3 月 26 日，Windows Azure 在中国正式商用。Windows Azure 提供云基础设施即服务（IaaS）和平台即服务（PaaS）两个层面的服务。在 IaaS 层面 Windows Azure 提供兼容主流操作系统（包括 Windows 和主流 Linux）的虚拟机；提供负载均衡服务，还可帮助用户按照自身要求构建虚拟网络拓扑；提供站到站及多站的 VPN 服务，打造可无缝迁移的混合云。在 PaaS 层面，Windows Azure 除了提供对多种开发语言和框架的支持外，还提供多种增值服务，包括媒体、身份识别、移动，Web 站点等。目前，Windows Azure 的数据中心建立在北京、上海两个城市，集成了优质的 BGP 网络，直接连接骨干网，彼此之间的数据可以进行互相备份，同时微软与蓝汛合作的 CDN 覆盖全国 90 多个城市，拥有 700 多个节点的物理服务器组。

国内用户可以通过其官网访问注册并定制 Windows Azure 的公有云服务，如图 2.5.1 所示。

Windows Azure 同样具备以下众多优良的特性。

① 安全性。安全性是云行业的先决条件，而 Azure 实现安全性、符合性以及隐私性的前瞻式和主动式方法是独具特色的。Azure 使用高度安全的云基础降低成本和复杂性。

② 全球性。借助多个区域的数据中心，Azure 为许多企业和组织同时提供全球和本地服务，使它们在满足本地数据驻留需求的同时，减少运营全球性基础结构的成本、时间和复杂性。

图 2.5.1 Windows Azure 国内公有云平台首页

③ Azure 可以使用任何开源 OS、语言和工具。Azure 在 2017 年对 GitHub 做出了最多的贡献，它是唯一对 Red Hat 提供集成支持的云。

④ 高可用性。Windows Azure 就单点故障问题做了平台设计优化，并在北京和上海的数据中心中实现了本地加异地的多点备份，从而确保了 Windows Azure 在服务等级协议（SLA）中承诺的业务和数据的高可用性。

⑤ 开放平台和优良的用户体验。Windows Azure 可以同时提供 Windows 和 Linux 虚拟机，特别对自家的 Windows 操作系统做了很多的优化，这点是其他公有云平台无法比拟的，虚拟机上支持 Java、.NET、PHP、Python 等语言。此外，Windows Azure 对 Hadoop、Wordpress、Dropbox 等开源框架都提供了良好的兼容性，并为安卓、iOS 及 Windows Phone 提供了 SDK。Windows Azure 的 PaaS 服务可以和底层的基础设施服务良好的结合，为用户提供从开发工具到开发环境的整合式服务，让用户专心开发和运行应用，而不用担心基础设施。

⑥ 混合云支持。通过 Windows Azure 的 VPN 服务，可以把用户的私有云和公有云通过同一个虚拟网络连接，使得计算任务能够无缝切换。目前全球 500 强中 57% 的企业采用了 Windows Azure 服务，在中国的企业用户数已经过万，主要客户包括有河北省廊坊市高新技术产业园、武汉经济开发区、CNTV、PPTV 等。

2. 国内独立运营的公有云

（1）百度云

百度云是百度公司旗下面向企业、开发者和政府机构的智能云计算服务商，是百度提供的公有云平台，于 2015 年正式开放运营。百度云不断将百度在云计算、大数据、人工智能的技术能力向社会输出，致力于为各行业提供以 ABC（人工智能、大数据、云计算）技术为一体的平台服务。百度智能云已发布超过 260 款产品和近 40 个解决方案，助力各行各业实现智能化转型。

在通过技术创新不断满足用户的移动搜索需求的同时，百度也在继续积极推动移动云生态系统的建设和发展，与产业实现共赢。2012 年 9 月，百度面向开发者全面开放包括云存储、大数据智能和云计算在内的核心云能力，为开发者提供更强大的技术运营支持与推广变现保障。2015 年，百度进一步开放其核心基础架构技术，为广大公有云需求者提供全系列可靠易用的高性能云计算产品。百度云将通过不断推出贴合生态需要的解决方案，为百度用户打造更为全面优质的生态服务，助力百度生态用户的业务价值最大化。

百度云主要产品如下。

① 计算与网络类产品。

• 云服务器 BCC（Baidu Cloud Compute）：支持弹性伸缩，镜像，快照，支持分钟级丰富灵活的计费模式，提供高性能云服务器服务。

• 负载均衡 BLB（Baidu Load Balance）：均衡应用流量，实现故障自动切换，消除故障节点，提高业务可用性。

• 专属服务器 DCC（Dedicated Cloud Compute），提供性能可控、资源独享、物理资源隔离的专属云计算服务，在满足超高性能及独占资源需求的同时，还可以与其他云产品自由互联，高效易用。

• 专线 ET（Express Tunnel）：是一种高性能、安全性极好的网络传输服务。专线服务避免了用户核心数据走公网线路带来的抖动、延时、丢包等网络质量问题，大大提升用户业务的性能与安全性。

• 应用引擎 BAE（Baidu App Engine）：提供弹性、便捷的应用部署服务，适于部署 App、公众号后台，以及电商、O2O、企业门户、博客、论坛、游戏等各种应用，极大简化运维工作。

② 存储和 CDN 产品。

• 对象存储 BOS（Baidu Object Storage）：提供稳定、安全、高效、高可扩展的云存储服务，支持最大 5TB 多媒体、文本、二进制等任意类型数据的存储。

• 云磁盘 CDS（Cloud Disk Service）：提供安全可靠、具备极高性能的块存储服务，为云服务器（BCC）提供高可用和高容量的数据存储服务。

• 内容分发网络 CDN（Content Delivery Network）：将网站内容发布到最接近用户的边缘节点，使用户可就近取得所需内容，提高用户访问的响应速度和成功率，同时能够保护源站。解决由于地域、带宽、运营商接入等问题带来的访问延迟问题，有效帮助站点提升访问速度。

③ 安全和管理类产品。

• 云安全服务 BSS（Baidu Security Service）：提供 DDoS 防护、云服务器防护、Web 漏洞检测等全方位的安全防护服务，实时发现用户的资源及业务系统的安全问题，保障用户的业务系统稳定运行。

- 云监控 BCM（Baidu Cloud Monitor）：提供 7×24 小时的实时监控服务，为用户的系统保驾护航。
- SSL 证书服务（SSL Certification Service）：百度云与全球知名的第三方数字证书认证和服务机构联合推出的 SSL 证书申请与管理一站式服务。无须繁杂流程，一键申请，轻松实现网站与 Web 应用的 HTTPS 加密部署。

此外还提供了数据库、数据分析、智能多媒体服务等各个类型的产品。

国内用户可以通过其官网访问注册并定制百度云的公有云服务，如图 2.5.2 所示。

图 2.5.2 百度云首页

（2）腾讯云

腾讯云有着深厚的基础架构，并且有着多年对海量互联网服务的经验，不管是社交、游戏还是其他领域，都有成熟产品来提供产品服务。腾讯在云端完成重要部署，为开发者及企业提供云服务、云数据、云运营等整体一站式服务方案。

具体包括云服务器、云存储、云数据库和弹性 Web 引擎等基础云服务；腾讯云分析（MTA）、腾讯云推送（信鸽）等腾讯整体大数据能力；以及 QQ 互联、QQ 空间、微云、微社区等云端链接社交体系。这些正是腾讯云可以提供给这个行业的差异化优势，造就了可支持各种互联网使用场景的高品质的腾讯云技术平台。

腾讯云包括云服务器、云数据库、CDN、云安全、万象图片和云点播等产品。

用户通过接入腾讯云平台，可降低初期创业的成本，能更轻松地应对来自服务器、存储以及带宽的压力。

国内用户可以通过其官网访问注册并定制腾讯云的公有云服务，如图 2.5.3 所示。

（3）阿里云

目前，阿里云在中国主要提供 IaaS 和 PaaS 两类服务。阿里云提供的 IaaS 服务包括云服务器（ECS）、负载均衡（SLB）和开放存储服务（OSS）。云服务器是一种处理能力可弹性伸缩的计算服务，其管理方式比物理服务器更简单高效；负载均衡是对多台云服务器

图 2.5.3　腾讯云首页

进行流量分发的负载均衡服务；开放存储服务是阿里云对外提供的海量、安全和高可靠的云存储服务。阿里云主要提供的 PaaS 服务是云引擎 ACE。云引擎是一款弹性、分布式的应用托管环境，支持 Java、PHP 多种语言。阿里云的国内数据中心建立在杭州、青岛、北京和中国香港 4 个城市。

　　国内用户可以通过其官网访问注册并定制阿里云的公有云服务，如图 2.5.4 所示。

图 2.5.4　阿里云首页

　　阿里云具备不少优良的特性。

　　首先阿里云前期积累丰厚。它是国内较早起步的云计算平台，在国内公有云市场参与者较少的时候就凭借阿里巴巴在中小站长和电子商务站长中的知名度和影响力，迅速聚集了一批客户。

其次阿里云针对不同行业定制不同的云应用策略。阿里云面向移动、游戏、金融等领域，根据各领域的特点，量身定制云计算服务，帮助各领域的企业创新业务、提升竞争力。目前阿里云的主要客户包括天弘基金、浙商证券、众安保险、特步官方商城等。

（4）华为云

华为云成立于2011年，隶属于华为公司，在北京、深圳、南京等多地设立有研发和运营机构，贯彻华为公司"云、管、端"的战略方针，汇集海内外优秀技术人才，专注于云计算中公有云领域的技术研究与生态拓展，致力于为用户提供一站式云计算基础设施服务，目标成为中国最大的公有云服务与解决方案供应商。

华为云主要产品如下。

① 弹性计算云（ECC—Elastic Computing Cloud）是整合了计算、存储与网络资源，按需使用、按需付费的一站式IT计算资源租用服务，以帮助开发者和IT管理员在不需要一次性投资的情况下，快速部署和管理大规模可扩展的IT基础设施资源。

② 对象存储服务（Object Storage Services）是一个基于对象的云存储服务，为客户提供海量、安全、高可靠、低成本的数据存储能力。客户可以通过REST接口或者基于Web浏览器的云管理平台界面对数据进行管理和使用。同时，提供了多种语言（如Java、PHP、C、Python）的SDK来简化编程。

华为对象存储服务可以为多种应用构建大规模的数据存储服务，如互联网海量内容（视频、图片、照片、图书、音像、杂志等）、网盘、数字媒体、备份、归档、BigData等服务。

③ 桌面云是采用最新的云计算技术开发出的一款智能终端产品，外表看起来是一个小盒子，但却可以代替普通计算机使用；同时用户也可以用PC和移动Pad等多种方式接入桌面云。华为桌面云改变了传统的PC办公模式，突破时间、地点、终端、应用的限制，随时随地办公介入，成就自由的现代办公时代，让客户更专注于核心业务的发展。

④ 云托管(CH—Cloud Hosting)是以应用为中心的公有云托管，以应用为单位整合计算、存储与网络资源，按需使用、按需付费的一站式IT计算资源租用服务，能够帮助企业在不需要一次性投资的情况下，快速部署和管理大规模可扩展的IT基础设施资源。云托管主要面向用户部署复杂应用，需要使用多台机器一起组网才能支撑一个应用的场景。

习题　云计算知名厂商及其产品

第 3 章 原生 OpenStack 云平台

本章导读：

本章介绍云计算产业界目前最大的开源技术项目 OpenStack 及其相关的功能组件，并在此基础上介绍 OpenStack 的版本更新历程及其主要差异。在后续章节的实训环节中，将介绍原生 OpenStack 云平台的构建及其管理和运维。

电子资源：

  电子教案 原生 Open Stack 云平台
  PPT 原生 Open Stack 云平台
  习题 原生 Open Stack 云平台

3.1 OpenStack 技术简介

OpenStack 起源于开源的亚马逊云平台,由 NASA(美国国家航空航天局)和 Rackspace 合作研发并发起的,以 Apache 许可证授权的自由软件和开放源代码项目。它既是一个开源社区,也是一个项目和一个开源软件(LOGO 如图 3.1.1 所示),它提供了一个部署云的操作平台或工具集,或者说整个 OpenStack 技术是一个开放的体系。其宗旨在于,使用开源的架构实现虚拟化计算或存储服务的云,为公有云、私有云,也为大云、小云提供可扩展的、灵活的云计算。据相关统计,当前市场上 70% 以上的公有云平台,以及私有云相关产品都是基于 OpenStack 这一开放体系构建的。

图 3.1.1 OpenStack 项目的 LOGO

OpenStack 是一个开源的云计算管理平台项目,由几个主要的组件组合起来完成具体工作。OpenStack 支持几乎所有类型的云环境,项目目标是提供实施简单、可大规模扩展、丰富、标准统一的云计算管理平台。OpenStack 通过各种互补的服务提供了基础设施即服务(IaaS)的解决方案,每个服务提供 API 以进行集成。

OpenStack 项目以 Python 语言编写,整合 Tornado 网页服务器、Nebula 运算平台,使用 Twisted 软件框架,并遵循 Open Virtualization Format、AMQP、SQLAlchemy 等标准,支持的虚拟化技术,包括 KVM、Xen、Vmware、LXC 等,并支持 VirtualBox、QEMU 等虚拟机管理软件。

然而 OpenStack 中所涉及的内容远非单纯的 Python 代码,想要真正理解并熟练部署维护 OpenStack 云计算平台,必须有大量的外围知识。OpenStack 社区十分强大,自成立以来,各门各派武林高手竞相亮招,不断地贡献各种各样的组件、模块。OpenStack 包含以下核心组件。

- KeyStone(身份认证):为 OpenStack 其他服务提供身份验证、服务规则和服务令牌的功能,管理 Domains、Projects、Users、Groups、Roles。自 Essex 版本集成到项目中。
- Nova(计算服务):一套控制器,用于为单个用户或使用群组管理虚拟机实例的整个生命周期,根据用户需求来提供虚拟服务。它负责虚拟机创建、开机、关机、挂起、暂停、调整、迁移、重启、销毁等操作,配置 CPU、内存等信息规格。自 Austin 版本集成到项目中。
- Glance(镜像支持):一套虚拟机镜像查找及检索系统,支持多种虚拟机镜像格式(如 AKI、AMI、ARI、ISO、QCOW2、Raw、VDI、VHD、VMDK),有创建上传镜像、删除镜像、编辑镜像基本信息的功能。自 Bexar 版本集成到项目中。
- Neutron(网络支持):提供云计算的网络虚拟化技术,为 OpenStack 其他服务提供网络连接服务。它为用户提供接口,可以定义 Network、Subnet、Router,配置 DHCP、DNS、负载均衡、L3 服务,网络支持 GRE、VLAN。插件架构支持许多主流的网络厂家和技术,如 OpenvSwitch,可包括 Linux 内核虚拟网络支持,如 Linuxbridge。自 Folsom 版本集成到项目中。

• Swift（对象存储）：利用一致性哈希算法构建了一个冗余的可扩展的分布式对象存储集群，包含 Proxy Server、Account Server、Container Server、Object Server 等组件，保证数据的可靠性以及完整性。

• Horizon（UI 界面支持）：OpenStack 中各种服务的 Web 管理门户，用于简化用户对服务的操作，如启动实例、分配 IP 地址、配置访问控制等。自 Essex 版本集成到项目中。

OpenStack 也包含了如 Placement（调度）、Ironic（裸金属服务）、Ceilometer（云测量）、Heat（编排）、Trove（数据库项目）等扩展组件，在社区中也能看到诸多第三方组件。每个组件都有各自的任务，看似独立负责自己的领域，但又互相衔接，相互协作，通过提供暴露的 API 接口，共同实现云计算服务。

OpenStack 作为一个开源项目，没有任何一家单独的公司在控制 OpenStack 的发展路线。OpenStack 本身具有巨大的市场动力，与此同时，很多大公司都在支持 OpenStack 发展。有了如此众多公司的资源投入，OpenStack 的发展是多元化的，它有全球大量的组织支持，大量的开发人员参与，发展迅速。国际上已经有很多使用 OpenStack 搭建的公有云、私有云、混合云，如 RackspaceCloud、惠普云、MercadoLibre 的 IT 基础设施云、AT&T 的 CloudArchitec、戴尔的 OpenStack 解决方案等。在国内，OpenStack 的热度也在逐渐升温，华胜天成、高德地图、京东、百度、中兴、华为等都对 OpenStack 产生了浓厚的兴趣并参与其中。自 2010 年创立以来，已发布 20 个版本，其中 Stein 版本有 139 个组织、1159 名代码贡献者参与，而最新的是 Train 版本。OpenStack 很可能在未来的基础设施即服务（IaaS）资源管理方面占据领导位置，成为公有云、私有云及混合云管理的"云操作系统"标准。

OpenStack 的蓬勃发展是多方努力的共同结果，包括 Canonical/Ubuntu、SUSE、VMware 和 Red Hat 等多家供应商，提供商业支持的 OpenStack 产品。此外，还有多个由 OpenStack 提供支持的云服务，包括 Oracle、Rackspace、Telefonica、OVH、vScaler 和 City Network。

OpenStack 发展之初，只有 Swift 存储和 Nova 计算两个项目。后续内容中最新版本 OpenStack Placement 服务是最初属于 Nova 的一部分，但现在已经被分离到它自己的项目中。根据发布说明，Placement 服务的目标是跟踪云资源清单和用法，以帮助其他服务有效地管理和分配其资源。

本书选用稳定的 Stein 版本，该版本中有一些关键亮点可用于帮助企业在灵活、可扩展的私有云上提供新的差异化应用程序和服务。随着组织希望从日益数字化的经济中提取更多的利益，Stein 版本增加了专注于支持新工作负载和用例的能力。

新兴的边缘计算概念也得到了 OpenStack Stein 版本的推动。借助边缘计算，计算可以扩展到网络的边缘，而不是位于中央核心的所有计算资源。

OpenStack 各版本主要更新功能见表 3.1.1。

表 3.1.1　OpenStack 各版本主要更新功能

版本	发布时间	主要更新功能
Train	2019/10/16	允许平台的 Nova 和 Cyborg 模块之间进行智能交互。Train 还包括硬件加速器的新的和改进的驱动程序以及大幅改进的服务来跟踪云资源的库存和使用情况

续表

版本	发布时间	主要更新功能
Stein	2019/4/10	加强容器功能，为 5G、边缘计算和 NFV 用例提供网络增强功能、改进了资源管理和跟踪功能
Rocky	2018/8/30	受人工智能、机器学习、NFV 和边缘计算等用例的驱动，提供增强的升级功能，并支持各种硬件架构（包括裸机）
Queens	2018/2/28	引入 Cinder 中的 Multi-Attach 功能，对虚拟 GPU（vGPU）支持和容器集成的改进
Pike	2017/8/30	Nova Cells v2 使得运营商能够分片部署，开始整合 Etcd，Ironic 增加了插入 Neutron 网络以实现多项目的能力等
Ocata	2017/2/22	Nova 调度策略使用 Placement API，并基于 CPU/RAM/Disk 的能力去选择计算节点，默认使用 Cell v2 部署，Ironic 裸机服务迎来网络与驱动程序增强，Cinder 块存储服务中的主动/主动高可用性如今可通过驱动程序实现，在网络层对基于容器的应用框架提供更为出色的支持能力
Newton	2016/10/6	Ironic 裸机开通服务，Magnum 容器编排集群管理器，此外，Kuryr 容器组网项目可将容器、虚拟和物理基础设施无缝集成为统一控制面板，解决扩展性和弹性问题
Mitaka	2016/4/7	统一的 API、更好的用户体验、更具可管理性，以及通过 Heat 可横向扩展能力为大型部署提供更高的性能与稳定性

3.2 体验原生 OpenStack 云平台

在本书后续章节中，将介绍原生 OpenStack 云平台的部署与维护，并基于此学习 OpenStack 中的各个组件，本节先体验一下该云平台，在云平台中启动一台虚拟机然后部署一个远程桌面服务，然后在平台上创建一个云硬盘并让启动的虚拟机使用到该云硬盘。

步骤 1：用浏览器登录云平台并创建一个 Windows Server 2012 虚拟机。

打开浏览器登录创建好的 OpenStack 云平台的 Web 界面，如图 3.2.1 所示。

图 3.2.1　云平台 Web 界面

用 admin 用户登录，密码为"000000"，进入云平台后，单击项目栏"计算"子栏中的"实例"项，界面如图 3.2.2 所示。

图 3.2.2　实例界面

单击界面中的"启动云主机"按钮,进入创建实例的子界面,创建一个名为 winserver2012 的实例,同时自定义一个云主机类型"win",具体参数参看右侧的"方案详情",选用创建的 winserver2012 的镜像,如图 3.2.3 和图 3.2.4 所示。

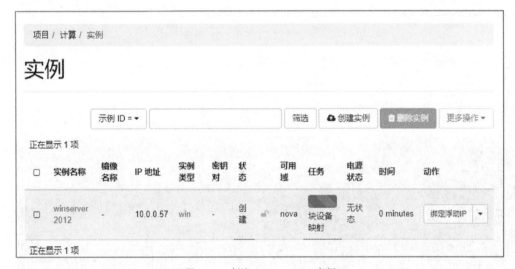

图 3.2.3　创建 winserver2012 实例

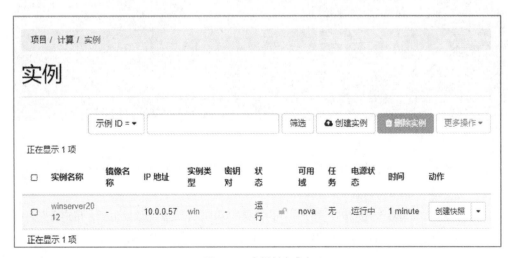

图 3.2.4　实例创建成功

单击右侧"更多"按钮▼,在弹出的下拉列表中选择"控制台"选项可以进入实例的控制台进行操作，如图 3.2.5 所示。

	实例名称	镜像名称	IP 地址	实例类型	密钥对	状态	可用域	任务	电源状态	时间	动作
☐	winserver2012	-	10.0.0.57	win	-	运行	🔓	nova	无	运行中	1 minute	创建快照 ▼

正在显示 1 项

动作菜单：
- 绑定浮动IP
- 连接接口
- 分离接口
- 编辑实例
- 连接卷
- 分离卷
- 更新元数据
- 编辑安全组
- 编辑端口安全组
- 控制台
- 查看日志

图 3.2.5　进入 winserver2012 实例控制台

选择"控制台"选项后，即可查看控制台，如图 3.2.6 所示。

图 3.2.6　winserver2012 控制台界面

单击界面上的"点击此处只显示控制台"超链接，即可在界面上操作此系统。

步骤 2：远程桌面服务的部署。

进入 Windows Server2012 系统后，为方便外部的主机连接到该虚拟机，用户需要关闭系统防火墙，调整 Internet 安全级别，允许下载，同时打开系统远程桌面。

首先关闭系统防火墙，如图 3.2.7 所示。

接着，需要更改 Internet 的安全级别，使系统允许下载文件。单击"自定义级别"按钮，在打开的对话框选中"下载"选项中"文件下载"的"启用"单选框，如图 3.2.8 和图 3.2.9 所示。

图 3.2.7 关闭系统防火墙

图 3.2.8 Internet 安全级别

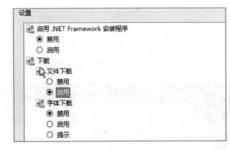

图 3.2.9 启用文件下载

  然后，打开系统的远程桌面功能，在"系统属性"对话框"远程"选项卡中选中"允许远程连接这台计算机"复选项，如图 3.2.10 所示。

  打开本地的远程桌面连接。在 Windows "运行"框中输入 mstsc 命令，虚拟机绑定的外网 IP 为 192.168.200.139，如图 3.2.11 所示。

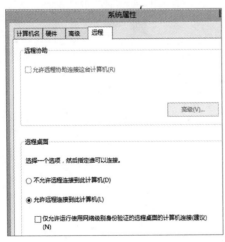

图 3.2.10 打开系统远程桌面

图 3.2.11 本地 Windows 的远程桌面连接

单击"连接"按钮,出现 Windows 安全认证界面,如图 3.2.12 所示。

输入创建的 winserver2012 实例的用户名"Administrator"和密码"000000",单击"确定"按钮,会出现证书错误提示,单击"是"按钮即可,如图 3.2.13 所示。

 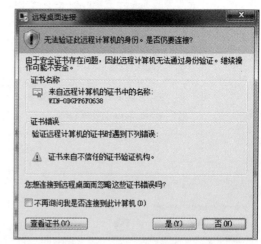

图 3.2.12　Windows 安全认证界面　　　　　图 3.2.13　证书错误

此时,可以看到成功连接到 OpenStack 云平台上的 winserver2012 虚拟机中,如图 3.2.14 所示。

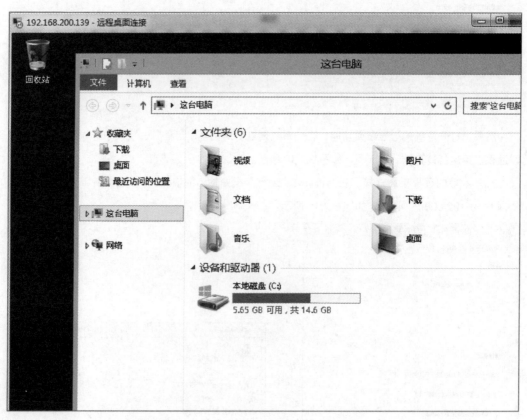

图 3.2.14　本地成功连接云平台中的虚拟机

步骤 3:创建云硬盘并在 Windows 虚拟机中使用。

首先在 OpenStack 云平台上创建一块名为"win"、大小 10GB 的空白云硬盘，创建详情如图 3.2.15 所示。

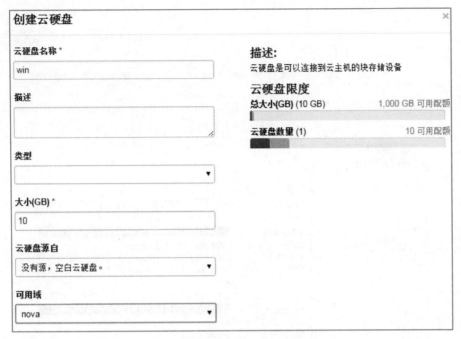

图 3.2.15　创建云硬盘界面

接着需要将云硬盘连接到 winserver 2012 实例中，单击"连接卷"按钮即可，如图 3.2.16 所示。

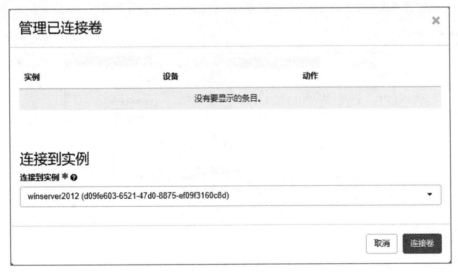

图 3.2.16　连接 winserver2012 实例

此时可以在界面看到连接成功的提示，如图 3.2.17 所示。

之后，进入 winserver2012 控制台，打开磁盘管理，此时可以看到有一块处于脱机状态的 10GB 未知磁盘，如图 3.2.18 所示。

图 3.2.17　界面显示连接实例成功

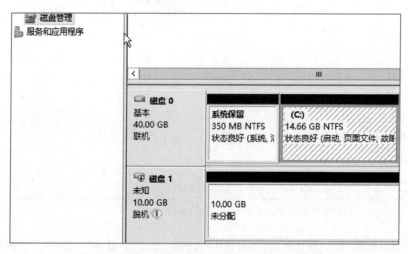

图 3.2.18　在 winserver 2012 中可以看到挂载的云硬盘

此时使用这块磁盘还需要对其进行联机、初始化和创建简单卷操作,如图 3.2.19~ 图 3.2.21 所示。

图 3.2.19　磁盘联机并初始化操作

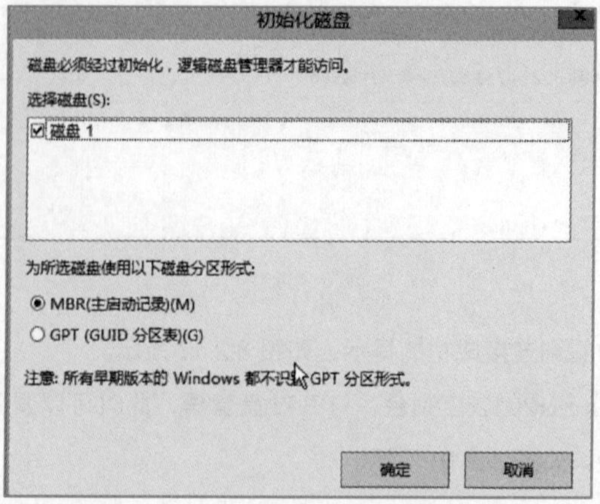

图 3.2.20　磁盘初始化为 MBR 形式

图 3.2.21　新建简单卷

上述步骤完成后,打开文件管理器,可以看到挂载的云硬盘已经可以使用,如图 3.2.22 所示。

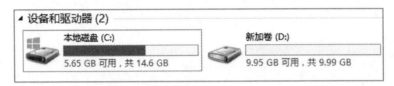

图 3.2.22　已经挂载的云硬盘

该云硬盘就可以像本地硬盘一样在虚拟机中使用,而且在后续内容中将看到该云硬盘可以根据使用情况进行动态扩容,实现云计算的按需分配、弹性的资源池功能,这样一台部署在云平台中的虚拟机可以根据需要进一步部署相关的服务,如 FTP、Web 服务等,读者可以自行尝试。

习题　原生 OpenStack 云平台

第二部分 原生OpenStack云平台基础环境的构建

第 4 章　原生 OpenStack 云平台的环境准备

 本章导读：

　　本章介绍原生 OpenStack 云平台的逻辑架构及其实现，包括云平台的软硬件架构实现的基本环境、实现方法。在此基础上，以控制节点为例详细介绍云平台操作系统 CentOS 7.6 的安装过程，还介绍两种主流终端仿真软件的使用方法，为后续实训项目安装运维云平台打下良好基础。本章最后的实训项目将详细介绍云平台正式开始安装前基本环境的准备。

电子资源：
　　电子教案　原生 OpenStack 云平台的环境准备
　　PPT　原生 OpenStack 云平台的环境准备
　　习题　原生 OpenStack 云平台的环境准备

第 4 章 原生 OpenStack 云平台的环境准备

4.1 原生 OpenStack 云平台的逻辑架构及其实现

电子教案 原生 OpenStack 云平台的逻辑架构及其实现

PPT 原生 OpenStack 云平台的逻辑架构及其实现

微课 1 原生 OpenStack 云平台的逻辑架构及其实现

本书以原生 OpenStack 云平台软件包为基础构建一个完整的基于 OpenStack 技术的云平台，通过后续的实训项目，读者可以逐步理解 OpenStack 技术各个组件模块的功能和作用，并且实现一个具备完整功能的 IaaS 云平台，实训项目逻辑架构如图 4.1.1 所示。云计算服务器基本参数见表 4.1.1。

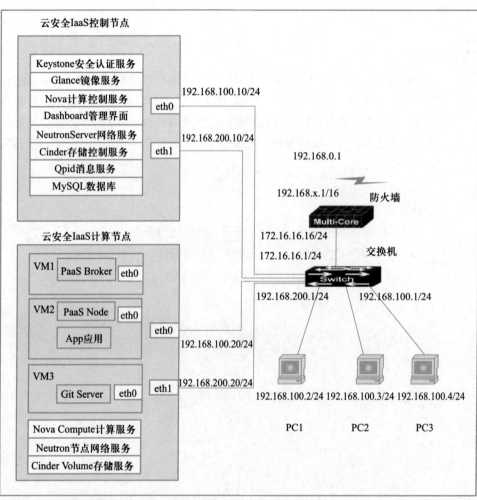

图 4.1.1 原生 OpenStack 云平台最小化逻辑架构

表 4.1.1 云计算服务器基本参数

型 号	CPU	内 存	存 储
其他品牌兼容服务器	Intel Xeon E5-2620 六核处理器	ECC REG 1600MHz 8GB	SAS 15000RPM 6.0Gbit/s 300GB X 2
其他品牌兼容服务器	Intel Xeon E5-2620 六核处理器	ECC REG 1600MHz 8GB	SATA 7200RPM 6.0Gbit/s 2TB 企业级 X 3

该架构也是实现原生 OpenStack 云平台的最小化逻辑架构，其中包括了一个控制节点（主机名为 controller）和一个计算节点（主机名为 compute）。每个节点配置两块网卡，一块用作内网间的通信以及对云平台的管理，其网段为 192.168.100.0/24，在后续实训

项目中该网卡简称为内网网卡，并且使用操作系统中名为 ens33 的网卡（其中 controller 节点的 ens33 网卡的 IP 地址为 192.168.100.10，compute 节点的 ens33 网卡的 IP 地址为 192.168.100.20）。另外一块用作外部网络对 IaaS 平台的访问以及平台对外部用户提供业务和服务，其网段为 192.168.200.0/24，在后续实训项目中该网卡简称为外网网卡，并且使用操作系统中名为 ens34 的网卡（由于后续实训中会清除两个节点第 2 块网卡的配置以保证 Neutron 服务的正常运行，所以这里不用给外网卡配置 IP 地址）。在后续实训项目中还会发现云平台构建成功后，创建的虚拟机实例的网络都在外网 192.168.200.0/24 的网段。

使用其他品牌的兼容服务器也可以完成本书中的实训项目，其中两台服务器 CPU 内核数在 4 个以上，控制节点内存大于 2GB，计算节点内存大于 4GB，两个节点的存储空间均在 100GB 以上，并且都最少配备两块网卡。

如果上述实验环境都不具备，可以通过 PC 上 Windows 7 环境下安装 VMware Workstation 15 或者 15 以上的版本，通过 VMware 中的虚拟机来模拟控制节点和计算节点，这样的操作环境对 PC 硬件要求较高，通常 PC 需配备 4 核 4 线程以上的 CPU，8GB 以上的内存，以及 300GB 以上的空余存储空间。推荐配置为 I7 处理器、16GB 内存、500GB 空余磁盘空间。

本书中实训项目用到两个软件包，如图 4.1.2 所示，其中第 1 个名为 CentOS-7-x86_64-DVD-1810 的 iso 镜像文件为 CentOS 7.6 版本的安装包，在控制节点和计算节点上均使用 CentOS 7.6 最小化安装作为原生 OpenStack 底层操作系统。OpenStack-Install-v1.0 的 iso 镜像文件为基于 OpenStack Stein 版本的原生 OpenStack 云平台软件包，该软件包用于构建本地 YUM 源并安装 IaaS 云平台的各个组件，如果读者在真实服务器上完成本书实训项目，则需要将第 1 个软件包（即 CentOS 7.6 版本的镜像文件）刻录到光盘上或 U 盘上，然后修改服务器的启动参数，从光盘或是 U 盘上启动最小化安装 CentOS 7.6。如果采用 PC 中 VMware Workstation 15 来模拟本书实训环境，则保证 PC 中存放有上述两个软件包即可。

名称	修改日期	类型	大小
CentOS-7-x86_64-DVD-1810	2019/6/5 10:33	UltraISO 文件	4,481,024...
OpenStack-Install-v1.0	2019/11/17 13:37	UltraISO 文件	1,847,892...

图 4.1.2　原生 OpenStack 云平台软件包

下面详细介绍通过 PC 中 VMware Workstation 15 来模拟本书所有实训项目逻辑架构的主要步骤，以控制节点虚拟机的构建过程为例进行介绍。首先在 PC 主板的 CMOS 设置中打开 CPU 的虚拟化（Virtualization）功能，之后在 Windows 操作系统中（推荐使用 Windows 7 64 位专业版）安装 VMware Workstation 15，安装完成后启动主界面，如图 4.1.3 所示。

在主界面中的"帮助"菜单中选择"关于"命令，查看当前的 VMware 版本号，确定为 15 及以上的版本，如图 4.1.4 所示。

接下来返回到主界面中，在"文件"菜单中选择"新建虚拟机"命令，在打开的对话框中选中"自定义（高级）"单选项，单击"下一步"按钮，如图 4.1.5 所示。

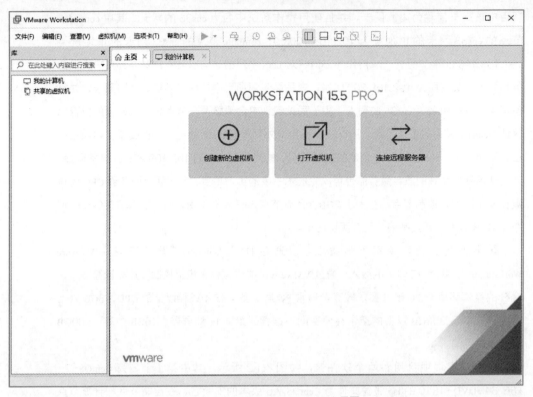

图 4.1.3　VMware Workstation 15 主界面

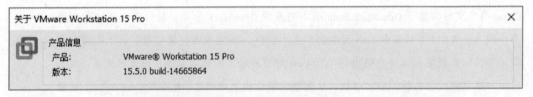

图 4.1.4　确认 VMware 的版本号

图 4.1.5　新建虚拟机

在选择虚拟机兼容性时，用户可以不做任何更改直接使用默认的兼容性，然后单击"下一步"按钮，如图 4.1.6 所示。

在"安装客户机操作系统"界面选择"稍后安装操作系统"单选项，如图 4.1.7 所示，之后继续单击"下一步"按钮。

在"选择客户机操作系统"界面时选择"Linux"单选项，具体版本为"CentOS 7 64 位"，如图 4.1.8 所示，之后继续单击"下一步"按钮。

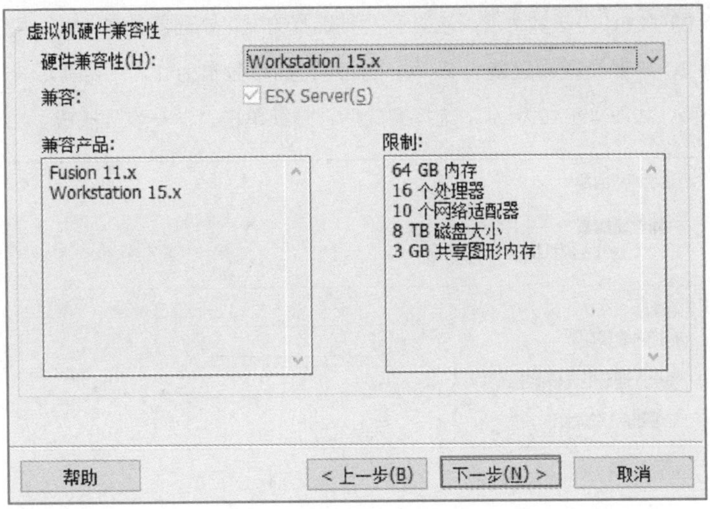

图 4.1.6　选择虚拟机兼容性

图 4.1.7　选择"稍后安装操作系统"单选项

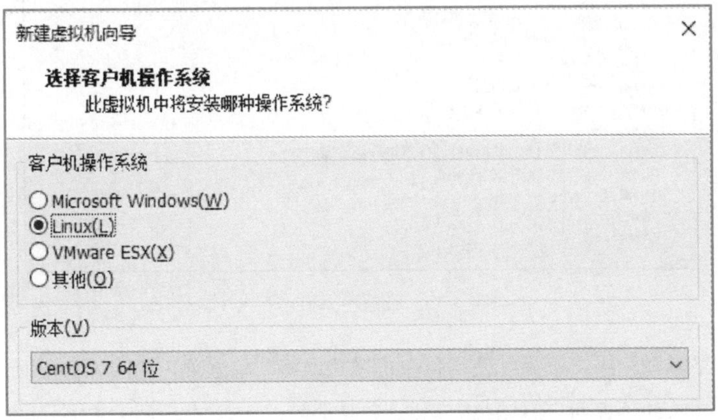

图 4.1.8　选择客户机操作系统

下面为虚拟机命名。这里将控制节点虚拟机命名为 controller，并选择合适的位置存储虚拟机镜像以及配置文件，本例中将控制节点虚拟机 controller 的所有文件存储在 D:\controller 目录下，如图 4.1.9 所示，之后继续单击"下一步"按钮。

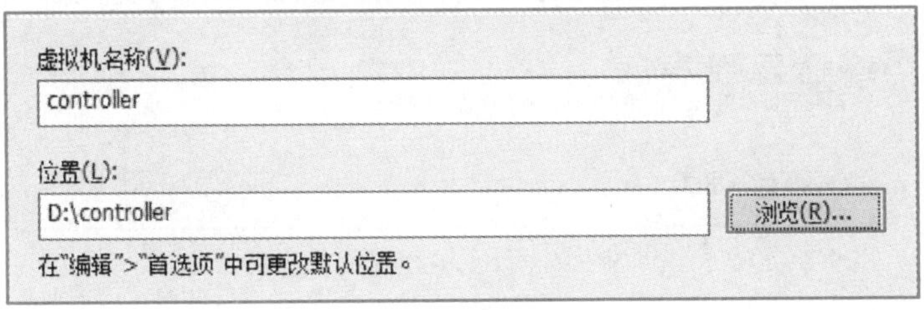

图 4.1.9　虚拟机名称和存放位置

控制节点的 CPU 内核数为 2，这里有两种选择方式：处理器数量为 1、每个处理器的内核数量为 2，或者处理器数量为 2、每个处理器内核数量为 1，只要满足总处理器内核总数为 2 即可，如图 4.1.10 所示，选择完成后，继续单击"下一步"按钮。

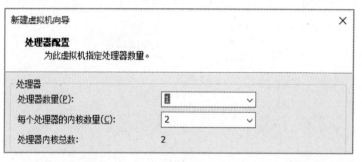

图 4.1.10 选择处理器以及核心数量

控制节点 controller 的内存为 4GB，即 4096MB，如图 4.1.11 所示，在界面中设置好即可单击"下一步"按钮。

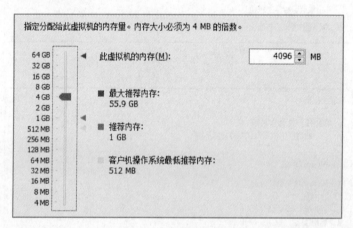

图 4.1.11 设置虚拟机内存大小

在"网络类型"中，选择"使用仅主机模式网络"单选项，如图 4.1.12 所示，这样使得每台用于实验的物理 PC 网络是相互独立和隔离的，之后继续单击"下一步"按钮。

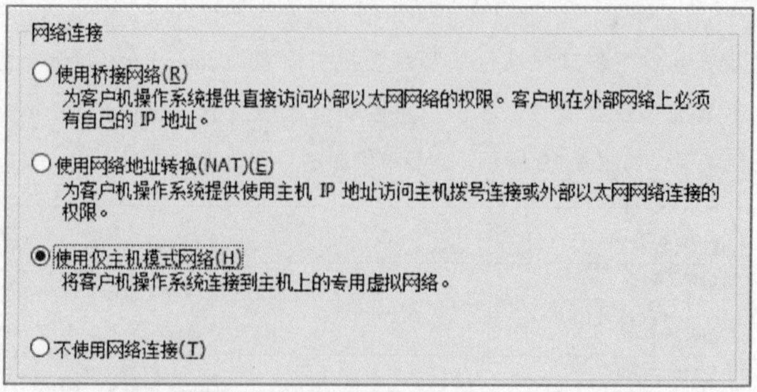

图 4.1.12 选择虚拟机网络工作模式

设置"I/O 控制器类型"为默认的"LSI Logic"类型，如图 4.1.13 所示，之后继续单击"下一步"按钮。

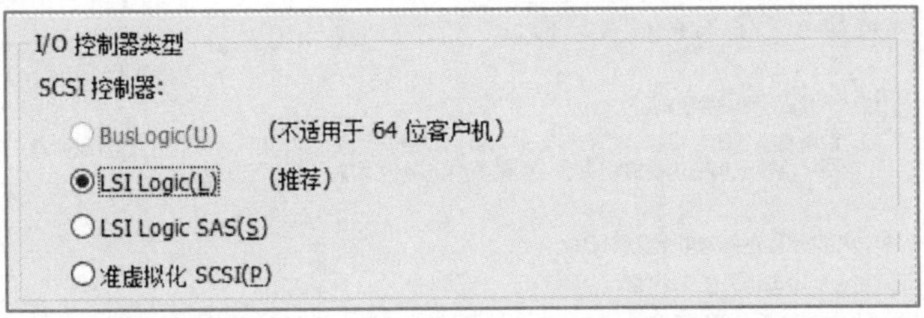

图 4.1.13　选择 I/O 控制器类型

设置"虚拟磁盘类型"为"SCSI",如图 4.1.14 所示,之后继续单击"下一步"按钮。

图 4.1.14　选择虚拟磁盘类型

在"选择磁盘"的步骤中,选择"创建新虚拟磁盘"单选项,如图 4.1.15 所示,并单击"下一步"按钮。

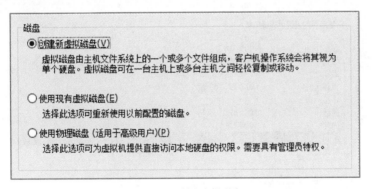

图 4.1.15　创建虚拟磁盘

控制节点的磁盘容量为 100GB 或者更大,可以直接在界面中输入,并选中"将虚拟磁盘存储为单个文件"单选项,如图 4.1.16 所示,单击"下一步"按钮。

在"指定磁盘文件"中使用默认的文件 controller.vmdk,不做任何更改,路径可以单击"浏览"按钮后自行定义,这里存储在 D:/controller 文件夹下,如图 4.1.17 所示,之后单击"下一步"按钮。

由于以上默认步骤为虚拟机只配备了一块网卡,无法满足原生 OpenStack 云平台逻辑架构中所使用的双网卡内外网环境,所以还需要为虚拟机添加第 2 块网卡。在"将使用下列设置创建虚拟机"界面中单击"自定义硬件"按钮,如图 4.1.18 所示。

之后单击"添加"按钮,在打开的对话框中选择"网络适配器"选项,如图 4.1.19 所示,继续单击"下一步"按钮。

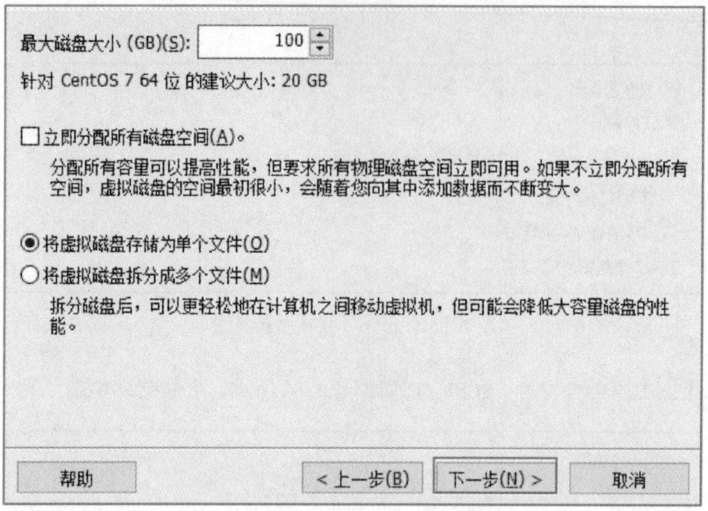

图 4.1.16　设置磁盘大小以及存储方式

图 4.1.17　指定磁盘文件以及存储位置

图 4.1.18　自定义硬件添加网卡

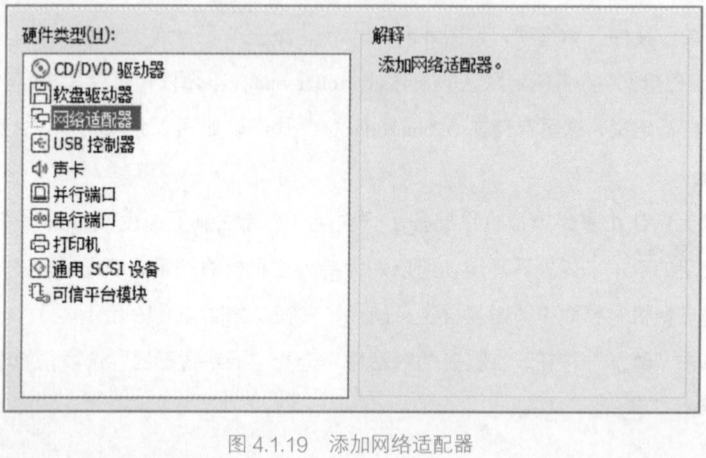

图 4.1.19　添加网络适配器

在"网络连接"区域中选中"NAT 模式：用于共享主机的 IP 地址"单选项，如图 4.1.20 所示，继续单击"下一步"按钮。

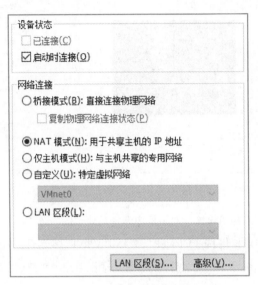

图 4.1.20 选择网络适配器类型

返回到"硬件"界面后，可以在设备框中看到有两块网卡，分别为网络适配器（类型为仅主机模式）以及网络适配器 2（类型为 NAT），在设备栏中选择"新 CD/DVD（IDE）"选项，此后单击右侧"使用 ISO 映像文件"右侧的"浏览"按钮，选择 CentOS-6.5-x86_64-bin 的 iso 映像文件（本书示例中该映像存放在 D 盘下，读者根据实际存放位置找到该文件并选择，确保路径正确），选中右上方"启动时连接"复选框，如图 4.1.21 所示。

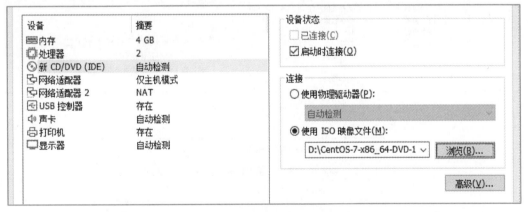

图 4.1.21 选择云平台镜像文件

完成上述所有步骤后单击界面右下方的"关闭"按钮，返回到"新建虚拟机"对话框，并单击"完成"按钮，之后在主界面中，就可以看到新建控制节点的虚拟机，如图 4.1.22 所示。

为了让以上控制节点的虚拟机能够顺利实现基于 OpenStack 的 IaaS 环境，还需要进一步开启虚拟机 CPU 的虚拟化功能，首先右击"我的计算机"标签中的"controller"虚拟机，然后在弹出的快捷菜单中选择"设置"命令，如图 4.1.23 所示。

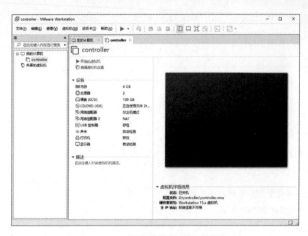
图 4.1.22　新建的控制节点虚拟机

图 4.1.23　设置控制节点虚拟机

之后在打开的"虚拟机设置"对话框中选择"处理器"选项，右下方将出现虚拟化引擎的各选项，选中"虚拟化 Intel VT-x/EPT 或 AMD V/RVI"复选框，如图 4.1.24 所示。

在启动虚拟机安装操作系统之前，用户还必须配置虚拟机的网络环境，保证启动的虚拟机和 PC 之间的连通性。在主界面菜单栏中选择"编辑"→"虚拟网络编辑器"命令，如图 4.1.25 所示。

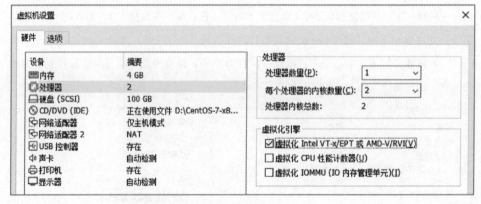
图 4.1.24　开启虚拟化引擎

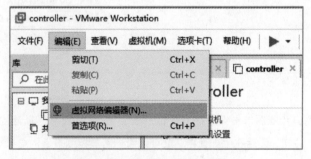
图 4.1.25　打开虚拟网络编辑器

在打开的"虚拟网络编辑器"对话框中将 VMnet1 类型为"仅主机模式"网络的网络地址改到控制节点内网网卡所在的网段，网络地址为 192.168.100.0，掩码为 255.255.255.0。将 VMnet8 类型为"NAT 模式"网络的网络地址改到控制节点外网网卡所在的网段

192.168.200.0，单击"应用"按钮，如图 4.1.26 所示。其后再单击"确定"按钮退出回到主界面，这里的 IP 地址段与之前规划的内网网卡以及外网网卡所在的地址段保持一致。

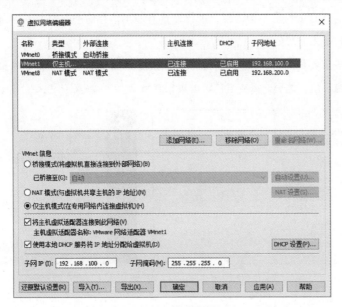

图 4.1.26 编辑虚拟网络

在主界面的 controller 虚拟机选项卡中单击"开启此虚拟机"按钮，即可顺利打开虚拟机并开始底层 CentOS 7.6 操作系统的安装过程，如图 4.1.27 和图 4.1.28 所示，此后 CentOS 7.6 的安装过程与在真实物理机中的安装过程完全一致。

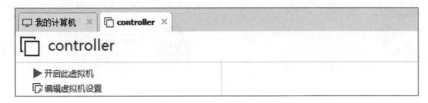

图 4.1.27 开启控制节点虚拟机

图 4.1.28 虚拟机中开启图 CentOS 7.6 的安装过程

计算节点虚拟机在 VMware Workstation 15 中的构建过程与以上控制节点的构建过程基本一致，上述所有使用到 controller 名称的地方均改为使用 compute。此外虚拟化的硬件配置略有不同，CPU 的核心数为 4 核，内存设置为 8GB，存储空间的大小为 200GB，最后构建的虚拟机基本配置如图 4.1.29 所示。

当出现安装界面时，选择第 1 个选项，直接按 Enter 键即可。下面，需要用户选择操作系统的语言环境，这里选择英语，如图 4.1.30 所示。

图 4.1.29　计算节点的虚拟机基本配置

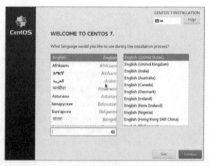
图 4.1.30　选择操作系统语言环境

单击"Continue"按钮后，在出现的界面中大部分保持默认即可。接下来需要配置磁盘存储，单击"INSTALLATION DESTINATION"按钮，如图 4.1.31 所示。

图 4.1.31　选择系统安装位置

在界面下方选中"I will configure partitioning"单选项来手动配置磁盘，之后单击左上角的"Done"按钮完成，如图 4.1.32 所示。

这里规划创建"/boot"分区、"swap"分区和"/"分区，"/boot"分区用来安装，设置为 200MB，通常"swap"分区是内存的两倍，然后将剩余的所有分区都分给"/"分区，如图 4.1.33 所示。

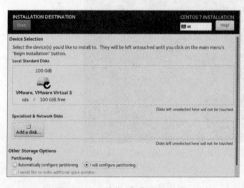
图 4.1.32　选择手动配置磁盘

图 4.1.33　划分分区主界面

单击该界面下方的"+"按钮来新建分区,按照规划,首先创建一个 200MB 的 /boot 分区,如图 4.1.34 所示。

其次,设置 8GB 大小的 swap 分区,如图 4.1.35 所示。

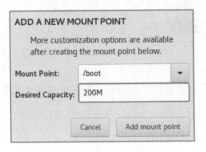 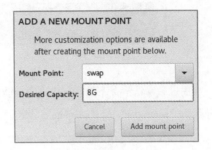

图 4.1.34 划分 /boot 分区　　　　　　　图 4.1.35 划分 swap 分区

将剩余的所有磁盘空间全部分配给根分区,不设置容量即表示剩余的所有空间,如图 4.1.36 所示。最后的磁盘分区情况如图 4.1.37 所示。

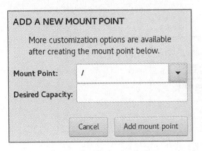 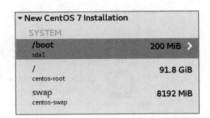

图 4.1.36 划分根分区　　　　　　　图 4.1.37 磁盘划分情况总览

单击左上角"Done"按钮,完成后出现如图 4.1.38 所示界面,单击"Accept Changes"按钮,这样磁盘分区就完成了。

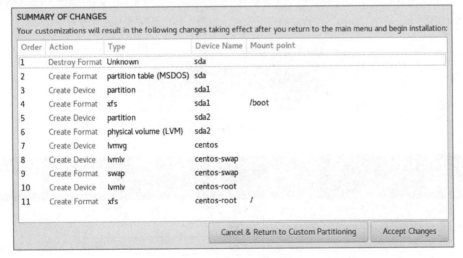

图 4.1.38 磁盘信息对照确认

存储配置完成后,再进入"Network & HOST NAME"网络和主机名配置界面,如图 4.1.39 所示。

图 4.1.39　配置网络和主机名

主机名默认情况下是"localhost.localdomain",本书选择在安装系统时直接配置主机名,分别为"controller"和"compute",这里仅以 controller 节点为例,如图 4.1.40 所示。

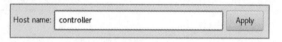

图 4.1.40　将主机名修改为 controller

需要选择系统的时区,可以根据实际位置更改系统的时区设置,这里选择默认时区,如图 4.1.41 所示。

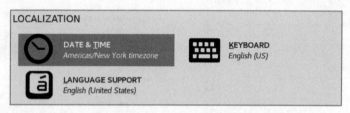

图 4.1.41　设置时间和日期

单击"DATE & TIME"按钮进入时区设置界面,设置时区和时间。

待上述步骤全部设置完成后,单击"Begin Install"按钮开始安装操作系统,在安装过程中,需要为系统 root 用户创建密码,本书中设置 root 用户的密码为"000000",系统提示为弱密码,此时需要单击两次"Done"按钮才能完成设置,如图 4.1.42 和图 4.1.43 所示。

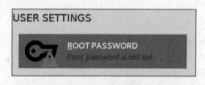

图 4.1.42　设置 root 密码

图 4.1.43　设置系统的 ROOT 密码

完成软件包安装之后，还需要操作系统进行额外的配置。单击"Finish configuration"按钮，一段时间后，系统会提示用户进行 Reboot（重启），这样安装就完成了，如图 4.1.44 所示。

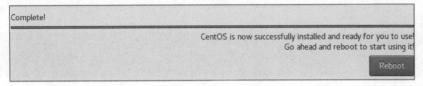

图 4.1.44　安装完成重启系统

单击"Reboot"按钮之后，系统就会进入重启阶段。重新开机后，会出现开机界面，在屏幕下方会出现一行进度条。同时需要说明，compute 节点的系统安装配置与 controller 节点基本一致，主机名设置为 copmute 即可，分区可以自行规划，这里不再赘述。

系统开机后，使用"root"用户和系统安装时为其设定的密码登录，如图 4.1.45 所示。

图 4.1.45　用 ROOT 密码登录 CentOS7.6

登录系统界面后，用户需要为每个节点的两张网卡分别进行配置。系统中，两张网卡分别为 ens33 和 ens34。配置系统网卡的方式有多种，本书采用编辑相应的网卡文本的方式进行配置，命令如下：

```
[root@controller ~]# vi /etc/sysconfig/network-scripts/ifcfg-ens33
```

这里首先配置控制节点的网卡参数，控制节点"controller"的内网网卡即 ens33 的 IP 地址为 192.168.100.10、掩码为 24 位、网关为 192.168.100.1，同时，还需要将"ONBOOT"的值改为"yes"即配置网卡为开机自启，将"BOOTPROTO"的值改为"static"，具体参数如图 4.1.46 所示。

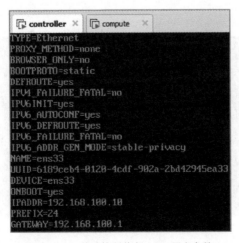

图 4.1.46　修改控制节点 ens33 网卡参数

其外网网卡即 ens34 先不做任何设置。

计算节点同样通过编辑网卡相应的文本文件来配置网卡参数，如图 4.1.47 所示。

图 4.1.47　打开计算节点的 ens33 网卡的配置文件

这里设定计算节点"compute"的内网网卡即 ens33 的 IP 地址为 192.168.100.20、掩码为 24 位、网关为 192.168.100.1，同时，还需要将"ONBOOT"的值改为"yes"即配置网卡为开机自启，将"BOOTPROTO"的值改为"static"，如图 4.1.48 所示。

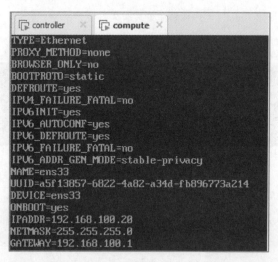

图 4.1.48　修改计算节点 ens33 网卡参数

同样，其外网网卡即 ens34 先不做任何设置。

配置好网卡信息后，用户需要通过"systemctl restart network"命令重启网络服务使配置生效，控制节点和计算节点的网络服务都需要重启，这里以控制节点的操作为例，配置成功如图 4.1.49 所示。

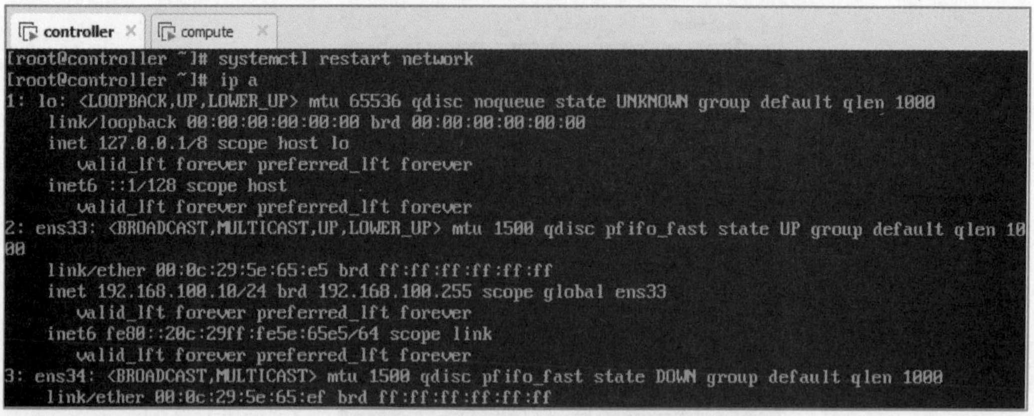

图 4.1.49　重启网络服务

在之后的 OpenStack 配置过程中，需要多次使用两个节点的 IP 地址，为方便使用，分别给两个节点配置相应的主机名进行解析，此时需要编辑“/etc/hosts”文件，使用 vi 命令编辑上述文件，控制节点和计算节点做相同的配置，如图 4.1.50 所示。

```
controller   compute
127.0.0.1    localhost localhost.localdomain localhost4 localhost4.localdomain4
::1          localhost localhost.localdomain localhost6 localhost6.localdomain6
192.168.100.10 controller
192.168.100.20 compute
```

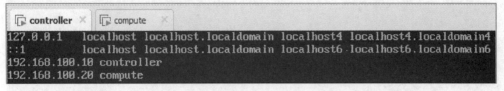

图 4.1.50 编辑 hosts 文件

通过两个节点之间互 ping 主机名的方式测试节点之间的网络连通性，如图 4.1.51 和图 4.1.52 所示。

```
controller   compute
[root@controller ~]# ping compute
PING compute (192.168.100.20) 56(84) bytes of data.
64 bytes from compute (192.168.100.20): icmp_seq=1 ttl=64 time=0.375 ms
64 bytes from compute (192.168.100.20): icmp_seq=2 ttl=64 time=0.411 ms
64 bytes from compute (192.168.100.20): icmp_seq=3 ttl=64 time=0.354 ms
64 bytes from compute (192.168.100.20): icmp_seq=4 ttl=64 time=0.340 ms
```

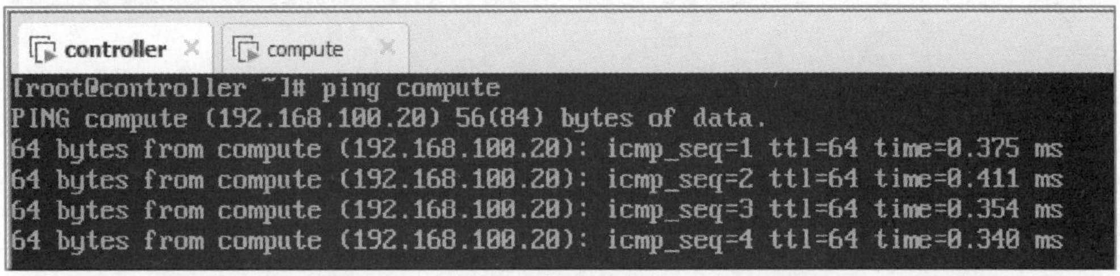

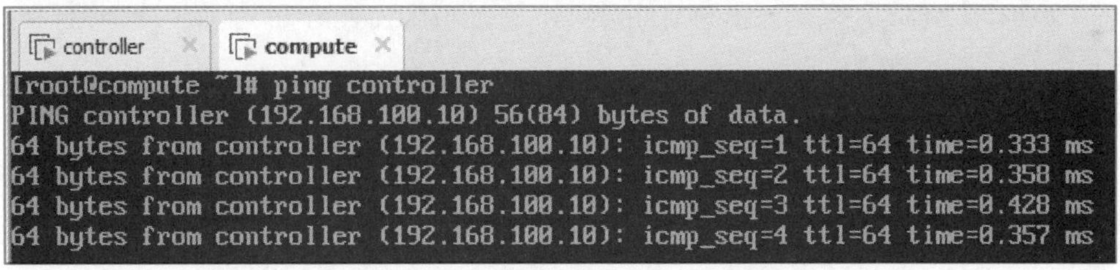

图 4.1.51 在控制节点测试主机名的连通性

```
controller   compute
[root@compute ~]# ping controller
PING controller (192.168.100.10) 56(84) bytes of data.
64 bytes from controller (192.168.100.10): icmp_seq=1 ttl=64 time=0.333 ms
64 bytes from controller (192.168.100.10): icmp_seq=2 ttl=64 time=0.358 ms
64 bytes from controller (192.168.100.10): icmp_seq=3 ttl=64 time=0.428 ms
64 bytes from controller (192.168.100.10): icmp_seq=4 ttl=64 time=0.357 ms
```

图 4.1.52 在计算节点测试主机名的连通性

最后，再通过本地主机分别 ping 两个节点内网 IP 地址的方式，来测试本机与控制节点以及计算节点之间的网络连通性，确保可以使用终端软件连接到控制节点以及计算节点进行操作，如图 4.1.53 和图 4.1.54 所示。

```
C:\Windows\system32\CMD.exe
Microsoft Windows [版本 10.0.18362.175]
(c) 2019 Microsoft Corporation。保留所有权利。

C:\Users\admin>ping 192.168.100.10

正在 Ping 192.168.100.10 具有 32 字节的数据:
来自 192.168.100.10 的回复: 字节=32 时间<1ms TTL=64
来自 192.168.100.10 的回复: 字节=32 时间<1ms TTL=64
来自 192.168.100.10 的回复: 字节=32 时间<1ms TTL=64
来自 192.168.100.10 的回复: 字节=32 时间<1ms TTL=64

192.168.100.10 的 Ping 统计信息:
    数据包: 已发送 = 4，已接收 = 4，丢失 = 0 (0% 丢失)，
往返行程的估计时间(以毫秒为单位):
    最短 = 0ms，最长 = 0ms，平均 = 0ms

C:\Users\admin>
```

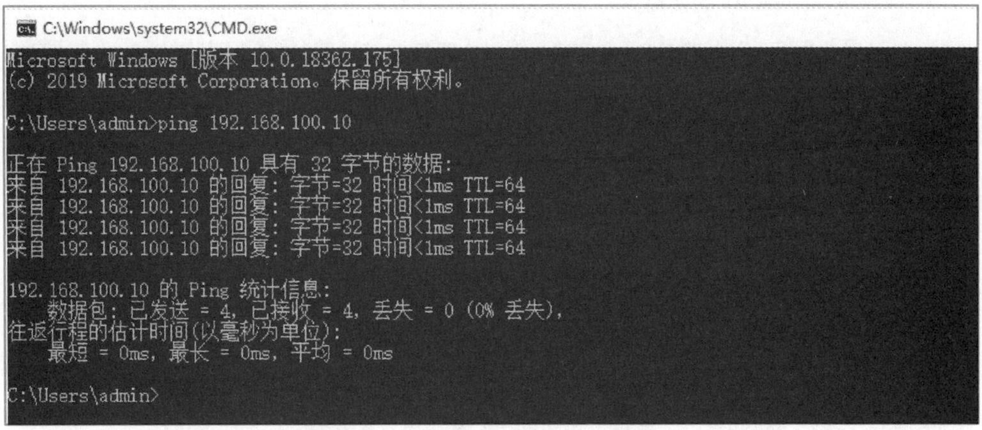

图 4.1.53 测试主机到控制节点的连通性

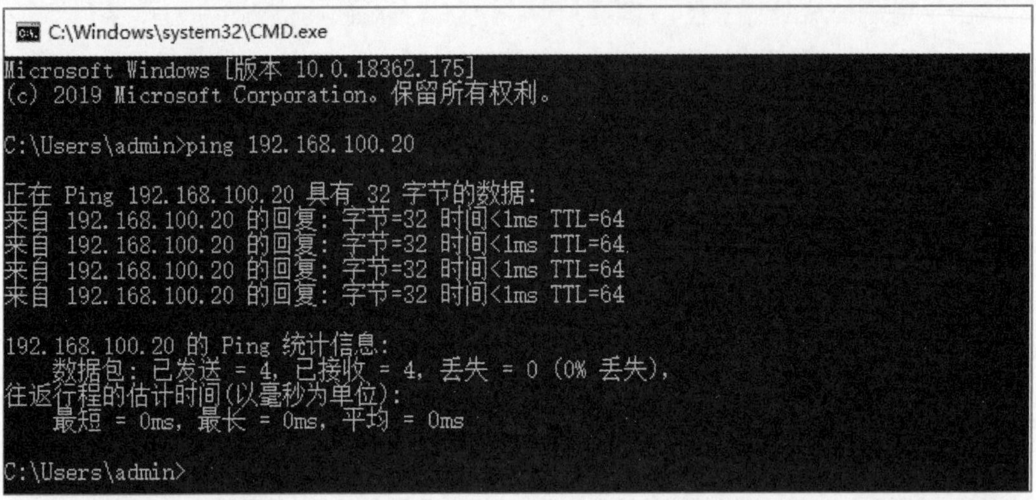

图 4.1.54 测试主机到计算节点的连通性

电子教案 终端仿真软件的使用

PPT 终端仿真软件的使用

4.2 终端仿真软件的使用

终端仿真软件，简单来讲就是在 PC 系统环境中，用程序仿真出一个远程控制台，作为一个远程终端，通过网络登录到远程 UNIX 或 Linux 系统。对于 Linux 系统，通常使用终端仿真软件进行远程管理，在终端仿真软件上用户可以更方便对命令进行操作。终端仿真软件有很多种类，这里介绍两种常用的终端仿真软件，SecureCRT 和 Xshell。

1. SecureCRT

（1）SecureCRT 简介

SecureCRT 是一款支持 SSH（SSH1 和 SSH2）的终端仿真程序，简单地说是 Windows 中登录 UNIX 或 Linux 服务器主机的软件。

SecureCRT 支持 SSH，同时支持 Telnet 和 Rlogin 协议。SecureCRT 是一款用于连接运行包括 Windows、UNIX 和 VMS 的理想工具。通过使用内含的 VCP 命令程序可以进行加密文件的传输。它具有流行 CRTTelnet 客户机的所有特点，包括自动注册、对不同主机保持不同的特性、打印功能、颜色设置、可变屏幕尺寸和用户定义的键位图等。

（2）SecureCRT 的使用

这里使用 7.0.0 版本的 SecureCRT，同时为方便通过当前登录窗口与主机之间进行文件传输，也安装了 ScureFX 工具，它可以与 SecureCRT 完美结合。本书通过使用 SecureCRT 登录 4.1 节中创建的 OpenStack 控制节点为例，讲解 SecureCRT 的常用功能。

① 安装完成 SecureCRT 后，双击图标，打开程序时，就会弹出"快速连接"对话框，如图 4.2.1 所示。

单击"协议"右侧下拉按钮，在下拉列表框中，可以看到 SecureCRT 支持的各种协议，这里使用默认的 SSH2 协议进行远程连接，同时在"主机名"文本框中输入要登录的主机 IP 地址，设置"端口"为默认的端口号 22。这里使用 root 用户登录，所以在"用户名"文本框中输入 root，单击"连接"按钮即可，如图 4.2.2 所示。

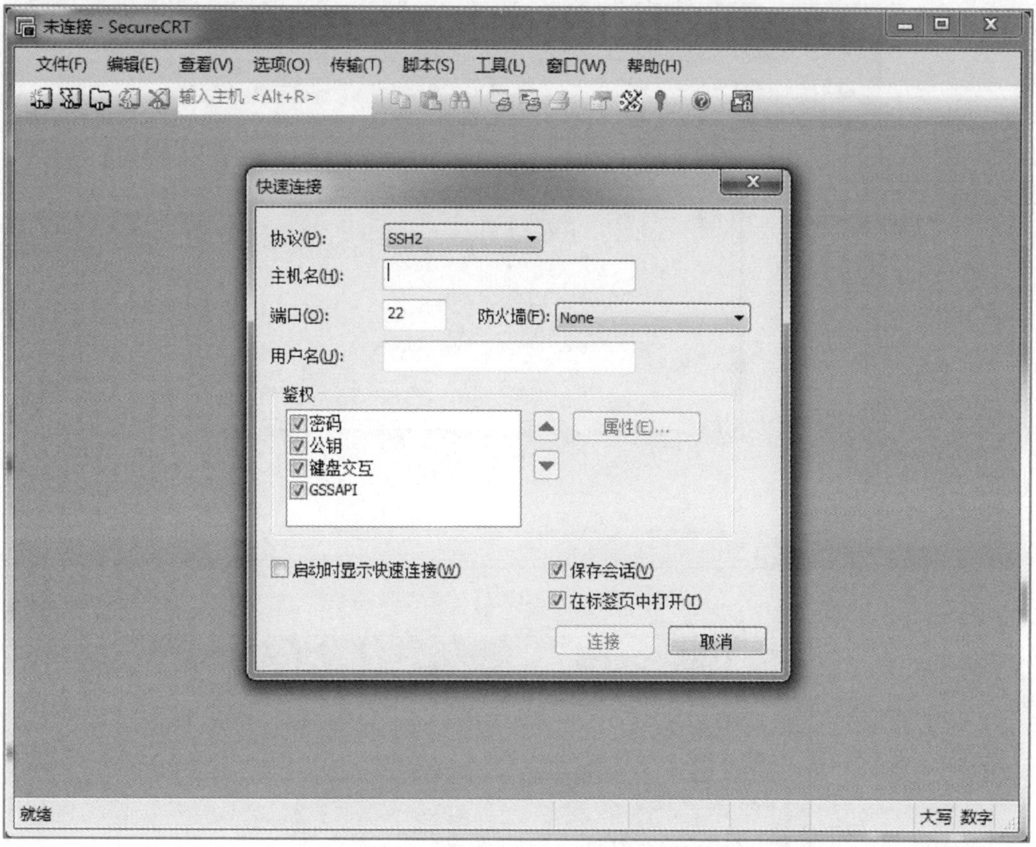

4.2 终端仿真软件的使用

图 4.2.1　SecureCRT "快速连接" 对话框

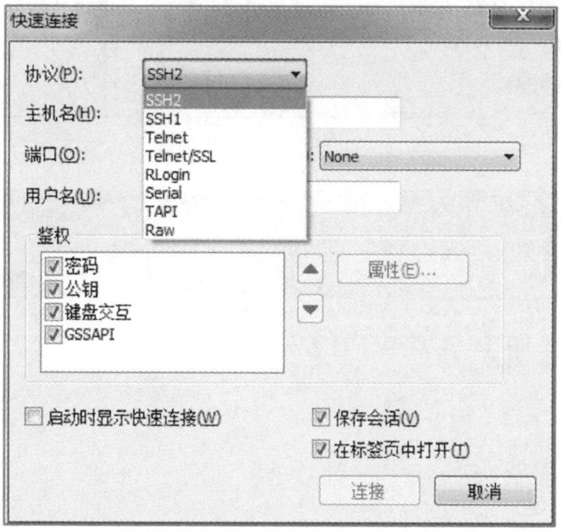

图 4.2.2　通过 "快速连接" 登录

第一次连接主机时，会弹出一个提示对话框，单击 "接受并保存" 按钮，如图 4.2.3 所示。

然后会出现需要输入登录主机密码的对话框，"用户名" 是用户之前创建连接时输入的 "root"，输入密码，并选中 "保存密码" 复选项，单击 "确定" 按钮即可，如图 4.2.4 所示。

连接主机成功后，标签页的左端出现绿色的勾，如图 4.2.5 所示。

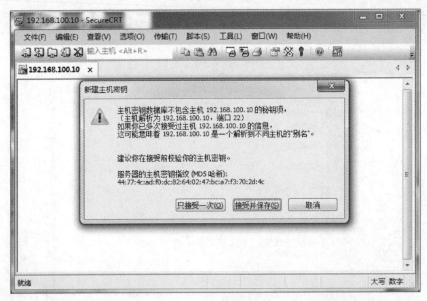

图 4.2.3　连接主机

图 4.2.4　输入登录主机的密码

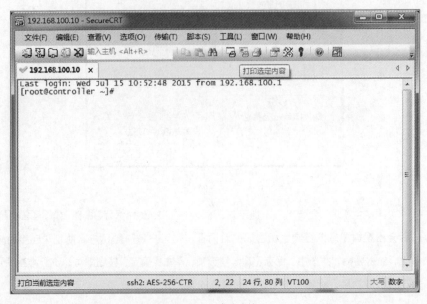

图 4.2.5　成功连接

②打开已经创建好的连接。单击工具栏中的"连接"按钮，会弹出一个"连接"对话框，其列表中是已经创建好的连接，如图 4.2.6 和图 4.2.7 所示。

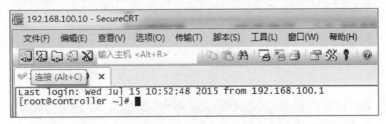

图 4.2.6　工具栏的"连接"按钮

图 4.2.7　"连接"对话框

③复制当前连接。有时用户对主机进行操作时，一个窗口可能不够，这时需要多打开几个相同的连接。除了再次打开创建好的连接这种方法，SecureCRT 还提供了复制连接的功能，右击已经打开的连接，在弹出的快捷菜单中选择"克隆会话"或"在新窗口中克隆"命令，即克隆出相同的连接，如图 4.2.8 和图 4.2.9 所示。

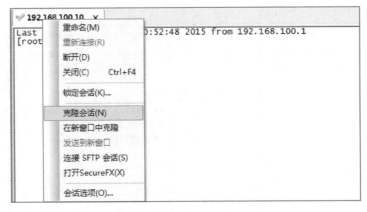

图 4.2.8　克隆会话

如果当前连接窗口需要断开，用户可以通过单击工具栏中"×"按钮（即"断开"按钮）来实现，同时断开的连接在标签页的左端出现一个红圈标记，如图 4.2.10 所示。

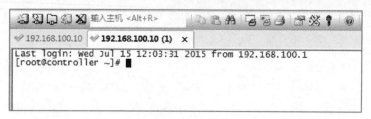

图 4.2.9　成功克隆会话

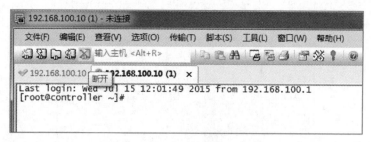

图 4.2.10　"断开"按钮和断开后的连接

当然，用户需要重新连接已断开的连接时，只需要单击工具栏中的"重新连接"按钮即可，如图 4.2.11 所示。

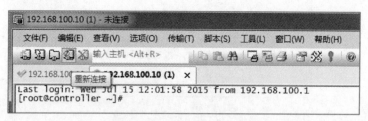

图 4.2.11　重新连接已断开的连接

④ 删除会话。当"连接"对话框列表中的失效连接过多时，有两种方法可以删除。方法不分优劣，读者可以根据自身习惯进行选择。

方法 1：先选中想要删除的连接，然后右击，在弹出的快捷菜单中选择"删除"命令即可，如图 4.2.12 所示。

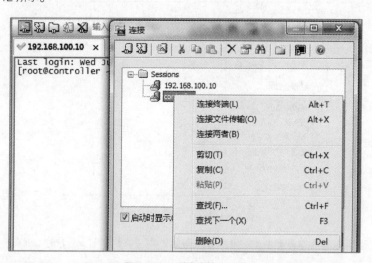

图 4.2.12　删除连接（1）

方法 2：同样先选中要删除的连接，然后单击"×"按钮（即"删除"按钮），即可进行删除，如图 4.2.13 所示。

图 4.2.13　删除连接（2）

通过"快速连接"创建的连接，其连接名称通常是创建连接时的主机 IP 地址，如果希望重命名为其他更方便识别的名称时，可以通过在"连接"对话框中右击想要重命名的连接，在弹出的快捷菜单中选择"重命名"命令，即可对连接进行重命名，如图 4.2.14 所示。

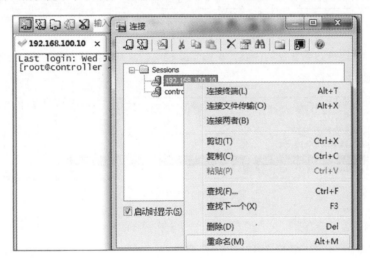

图 4.2.14　重命名已创建的连接

⑤ 在终端中复制。使用终端仿真软件的好处之一就是可以很方便地复制和粘贴命令中的内容。

● 选择。在介绍复制之前，先了解如何在 SecureCRT 中选择仿真终端的内容。有以下 3 种方法选择相关内容，方法不分优劣，读者可以根据自身习惯进行选择。

方法 1：单击然后移动鼠标选择想选择的内容。

方法 2：双击要选择的单词即可选择相关单词。

方法 3：点击 3 次想要选择的行即可选择整行。

● 复制功能。在 SecureCRT 中有 5 种常用的复制终端内容的方法。读者可以根据自己的习惯选择。

方法 1：选择想要复制的内容，在菜单栏选择"编辑"→"复制"命令，即可完成对所选内容的复制，如图 4.2.15 所示。

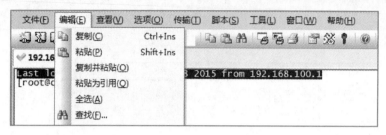

图 4.2.15　复制终端内容（1）

方法 2：选择想要复制的内容并右击，在弹出的快捷菜单中选择"复制"命令，即可完成对所选内容的复制，如图 4.2.16 所示。

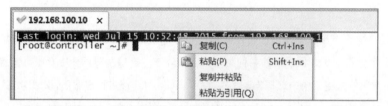

图 4.2.16　复制终端内容（2）

方法 3：选择想要复制的内容，单击工具栏的"复制"按钮，即可完成对所选内容的复制，如图 4.2.17 所示。

图 4.2.17　复制终端内容（3）

方法 4：选择"选项"→"全局选项"命令，在打开的"全局选项"对话框中，在"终端"组的"鼠标"区域中选中"选中时复制"复选框，即可实现通过鼠标选中即复制的功能，如图 4.2.18 所示。

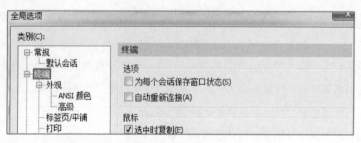

图 4.2.18　复制终端内容（4）

方法 5：在通过按钮进行复制时，可以看到提示的快捷键，即同时按住 **Ctrl+Insert** 快捷键同样可以完成复制。

- 粘贴。粘贴同复制一样，同样有 5 种常用方法供读者选择。

方法 1：选择需要粘贴的位置，单击工具栏的"粘贴"按钮，如图 4.2.19 所示。

方法 2：选择粘贴的位置并右击，在弹出的快捷菜单中选择"粘贴"命令，如图 4.2.20

所示。

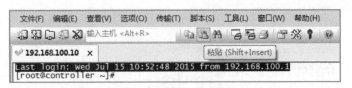

图 4.2.19 粘贴复制的内容（1）

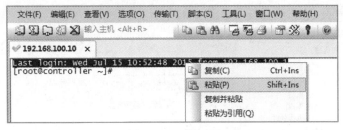

图 4.2.20 粘贴复制的内容（2）

方法 3：在菜单栏选择"编辑"→"粘贴"命令，如图 4.2.21 所示。

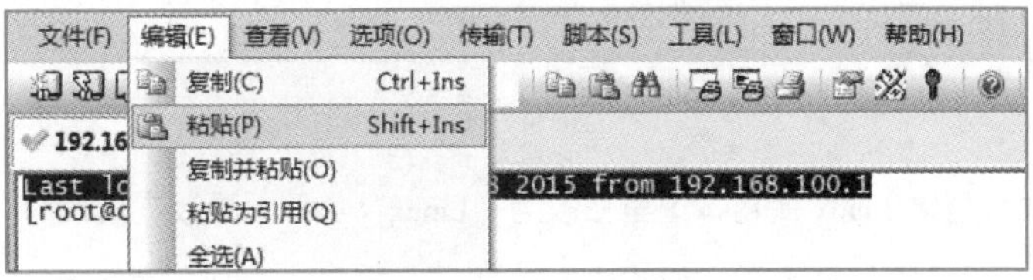

图 4.2.21 粘贴复制的内容（3）

方法 4：在菜单栏选择"选项"→"全局选项"命令，在打开的"全局选项"对话框中，在"终端"组的"鼠标"区域中选中"粘贴用右按钮"复选项，然后单击"确定"按钮，之后在需要粘贴的位置，直接按鼠标右键或中键即可完成粘贴，如图 4.2.22 所示。

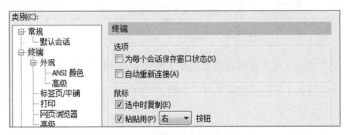

图 4.2.22 粘贴复制的内容（4）

方法 5：在需要粘贴的位置，通过组合键 Shift+Insert 同样可以完成粘贴。

以上介绍了 SecureCRT 的常用功能，下面介绍终端窗口属性的一些简单设置。

⑥ 窗口属性设置分为当前连接窗口设置和全局设置两种。人们通常通过它们设置仿真终端的类型、颜色方案、字体以及编码等属性。在菜单栏中打开"选项"菜单，如图 4.2.23 所示。

⑦ 更改仿真终端的类型。选择"会话选项"命令，打开"会话选项"对话框，选择"终端"→"仿真"类别，在右侧"终端"下拉列表中会列出当前版本 SecureCRT 支持的所有

终端类型，默认是"VT100"类型，如图 4.2.24 所示。

图 4.2.23 "选项"菜单中的"会话选项"和"全局选项"

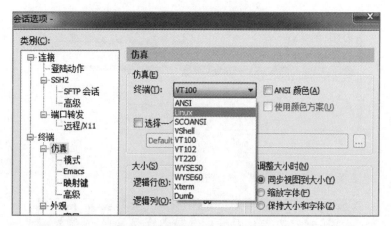

图 4.2.24 仿真终端的类型列表

为了更方便登录 Linux 使用，这里修改为"Linux"类型，更改完成后，会发现程序自动选中了"ANSI 颜色"复选框，如图 4.2.25 所示，同时也可以使用 SecureCRT 提供的颜色方案，这需要选中"使用颜色方案"复选框。

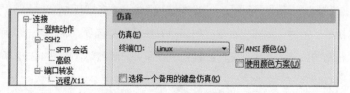

图 4.2.25 使用"颜色方案"复选框

选择"外观"类别，用户可以编辑窗口文本的外观，包括"当前颜色方案""字体""编码"等。SecureCRT 提供了 8 种颜色方案可供选择，同时允许用户自定义颜色方案，如图 4.2.26 所示。

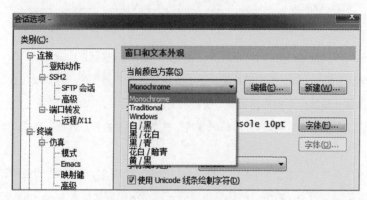

图 4.2.26 当前颜色方案

单击图 4.2.26 中的"字体"按钮,可以在打开的对话框中配置文字的"字体""字形""大小"选项,选择合适的内容,单击"确定"按钮即可,如图 4.2.27 所示。

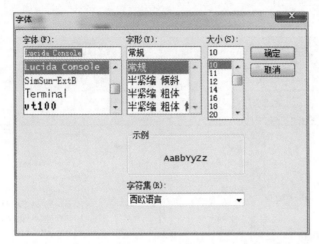

图 4.2.27　终端字体

有时由于登录系统语言编码或者是用户需要在终端中查看的文本编码问题,显示到终端中的内容可能会出现乱码的现象,这时就需要对终端仿真软件的编码进行设置。在如图 4.2.28 所示的"字符编码"下拉列表中列出了 SecureCRT 支持的所有字符编码,通常为方便显示选择"UTF-8"编码。

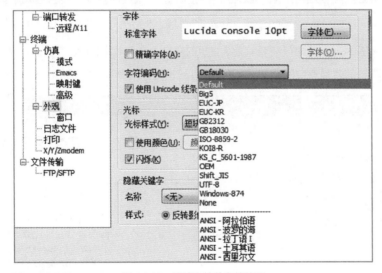

图 4.2.28　终端支持的字符编码

⑧ 用仿真终端与本地主机进行文件传输。这里需要预先安装 SecureFX,它与 SecureCRT 是配套软件,单击 SecureCRT 窗口的"SecureFX"按钮,如图 4.2.29 所示。

图 4.2.29　"SecureFX"按钮

SecureFX 窗口主要由本地窗口和当前登录的主机窗口组成，用户可以在两个窗口中通过鼠标拖曳方便地进行文件及文件夹的传输，SecureFX 窗口的下半部分可以显示文件传输过程的具体细节，如图 4.2.30 所示。

图 4.2.30　SecureFX 的窗口

如果打开 SecureFX 时只有当前登录主机的窗口，没有本地窗口，可以通过在菜单栏选择"视图"→"本地窗口"命令，本地窗口即可出现，如图 4.2.31 所示。

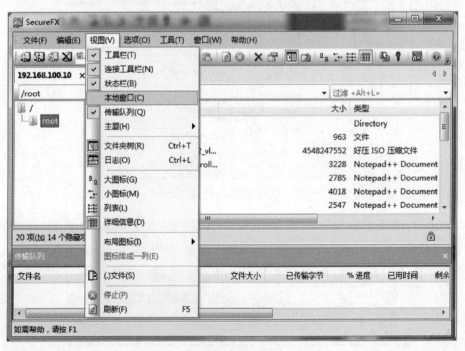

图 4.2.31　恢复本地窗口

2. Xshell 基本使用操作

（1）Xsehll 简介

Xshell 是在 Windows 环境下使用的强大的 SSH、Telent、Rlogin 终端仿真软件。Windows 用户通过 Xshell 可以便捷安全地登录 UNIX/Linux 主机。SSH（Secure Shell）协议支持加密及用户身份验证功能以安全连接网络并可替代 Telent、Rlogin 等原协议。

（2）Xshell 的使用

本次使用的 Xshell 为第 4 版，同时为方便通过当前 Xshell 与登录中的主机建立快速便利的 FTP/SFTP 连接，安装了 Xftp 4。本书通过使用 Xshell 登录 4.1 节中创建的 OpenStack 控制节点为例讲解 Xshell 的常用功能。

安装完成 Xshell 之后，双击图标，会出现一个没有进行连接的本地 Shell 的对话框，如图 4.2.32 所示。

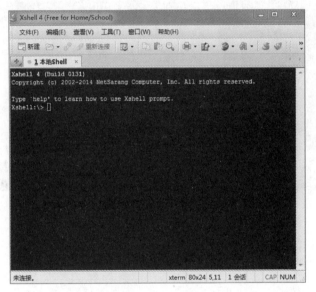

图 4.2.32　打开 Xshell

① 新建会话。单击工具栏中的"新建"按钮，弹出一个"新建会话（2）属性"的对话框，在"连接"选项中分别配置"名称""协议""主机"，"端口号"直接选择默认，"协议"选择"SSH"协议，"主机"输入要连接登录的主机地址，这里是虚拟机中控制节点的内网 IP，其他默认即可，如图 4.2.33 所示。

图 4.2.33　"新建会话"对话框

配置好新的会话后，单击"确定"按钮后，新创建的会话就被加入"会话"对话框中，然后单击"连接"按钮即可，如图 4.2.34 所示。

单击"连接"按钮后，会弹出一个"SSH 安全警告"对话框，为方便以后使用，这里单击"接受并保存"按钮，如图 4.2.35 所示。

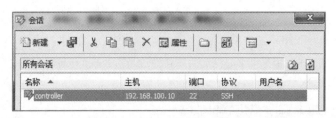

图 4.2.34 "会话"对话框

图 4.2.35 SSH 安全警告

紧接着会弹出"SSH 用户名"对话框，这里使用"root"用户登录，同时选中"记住用户名"复选框，单击"确定"按钮，如图 4.2.36 所示。

图 4.2.36 登录主机用户名

弹出"SSH 用户身份验证"对话框，此时输入 root 用户的密码，然后单击"确定"按钮，如图 4.2.37 所示。

现在从 Xshell 界面的提示信息来看，已经成功登录到所需的控制节点上了，如图 4.2.38 所示。

② 打开已创建的会话。这里介绍 3 种打开会话框的方式。以下方法不分优劣，读者根据自己的习惯使用。

图 4.2.37　用户登录密码

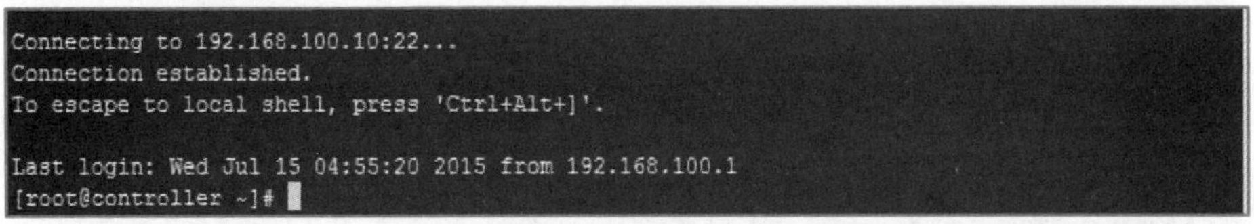

图 4.2.38　成功登录远程主机

方法 1：在菜单栏选择"文件"→"打开"命令，弹出"会话"对话框，如图 4.2.39 所示。

图 4.2.39　打开已创建的会话（1）

方法 2：单击工具栏中的"文件夹"按钮的倒三角按钮，弹出的下拉列表中包含多个选项，如"打开""本地 Shell"和已经创建的所有会话。若直接单击"文件夹"按钮默认是打开"会话"对话框，如图 4.2.40 所示。

图 4.2.40　打开已创建的会话（2）

方法 3：单击标签页最左边的"+"按钮会同方法 2 一样弹出多个选项，如图 4.2.41 所示。

图 4.2.41　打开已创建的会话（3）

③关闭会话。关闭会话的方法相对简单，单击打开的会话标签页右边的"×"按钮，

就会弹出是否断开连接的提示,单击"是"按钮即可关闭当前会话。当仅有一个会话时,程序还会提示是否断开将连接并退出程序,如图 4.2.42 所示。

图 4.2.42　断开连接提示

④ 编辑会话。用户可以通过编辑会话更改已创建会话的相关信息。打开"会话"对话框,选择要编辑的会话,然后单击"属性"按钮,就可以进行相关内容的编辑了,编辑完成后单击"确定"按钮,如图 4.2.43 所示。

图 4.2.43　编辑会话

⑤ 重命名会话。当用户对已经创建的会话名称不满意时,可以通过重命名的方式更改会话的名称。首先打开"会话"对话框,然后选择需要重命名的会话,右击,在弹出的快捷菜单选择"重命名"命令,即可进行更改,如图 4.2.44 所示。

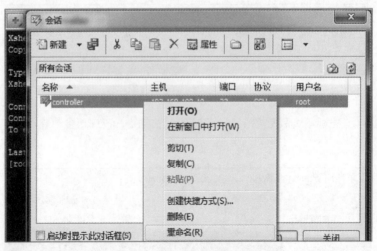

图 4.2.44　重命名会话

⑥ 删除会话。当用户的"会话"对话框中存在已经失效的会话,或者想要删除某会话时,可以通过下面两种方式删除会话。

方法 1：在打开的"会话"对话框中，选择需要删除的会话，单击"×"按钮，即可删除会话，如图 4.2.45 所示。

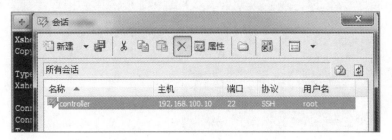

图 4.2.45　删除会话（1）

方法 2：在打开的"会话"对话框中，选中想要删除的会话并右击，然后在弹出的快捷菜单中选择"删除"命令，如图 4.2.46 所示。

图 4.2.46　删除会话（2）

⑦ 在终端中复制。使用终端仿真软件的好处之一就是可以很方便地复制和粘贴命令中的内容。

● 选择方式。在介绍复制之前，先了解如何在 Xshell 中选择仿真终端的内容。有 3 种方法选择相关内容，以下方法不分优劣，读者根据自身习惯进行选择。

方法 1：单击然后移动鼠标选择想选择的内容。

方法 2：双击要选择的单词即可选择相关单词。

方法 3：点击 3 次想要选择的行即可选择整行。

● 复制功能。在 Xshell 中有 3 种常用的复制终端中内容的方法，读者可以根据自己的习惯选择。

方法 1：选择想要复制的内容并右击，在弹出的快捷菜单中鼠标选择"复制"命令，即可实现复制功能，如图 4.2.47 所示。

方法 2：选择想要复制的内容，在菜单栏中选择"编辑"→"复制"命令，实现复制，如图 4.2.48 所示。

方法 3：选择"工具"→"选项"命令，在"键盘和鼠标"选项"选择"区域中选中"将选定的文本自动复制到剪贴板"复选项，然后单击"确定"按钮，如图 4.2.49 所示。

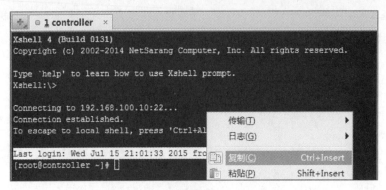

图 4.2.47　复制内容（1）

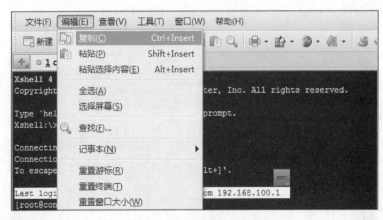

图 4.2.48　复制内容（2）

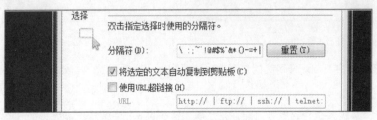

图 4.2.49　复制内容（3）

• 粘贴功能。复制完成后需要将复制的内容进行粘贴，可以将复制的内容在终端上进行粘贴，同时方便以后还可以将内容粘贴在本地文档中。粘贴同样有以下 3 种方法。

方法 1：在需要粘贴的位置右击，在弹出的快捷菜单中选择"粘贴"命令即可完成粘贴，如图 4.2.50 所示。

图 4.2.50　粘贴复制的内容（1）

方法 2：在菜单栏选择"编辑"→"粘贴"命令，如图 4.2.51 所示。

图 4.2.51　粘贴复制的内容（2）

方法 3：在菜单栏选择"工具"→"选项"命令，在打开对话框的"键盘和鼠标"选项卡中，可以将鼠标的中键或右键设置为"Paste the clipboard contents"即粘贴所选择的内容，这样每次复制完直接在终端直接按中键或右键即可完成粘贴，如图 4.2.52 所示。

图 4.2.52　粘贴复制的内容（3）

有时由于登录系统语言编码或者是需要在终端中查看的文本编码问题，显示到终端中的内容可能会出现乱码现象，这时就需要对终端仿真软件的编码进行设置，在 Xshell 中最简单的方法是单击工具栏的地球图标（即"编码"按钮），在弹出的下拉列表中列出了各种编码的选择项，从中选择自己需要的即可，通常选择"Unicode（UTF-8）"选项，如图 4.2.53 所示。

图 4.2.53　修改字符编码

⑧ Xshell 与 Xftp 联动的文件传输功能，这是一项非常方便实用的功能，需要预先安装 Xftp，当用户安装完 Xftp 后，它会自动加入到 Xshell 的工具栏中，如图 4.2.54 所示。

图 4.2.54　Xftp

单击"新建文件传输"按钮后，会出现一个有两个窗口的界面，一般左侧为本地窗口，右侧为当前通过 Xshell 登录的主机界面。界面的下方显示文件在传输过程中的详细信息。在 Xftp 中用户可以直接通过鼠标的拖曳进行文件的传输，也可以直接在窗口中对文件进行删除、重命名等基本操作，如图 4.2.55 所示。

第 4 章 原生 OpenStack 云平台的环境准备

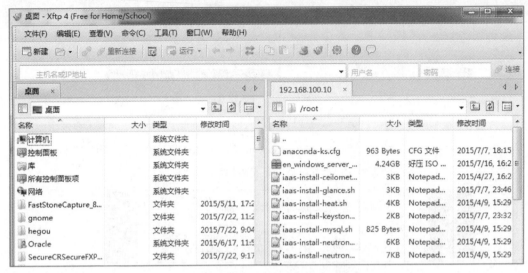

图 4.2.55　Xftp 窗口

4.3　实训项目 1　原生 OpenStack 云平台基本环境配置

电子教案　原生 OpenStack 云平台基本环境配置

PPT　原生 OpenStack 云平台基本环境配置

微课 2　原生 OpenStack 云平台基本环境配置

1. 实训前提环境

本地主机需准备好终端软件，通过终端软件连接安装好 CentOS 7.6 最小化系统的控制节点以及计算节点，上述节点可以是物理机，也可以是 VM 环境下的虚拟机，本地主机中还需存放 CentOS-7-x86_64-DVD-1810.iso 和 OpenStack-Install-v1.0.iso 两个用于安装原生 OpenStack 云平台的镜像。

2. 实训涉及节点

controller 节点和 compute 节点。

3. 实训目标

① 完成 hosts 文件的修改。
② 完成 selinux 的修改。
③ 完成 yum 源的配置。
④ 完成 NTP 服务的安装。
⑤ 完成 RabbitMQ 服务的安装。
⑥ 完成 Memcached 服务的安装。
⑦ 完成 OpenStack 环境准备包的安装。

4. 实训步骤及其详解

步骤 1：修改 hosts 文件。

在 controller 节点和 compute 节点，分别执行 vi 命令修改 /etc/hosts 文件，下面以 controller 节点上的操作为例，命令如下。

```
[root@controller ~]# vi /etc/hosts
```

4.3 实训项目 1 原生 OpenStack 云平台基本环境配置

打开 /etc/hosts，添加 controller 和 compute 主机名对应 IP 地址的解析，将 192.168.100.10 解析为 controller、192.168.100.20 解析为 compute。配置结果，如图 4.3.1 所示。

```
127.0.0.1       localhost localhost.localdomain localhost4 localhost4.localdomain4
::1             localhost localhost.localdomain localhost6 localhost6.localdomain6
192.168.100.10 controller
192.168.100.20 compute
```

图 4.3.1 host 文件中添加主机名解析

步骤 2：修改 selinux 配置文件。

在 controller 节点和 compute 节点，分别执行 vi 命令修改 /etc/sysconfig/selinux 文件，下面以 controller 节点上的操作为例，命令如下。

```
[root@controller ~]# vi /etc/sysconfig/selinux
```

在文件中修改参数 SELINUX=permissive，配置结果如图 4.3.2 所示。

```
# This file controls the state of SELinux on the system.
# SELINUX=permissive
#     enforcing - SELinux security policy is enforced.
#     permissive - SELinux prints warnings instead of enforcing.
#     disabled - No SELinux policy is loaded.
SELINUX=permissive
# SELINUXTYPE= can take one of three values:
#     targeted - Targeted processes are protected,
#     mls - Multi Level Security protection.
SELINUXTYPE=targeted
```

图 4.3.2 设置 selinux 配置文件中的参数

selinux 配置文件内容修改后重启生效，可以通过 setenforce 命令临时生效配置，命令如下。

```
[root@controller ~]# setenforce 0
```

配置完成后可以通过 getenforce 命令查看 selinux 的规则，命令如下。

```
[root@controller ~]# getenforce
```

如果上述的命令输出结果为 permissive，表示修改成功。

步骤 3：在 controller 节点配置 yum 源。

首先，将 CentOS-7-x86_64-DVD-1810.iso 和 OpenStack-Install-v1.0.iso 两个镜像文件上传至 controller 节点 opt 目录下，这里使用 SecureFX 终端软件直接登录 controller 节点，将本地文件拖入上述目录即可，如图 4.3.3 所示。

上传完毕后，需要再确认 /opt/ 目录下是否有这两个镜像文件，命令如下。

```
[root@controller ~]# ll /opt/
```

确认结果如图 4.3.4 所示。

接下来在 /opt/ 目录下创建 centos 和 iaas 目录，命令如下。

```
[root@controller ~]# mkdir /opt/centos
[root@controller ~]# mkdir /opt/iaas
```

目录创建完毕后，将 CentOS-7-x86_64-DVD-1810.iso 镜像挂载到 centos 目录，命令如下。

```
[root@controller ~]# mount -o loop /opt/CentOS-7-x86_64-DVD-1810.iso /opt/centos/
```

之后再将 OpenStack-Install-v1.0.iso 镜像挂载到 iaas 目录，命令如下。

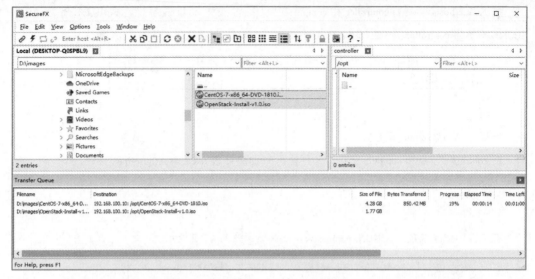

图 4.3.3　使用 SecureFX 上传两个镜像文件

```
[root@controller ~]# ll /opt/
total 6328916
-rw-r--r--. 1 root root 4588568576 Jun  4 22:33 CentOS-7-x86_64-DVD-1810.iso
-rw-r--r--. 1 root root 1892241408 Nov 17 00:37 OpenStack-Install-v1.0.iso
```

图 4.3.4　确认两个镜像文件已上传

```
[root@controller ~]# mount -o loop /opt/OpenStack-Install-v1.0.iso /opt/iaas/
```

挂载完成后可以通过 ll 命令查看 centos 和 iaas 目录的内容，命令如下。

```
[root@controller ~]# ll /opt/centos/
[root@controller ~]# ll /opt/iaas/
```

其结果如图 4.3.5 和图 4.3.6 所示。

图 4.3.5 中 Packages 目录为 CentOS 系统软件包存放目录，repodata 存放 yum 索引文件。yum 源配置文件内 baseurl 指定目录应该为索引目录的上级目录（即 centos 目录）。

```
-rw-rw-r--. 1 root root     14 Nov 25 2018 CentOS_BuildTag
drwxr-xr-x. 3 root root   2048 Nov 25 2018 EFI
-rw-rw-r--. 1 root root    227 Aug 30 2017 EULA
-rw-rw-r--. 1 root root  18009 Dec  9 2015 GPL
drwxr-xr-x. 3 root root   2048 Nov 25 2018 images
drwxr-xr-x. 2 root root   2048 Nov 25 2018 isolinux
drwxr-xr-x. 2 root root   2048 Nov 25 2018 LiveOS
drwxrwxr-x. 2 root root 663552 Nov 25 2018 Packages
drwxrwxr-x. 2 root root   4096 Nov 25 2018 repodata
-rw-rw-r--. 1 root root   1690 Dec  9 2015 RPM-GPG-KEY-CentOS-7
-rw-rw-r--. 1 root root   1690 Dec  9 2015 RPM-GPG-KEY-CentOS-Testing-7
-r--r--r--. 1 root root   2883 Nov 25 2018 TRANS.TBL
```

图 4.3.5　确认镜像文件正确挂载（1）

```
drwxrwxr-x. 6 nobody nobody 2048 Mar  6 01:12 iaas-repo
drwxrwxr-x. 2 nobody nobody 2048 Mar  6 01:45 images
```

图 4.3.6　确认镜像文件正确挂载（2）

在上述所有操作成功之后，删除 /etc/yum.repo.d/ 内所有原有的 yum 源配置文件，命令如下。

```
[root@controller ~]# rm -rf /etc/yum.repos.d/*
```

4.3 实训项目 1 原生 OpenStack 云平台基本环境配置

删除完成后，手动创建 yum 源配置文件 local.repo，建立本地的 yum 源，让之后 controller 节点所有用到的软件包从本地安装，命令如下。

```
[root@controller ~]# vi /etc/yum.repos.d/local.repo
```

在 local.repo 内写入以下内容：

```
[centos]
name=centos
baseurl=file:///opt/centos
enbaled=1
gpgcheck=0
[iaas]
name=iaas
baseurl=file:///opt/iaas/iaas-repo
enbaled=1
gpgcheck=0
```

上述文件完成后保存退出，还要使用 yum list 命令确认 yum 源是否配置成功，命令如下。

```
[root@controller ~]# yum list
```

其结果如图 4.3.7 所示。

```
zsh-html.x86_64                    4.3.10-7.el6                centos
zsh-lovers.noarch                  0.9.0-1.el6                 iaas
zvbi.x86_64                        0.2.33-6.el6                iaas
zvbi-devel.x86_64                  0.2.33-6.el6                iaas
zvbi-fonts.noarch                  0.2.33-6.el6                iaas
zziplib.x86_64                     0.13.62-1.el6               iaas
zziplib-devel.x86_64               0.13.62-1.el6               iaas
zziplib-utils.x86_64               0.13.62-1.el6               iaas
[root@controller ~]#
```

图 4.3.7 确认本地 yum 源配置成功

完成上述配置之后，在 controller 节点安装 vsftpd 服务，使得 compute 节点能够访问 controller 节点，通过 ftp 构建的 yum 源并安装软件包，命令如下。

```
[root@controller ~]# yum -y install vsftpd
```

其安装成功的结果如图 4.3.8 所示。

```
Running transaction
  Installing : vsftpd-3.0.2-25.el7.x86_64
  Verifying  : vsftpd-3.0.2-25.el7.x86_64

Installed:
  vsftpd.x86_64 0:3.0.2-25.el7

Complete!
[root@controller ~]#
```

图 4.3.8 确认 vsftpd 安装成功

接下来，通过 vi 命令修改 vsftpd 服务配置文件，命令如下。

```
[root@controller ~]# vi /etc/vsftpd/vsftpd.conf
```

在上述配置文件的第 1 行下插入参数 anon_root=/opt，如图 4.3.9 所示。

```
# Example config file /etc/vsftpd/vsftpd.conf
anon_root=/opt
#
# The default compiled in settings are fairly paranoid. This sample file
# loosens things up a bit, to make the ftp daemon more usable.
```

图 4.3.9 修改 vsftpd.conf 文件中的参数

修改完毕后保存退出，之后启动 vsftpd 服务，并设置为开机自动启动，命令如下。

```
[root@controller ~]# systemctl restart vsftpd
[root@controller ~]# systemctl enable vsftpd
```

为了保证之后 compute、controller 节点云平台组件之间能够正常通过 IP 网络通信，以及 compute 的云平台组件可以通过 ftp 安装，还需要同时关闭上述两个节点的防火墙，命令如下。

```
[root@controller ~]# systemctl stop firewalld
[root@controller ~]# systemctl disable firewalld
[root@compute ~]# systemctl stop firewalld
[root@compute ~]# systemctl disable firewalld
```

和 controller 节点一样，需要在 compute 节点删除 /etc/yum.repo.d/ 目录下原有的 yum 源配置文件。

```
[root@compute ~]# rm -rf /etc/yum.repos.d/*
```

之后通过 vi 命令在 compute 节点手动创建 yum 源配置文件 local.repo，建立 ftp 远程的 yum 源，其路径指向之前在 controller 节点下构建的 ftp，命令如下。

```
[root@compute ~]# vi /etc/yum.repos.d/local.repo
```

在文件内输入以下内容：

```
[centos]
name=centos
baseurl=ftp://192.168.100.10/centos
enabled=1
gpgcheck=0
[iaas]
name=iaas
baseurl=ftp://192.168.100.10/iaas/iaas-repo
enabled=1
gpgcheck=0
```

完成后保存退出，并使用 yum list 命令确认 compute 的 yum 源是否配置成功，命令如下。

```
[root@compute ~]# yum list
```

其结果如图 4.3.10 所示。

```
zsh-html.x86_64                 4.3.10-7.el6        centos
zsh-lovers.noarch               0.9.0-1.el6         iaas
zvbi.x86_64                     0.2.33-6.el6        iaas
zvbi-devel.x86_64               0.2.33-6.el6        iaas
zvbi-fonts.noarch               0.2.33-6.el6        iaas
zziplib.x86_64                  0.13.62-1.el6       iaas
zziplib-devel.x86_64            0.13.62-1.el6       iaas
zziplib-utils.x86_64            0.13.62-1.el6       iaas
[root@compute ~]#
```

图 4.3.10　确认 compute 节点的 yum 配置成功

步骤 4：配置 NTP 时钟同步服务。

controller 节点和 compute 节点需要同时被外部主机访问，所以两个节点主机时钟必须同步。可以在 controller 节点安装 NTP 服务端，让 compute 节点与 controller 节点的时钟同步。

在 controller 节点和 compute 节点，分别安装 NTP 服务，命令如下。

```
[root@controller ~]# yum -y install chrony
[root@compute ~]# yum -y install chrony
```

安装成功的结果如图 4.3.11 所示。

```
Running transaction
  Updating    : chrony-3.4-1.el7.x86_64
  Cleanup     : chrony-3.2-2.el7.x86_64
  Verifying   : chrony-3.4-1.el7.x86_64
  Verifying   : chrony-3.2-2.el7.x86_64

Updated:
  chrony.x86_64 0:3.4-1.el7

Complete!
```

图 4.3.11　chrony 服务安装成功

之后在 controller 节点修改 chrony.conf 配置文件，将 controller 节点作为服务端，使用 vi 命令打开 chrony 服务的主配置文件，命令如下。

```
[root@controller ~]# vi /etc/chrony.conf
```

删除文件中的以下参数：

```
server 0.centos.pool.ntp.org iburst
server 1.centos.pool.ntp.org iburst
server 2.centos.pool.ntp.org iburst
server 3.centos.pool.ntp.org iburst
```

并添加如下内容：

```
server 127.127.1.0 iburst
local stratum 10
allow 127/8
allow 192.168.100.0/24
```

结果如图 4.3.12 所示。

```
# Use public servers from the pool.ntp.org project.
# Please consider joining the pool (http://www.pool.ntp.org/join.html).
server 127.127.1.0 iburst
local stratum 10
allow 127/8
allow 192.168.100.0/24

# Record the rate at which the system clock gains/losses time.
driftfile /var/lib/chrony/drift

# Allow the system clock to be stepped in the first three updates
# if its offset is larger than 1 second.
makestep 1.0 3

# Enable kernel synchronization of the real-time clock (RTC).
rtcsync

# Enable hardware timestamping on all interfaces that support it.
#hwtimestamp *
```

图 4.3.12　修改 NTP 主配置文件中的参数

完成后保存退出，在 controller 节点启动 NTP 服务并设置为开机自动启动，命令如下。

```
[root@controller ~]# systemctl restart chronyd
[root@controller ~]# systemctl enable chronyd
```

在 compute 节点，也需要修改 chrony.conf 配置文件，将 compute 节点作为客户端，命令如下。

```
[root@controller ~]# vi /etc/chrony.conf
```

删除文件中的以下参数：

```
server 0.centos.pool.ntp.org iburst
server 1.centos.pool.ntp.org iburst
server 2.centos.pool.ntp.org iburst
server 3.centos.pool.ntp.org iburst
```

并添加如下内容：

```
server controller iburst
```

完成后保存退出，在 compute 节点启动 NTP 服务并设置为开机自动启动（务必确保 controller 节点的防火墙为关闭状态），命令如下。

```
[root@controller ~]# systemctl restart chronyd
[root@controller ~]# systemctl enable chronyd
```

在 compute 节点执行 chronyc sources 命令来测试，看到如图 4.3.13 所示的输出信息表示时钟同步成功。

```
[root@compute ~]# chronyc sources
210 Number of sources = 1
MS Name/IP address         Stratum Poll Reach LastRx Last sample
===============================================================================
^* controller                   12   6     7     1   -672ns[  -45us] +/-  228us
```

图 4.3.13　验证 NTP 时间同步

自此，NTP 服务已经全部安装完成。

步骤 5：RabbitMQ 服务的安装。

在 controller 节点安装 RabbitMQ 软件包，命令如下。

```
[root@controller ~]# yum -y install rabbitmq-server
```

安装完成后，在 controller 节点启动 RabbitMQ 服务，并设置为开机自启动，命令如下。

```
[root@controller ~]# systemctl start rabbitmq-server.service
[root@controller ~]# systemctl enable rabbitmq-server.service
```

接下来创建一个名为 openstack 的 RabbitMQ 用户，密码为 000000，并为其设置权限，命令如下。

```
[root@controller ~]# rabbitmqctl add_user openstack 000000
[root@controller ~]# rabbitmqctl set_permissions openstack ".*" ".*" ".*"
```

自此，RabbitMQ 服务就已全部安装完成。

步骤 6：Memcached 服务的安装。

在 controller 节点，安装 Memcached 服务，命令如下。

```
[root@controller ~]# yum -y install memcached python-memcached
```

安装完成后，将配置文件里面 OPTIONS 字段后加上控制节点的主机名即 controller。

```
[root@controller ~]# sed -i '/OPTIONS/d' /etc/sysconfig/memcached
[root@controller ~]# echo OPTIONS=\"-l 127.0.0.1,::1,controller\" >> /etc/sysconfig/memcached
```

上述第 1 个命令是使用 sed 命令行工具将匹配到 OPTIONS 字段的行删除，-i 参数是使修改立即对文件生效，单引号中 d 为删除；第 2 个命令使用 echo 向配置文件中追加带有主机名的一行内容，符号 ">>" 是向文件内追加内容而不是覆盖。

接下来启动 Memcached 服务并设置为开机自启的，命令如下。

```
[root@controller ~]# systemctl start memcached.service
[root@controller ~]# systemctl enable memcached.service
```

步骤 7：OpenStack 环境准备包的安装。

在 controller 节点和 compute 节点，分别安装 OpenStack 环境准备包，命令如下。

```
[root@controller ~]# yum -y install openstack-utils openstack-selinux python-openstackclient crudini
[root@compute ~]# yum -y install openstack-utils openstack-selinux python-openstackclient crudini
```

自此，OpenStack 软件包已经安装完成，所有基本环境配置已经完成。

习题 原生 OpenStack 云平台的环境准备

第 5 章 MySQL 数据库的安装及其配置

本章导读：

本章介绍 MySQL 数据库的基本功能、基本操作，MySQL 作为云平台后台的数据库服务对整个云平台的正常运行起着至关重要的作用，在实训项目 2 中重点介绍如何通过命令行在底层 CentOS 操作系统上安装部署 MySQL 数据库服务，同时解读 MySQL 数据库在安装和使用过程中可能出现的问题，最后将进一步介绍 MySQL 数据库安装脚本的制作和使用。

电子资源：

  电子教案　MySQL 数据库的安装及其配置
  PPT　MySQL 数据库的安装及其配置
  习题　MySQL 数据库的安装及其配置

5.1 MySQL 数据库功能简介

电子教案 MySQL 数据库功能简介

PPT MySQL 数据库功能简介

MySQL 是目前最流行的关系型数据库管理系统之一，由瑞典 MySQL AB 公司开发，目前属于美国 Oracle 公司。MySQL 是一种关系型数据库管理系统，关系型数据库将数据保存在不同的表中，而不是将所有数据放在一个大仓库内，这样就增加了速度并提高了灵活性。MySQL 是开源的，用户不需要支付额外的费用。MySQL 支持大型的数据库，可以处理拥有上千万条记录的大型数据库。MySQL 使用标准的 SQL 数据语言形式，它可以运行于多个系统上，并支持多种语言，这些编程语言包括 C、C++、Python、Java、Perl、PHP、Eiffel、Ruby 等。MySQL 对 PHP 有很好的支持，PHP 是目前最流行的 Web 开发语言之一。MySQL 支持大型数据库，支持 5000 万条记录的数据仓库，32 位系统表文件最大可支持 4GB，64 位系统支持最大的表文件为 8TB。MySQL 是可以定制的，采用了 GPL 协议，用户可以修改源码来开发自己的 MySQL 系统。

在 Web 应用方面，MySQL 是最好的关系数据库管理系统（Relational Database Management System，RDBMS）应用软件之一。

对于 OpenStack 来说，MySQL 是在组件中应用最流行的数据库，后面的实训内容会详细介绍如何安装使用 MySQL 数据库为 OpenStack 的组件提供服务。

1. MySQL 数据库的特点以及主要术语

数据库（Database）是按照数据结构来组织、存储和管理数据的仓库，每个数据库都有一个或多个不同的 API 用于创建、访问、管理、搜索和复制所保存的数据。也可以将数据存储在文件中，但是在文件中读写数据速度相对较慢。所以，使用关系型数据库管理系统（RDBMS）来存储和管理的大数据量。所谓的关系型数据库，是建立在关系模型基础上的数据库，借助于集合代数等数学概念和方法来处理数据库中的数据，它具有如下的特点。

① 数据以表格的形式出现。
② 每行为各种记录名称。
③ 每列为记录名称所对应的数据域。
④ 许多的行和列组成一张表单。
⑤ 若干的表单组成数据库。

在使用 MySQL 数据库前，首先了解下关系型数据库管理系统（RDBMS）的一些主要术语。

- 数据库：数据库是一些关联表的集合。。
- 数据表：表是数据的矩阵。在一个数据库中的表看起来像一个简单的电子表格。
- 列：一列（数据元素）包含了相同的数据，如邮政编码的数据。
- 行：一行（元组或记录）是一组相关的数据，如一条用户订阅的数据。
- 冗余：存储两倍数据，冗余可以使系统速度更快。
- 主键：主键是唯一的。一个数据表中只能包含一个主键，可以使用主键来查询数据。
- 外键：外键用于关联两个表。
- 复合键：复合键（组合键）将多个列作为一个索引键，一般用于复合索引。

- 索引：使用索引可快速访问数据库表中的特定信息。索引是对数据库表中一列或多列的值进行排序的一种结构，类似于书籍的目录。
- 参照完整性：参照的完整性要求关系中不允许引用不存在的实体。与实体完整性是关系模型必须满足的完整性约束条件，目的是保证数据的一致性。

2. MySQL 数据库的基本操作

在已经安装好 MySQL 数据库的 Linux 系统中，可以使用 MySQL 二进制方式进入到 MySQL 命令提示符下，连接 MySQL 数据库。使用命令连接 MySQL 服务器，命令如下。

```
[root@controller ~]# mysql -u root -p000000
```

其中 -u 后面是数据库的用户名，默认的管理员用户为 root，-p 后面是该用户的密码，也可以在 -p 后不输入密码，系统出现提示输入时再输入。

```
[root@controller ~]# mysql -u root -p
Enter password: ******
```

在登录成功后会出现 mysql> 命令提示窗口，用户可以在上面执行任何 SQL 语句，如图 5.1.1 所示。

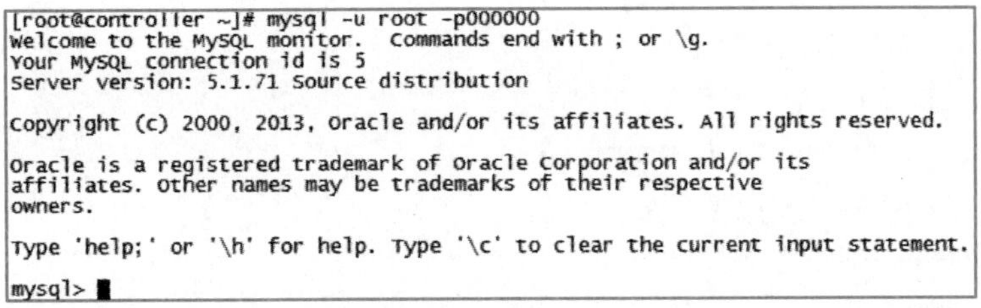

```
[root@controller ~]# mysql -u root -p000000
Welcome to the MySQL monitor.  Commands end with ; or \g.
Your MySQL connection id is 5
Server version: 5.1.71 Source distribution

Copyright (c) 2000, 2013, Oracle and/or its affiliates. All rights reserved.

Oracle is a registered trademark of Oracle Corporation and/or its
affiliates. Other names may be trademarks of their respective
owners.

Type 'help;' or '\h' for help. Type '\c' to clear the current input statement.

mysql>
```

图 5.1.1 成功登录 MySQL 数据库

在上述命令中使用了 root 用户登录到 MySQL 服务器，也可以使用其他 MySQL 用户登录。如果用户权限足够，任何用户都可以在 MySQL 的命令提示窗口中进行 SQL 操作。

退出 mysql> 命令提示窗口可以使用 exit 命令，命令如下。

```
mysql> exit;
```

使用普通用户登录，可能需要特定的权限来创建或者删除 MySQL 数据库。这里使用 root 用户登录，root 用户拥有最高权限，可以使用 mysqladmin 命令来创建数据库。

下面创建一个名为 demo 的数据库，命令如下。

```
[root@controller ~]# mysqladmin -u root -p create demo
Enter password: ******
```

也可以先登录数据库，再进行创建数据库，命令如下。

```
[root@controller ~]# mysql -uroot -p000000
mysql> create database demo;
```

以上两种命令执行后会创建 MySQL 数据库 demo，在登录后用语句创建数据库时要注意语句最后的分号。

在连接到 MySQL 数据库后，可能有多个可以操作的数据库，要通过命令选择相关的数据库进行操作。

下面登录数据库,然后选择数据库 demo,命令如下。

```
[root@controller ~]# mysql -u root -p
Enter password: ******
mysql> use demo;
Database changed
mysql>
```

可以在一个数据库中继续创建属于该库的 MySQL 数据表,以下为创建 MySQL 数据表的 SQL 通用语法。

```
CREATE TABLE table_name ( column_name column_type );
```

通过 mysql> 命令窗口可以很简单地创建 MySQL 数据表,可以使用 SQL 语句 CREATE TABLE 来创建数据表。

下面创建一个名为 demo_tbl 的数据表,命令如下:

```
[root@controller ~]# mysql -u root -p
Enter password: *******
mysql> use demo;
Database changed
mysql> CREATE TABLE demo_tbl(
    -> demo_id INT NOT NULL AUTO_INCREMENT,
    -> demo_title VARCHAR(100) NOT NULL,
    -> demo_author VARCHAR(40) NOT NULL,
    -> submission_date DATE,
    -> PRIMARY KEY ( demo_id )
    -> );
Query OK, 0 rows affected (0.16 sec)
mysql>
```

MySQL 表中使用 INSERT INTO SQL 语句来插入数据。

以下是向 MySQL 数据表插入数据通用的 INSERT INTO SQL 语法。

```
INSERT INTO table_name ( field1, field2, ...fieldN )
                VALUES
                    ( value1, value2, ...valueN );
```

如果数据是字符型,必须使用单引号或者双引号,如 "value"。

下面将向 demo_tbl 表中插入 3 条数据,命令如下。

```
[root@controller ~]# mysql -u root -p;
Enter password: *******
mysql> use demo;
Database changed
mysql> INSERT INTO demo_tbl
    ->( demo_title, demo_author, submission_date )
    ->VALUES
```

```
    ->("Learn PHP", "John Poul", NOW( ) );
Query OK, 1 row affected (0.01 sec)
mysql> INSERT INTO demo_tbl
    ->( demo_title, demo_author, submission_date )
    ->VALUES
    ->("Learn MySQL", "Abdul S", NOW( ) );
Query OK, 1 row affected (0.01 sec)
mysql> INSERT INTO demo_tbl
    ->( demo_title, demo_author, submission_date )
    ->VALUES
    ->("JAVA Demo", "Sanjay", '2019-11-10');
Query OK, 1 row affected (0.01 sec)
mysql>
```

需要注意的是，使用箭头标记（->）不是SQL语句的一部分，它仅仅表示一个新行，如果一条SQL语句太长，可以通过Enter键来创建一个新行来编写SQL语句，SQL语句的命令结束符为分号（；）。

在上述命令中，并没有提供demo_id的数据，因为该字段在创建表时已经设置它为AUTO_INCREMENT（自动增加）属性。所以，该字段会自动递增而不需要用户去设置。实例中NOW( )是一个MySQL函数，该函数返回日期和时间。

MySQL数据库使用SELECT语句来查询数据。以下是在MySQL数据库中查询数据通用的SELECT语法。

```
SELECT field1, field2, ...fieldN table_name1, table_name2...
[WHERE Clause]
[OFFSET M ][LIMIT N]
```

查询语句中可以使用一个或者多个表，表之间使用逗号（,）分割，并使用WHERE语句来设定查询条件。SELECT命令可以读取一条或者多条记录。可以使用星号（*）来代替其他字段，SELECT语句会返回表的所有字段数据，可以使用WHERE语句来包含任何条件。可以通过OFFSET指定SELECT语句开始查询的数据偏移量，默认情况下偏移量为0。可以使用LIMIT属性来设定返回的记录数。

下面查询数据表demo_tbl的所有记录，命令如下。

```
[root@controller ~]# mysql -u root -p
Enter password: *******
mysql> use demo;
Database changed
mysql> SELECT * from demo_tbl;
+---------+-------------+-------------+-----------------+
| demo_id | demo_title  | demo_author | submission_date |
+---------+-------------+-------------+-----------------+
|       1 | Learn PHP   | John Poul   | 2019-11-10      |
|       2 | Learn MySQL | Abdul S     | 2019-11-10      |
|       3 | JAVA Demo   | Sanjay      | 2019-11-10      |
+---------+-------------+-------------+-----------------+
```

```
3 rows in set (0.01 sec)
mysql>
```

MySQL 中 WHERE 的使用非常灵活，知道从 MySQL 表中使用 SELECT 语句来读取数据。如需有条件地从表中选取数据，可将 WHERE 子句添加到 SELECT 语句中。以下是 SELECT 语句使用 WHERE 子句从数据表中读取数据的通用语法。

```
SELECT field1,field2,...fieldN FROM table_1,table_1...
[WHERE condition1 [AND [OR]] condition2...
```

field1、field2 为需要查询的字段，table_1、tabled_2 为查询的表名，WHERE 后面为查询条件。

如果需要修改或更新 MySQL 中的数据，可以使用 UPDATE 命令来操作。以下是 UPDATE 命令修改 MySQL 数据表数据的通用 SQL 语法。

```
UPDATE table_name SET field1=new-value1,field2=new-value2
[WHERE Clause]
```

可以同时更新一个或多个字段。还可以在 WHERE 子句中指定任何条件。当需要更新数据表中指定行的数据时，WHERE 子句是非常有用的。

以下将在 UPDATE 命令中使用 WHERE 子句来更新 demo_tbl 表中指定的数据。

下面将更新数据表中 demo_id 为 3 的记录中 demo_title 的字段值，命令如下。

```
[root@controller ~]# mysql -u root -p
Enter password: *******
mysql> use demo;
Database changed
mysql> UPDATE demo_tbl
    -> SET demo_title='Learning JAVA'
    -> WHERE demo_id=3;
Query OK, 1 row affected (0.04 sec)
Rows matched: 1  Changed: 1  Warnings: 0
mysql>
```

在 MySQL 可以使用 SQL 的 DELETE FROM 命令来删除 MySQL 数据表中的记录。以下是 DELETE 语句从 MySQL 数据表中删除数据的通用语法。

```
DELETE FROM table_name [WHERE Clause]
```

如果没有指定 WHERE 子句，MySQL 表中的所有记录将被删除。

可以在 WHERE 子句中指定任何条件，还可以在单个表中一次性删除记录。例如，当删除数据表中指定的记录时 WHERE 以下实例将删除 demo_tbl 表中 demo_id 为 3 的记录，WHERE 子句是非常有用的。

```
[root@controller ~]# mysql -u root -p
Enter password: *******
mysql> use demo;
Database changed
```

```
mysql> DELETE FROM demo_tbl WHERE demo_id=3;
Query OK, 1 row affected (0.23 sec)
mysql>
```

MySQL 删除数据表：MySQL 中删除数据表是非常容易操作的，但在进行删除表操作时要非常小心，因为执行删除命令后所有数据都会消失。

以下为删除 MySQL 数据表的通用语法。

```
DROP TABLE table_name ;
```

在 mysql> 命令提示窗口中删除数据表的 SQL 语句为 DROP TABLE。

下面删除刚刚创建的数据表 demo_tbl，命令如下。

```
[root@controller ~]# mysql -u root -p
Enter password: *******
mysql> use demo;
Database changed
mysql> DROP TABLE demo_tbl;
Query OK, 0 rows affected (0.8 sec)
mysql>
```

以上是对 MySQL 数据库基本操作的介绍。

5.2 实训项目 2　MySQL 数据库的手动安装与配置

1. 实训前提环境

完成实训项目 1 中所有内容，即原生 OpenStack 云平台基本环境配置，或者从已完成实训项目 1 的镜像开始，继续完成本实训内容。

2. 实训涉及节点

controller。

3. 实训目标

① 完成 MySQL 数据库包的安装。
② 完成 MySQL 数据库配置文件的修改。
③ 完成 MySQL 数据库的启动以及各项配置。

4. 实训步骤及其详解

步骤 1：MySQL 基本组件的安装。

在 controller 节点执行 yum 源安装命令，命令如下。

```
[root@controller ~]# yum -y install mariadb mariadb-server python2-PyMySQL
```

执行上述安装命令成功后，可以看到成功标志，所有 Keystone 依赖包都安装完成，结果如图 5.2.1 所示。

步骤 2：修改 MySQL 数据库的配置文件。

首先使用 vi 进入 /etc/my.cnf.d/openstack.cnf 配置文件内，命令如下。

电子教案　MySQL 数据库的手动安装与配置

PPT　MySQL 数据库的手动安装与配置

微课 3　MySQL 数据库的手动安装与配置

```
Installed:
  mariadb.x86_64 3:10.3.10-1.el7.0.0.rdo2
Dependency Installed:
  mariadb-common.x86_64 3:10.3.10-1.el7.0.0.rdo2
  perl-Carp.noarch 0:1.26-244.el7
  perl-File-Temp.noarch 0:0.23.01-3.el7
  perl-PathTools.x86_64 0:3.40-5.el7
  perl-Pod-Usage.noarch 0:1.63-3.el7
  perl-Text-ParseWords.noarch 0:3.29-4.el7
  perl-libs.x86_64 4:5.16.3-294.el7_6
  perl-threads.x86_64 0:1.87-4.el7
  python-ply.noarch 0:3.4-11.el7
  python2-cryptography.x86_64 0:2.5-1.el7

Dependency Updated:
  mariadb-libs.x86_64 3:10.3.10-1.el7.0.0.rdo2

Complete!
[root@controller ~]#
```

图 5.2.1　MySQL 数据库安装成功

```
[root@controller ~]# vi /etc/my.cnf.d/openstack.cnf
```

然后在该文件中输入以下 7 行配置文件。

```
[mysqld]
bind-address = 192.168.100.10
default-storage-engine = innodb
innodb_file_per_table = on
max_connections = 4096
collation-server = utf8_general_ci
character-set-server = utf8
```

修改完成后保存退出。

上述配置文件中"bind-address = 192.168.100.10"作用是绑定数据库的 IP 地址为 192.168.100.10，可以通过绑定的 IP 地址来访问数据库。

步骤 3：启动 MySQL 数据库。

首先在 controller 节点执行 service 命令开启数据库服务，并通过 systemctl 命令设置数据库服务为开机自动启动，命令如下。

```
[root@controller ~]# systemctl start mariadb
[root@controller ~]# systemctl enable mariadb
```

步骤 4：对数据库进行配置。

初始化数据库，命令如下。

```
[root@controller ~]# mysql_secure_installation
```

执行上述命令后，出现如图 5.2.2 所示结果。

```
NOTE: RUNNING ALL PARTS OF THIS SCRIPT IS RECOMMENDED FOR ALL MySQL
      SERVERS IN PRODUCTION USE!    PLEASE READ EACH STEP CAREFULLY!

In order to log into MySQL to secure it, we'll need the current
password for the root user.  If you've just installed MySQL, and
you haven't set the root password yet, the password will be blank,
so you should just press enter here.

Enter current password for root (enter for none):
```

图 5.2.2　MySQL 数据库安全配置

提示输入当前的数据库密码，目前数据库没有设置密码，直接按 Enter 键即可，出现如图 5.2.3 所示结果。

```
Enter current password for root (enter for none):
OK, successfully used password, moving on...
Setting the root password ensures that nobody can log into the MySQL
root user without the proper authorisation.
Set root password? [Y/n]
```

图 5.2.3 提示设置 MySQL 数据库的密码

提示是否设置数据库密码后，输入"y"后按 Enter 键，接着输入新密码，如图 5.2.4 所示。

```
Setting the root password ensures that nobody can log into the MySQL
root user without the proper authorisation.
Set root password? [Y/n] y
New password:
```

图 5.2.4 设置 MySQL 数据库的密码

提示输入新的密码，这里设置密码为"000000"。输入"000000"后按 Enter 键，提示确认输入新密码，再次输入"000000"并按 Enter 键，如图 5.2.5 所示。

```
Setting the root password ensures that nobody can log into the MySQL
root user without the proper authorisation.
Set root password? [Y/n] y
New password:
Re-enter new password:
```

图 5.2.5 提示确认数据库密码

提示密码设置成功，并且提示是否要删除匿名用户。输入"y"并按 Enter 键，如图 5.2.6 所示，提示成功删除匿名用户，并且提示是否不允许 root 用户远程登录。输入"n"并按 Enter 键，如图 5.2.6 所示。

```
Remove anonymous users? [Y/n] y
 ... Success!
Normally, root should only be allowed to connect from 'localhost'.  This
ensures that someone cannot guess at the root password from the network.
Disallow root login remotely? [Y/n] n
 ... skipping.
```

图 5.2.6 删除匿名用户并禁用 root 远程登陆

提示是否删除"test"数据库，输入"y"并按 Enter 键，提示是否现在重载权限表，输入"y"并按 Enter 键，如图 5.2.7 所示。

```
Remove test database and access to it? [Y/n] y
 - Dropping test database...
 ... Success!
 - Removing privileges on test database...
 ... Success!
Reloading the privilege tables will ensure that all changes made so far
will take effect immediately.
Reload privilege tables now? [Y/n] y
 ... Success!

Cleaning up...


All done!  If you've completed all of the above steps, your MySQL
installation should now be secure.

Thanks for using MySQL!

[root@controller ~]#
```

图 5.2.7 删除 test 数据库并重载权限表

出现上述提示，说明对数据库的安全配置已经完成。

至此，已经完成所有数据库的安装。

步骤 5：常见错误及调试排错。

① 使用 MySQL 命令登录数据库时登录失败，提示以下错误：

```
ERROR 1045 (28000): Access denied for user 'root'@'localhost' (using password: YES)
```

解决方法：提示 root 用户使用密码访问本地数据库被拒绝，一般是密码输入有误。重新输入正确的密码即可。

② 启动数据库服务失败。

在启动 MySQL 服务时，提示失败，命令如下。

```
[root@controller ~]# systemctl status mariadb
```

反馈结果如图 5.2.8 所示。

```
Another MySQL daemon already running with the same unix soc
Starting mysqld:
```

图 5.2.8　MySQL 服务启动失败反馈结果

解决方法如下。

● 删除 /var/lib/mysql/ 目录下的 mysql.sock 文件。

```
[root@controller ~]# rm -rf /var/lib/mysql/mysql.sock
```

● 然后重新启动即可，命令如下。

```
[root@controller ~]# systemctl restart mariadb
```

5.3　MySQL 数据库安装脚本及其解读

在了解了如何手动安装 MySQL 数据库后，下面介绍另一种安装 MySQL 数据库的方法，即利用已经写好的脚本自动安装 MySQL 数据库。

1. 需要安装的脚本内容

```bash
#!/bin/bash
yum -y install mariadb mariadb-server python2-PyMySQL expect
cat > /etc/my.cnf.d/openstack.cnf << EOF
[mysqld]
bind-address = 192.168.100.10
default-storage-engine = innodb
innodb_file_per_table = on
max_connections = 4096
collation-server = utf8_general_ci
character-set-server = utf8
EOF
systemctl enable mariadb.service
systemctl start mariadb.service
expect -c "
spawn /usr/bin/mysql_secure_installation
```

```
expect \"Enter current password for root (enter for none): \"
send \"\r\"
expect \"Set root password?\"
send \"y\r\"
expect \"New password: \"
send \"000000\r\"
expect \"Re-enter new password: \"
send \"000000\r\"
expect \"Remove anonymous users?\"
send \"y\r\"
expect \"Disallow root login remotely?\"
send \"n\r\"
expect \"Remove test database and access to it?\"
send \"y\r\"
expect \"Reload privilege tables now?\"
send \"y\r\"
expect eof
"
```

上面内容与手动安装 MySQL 数据库的过程基本一致，这里讲解在上述脚本中出现的特殊命令。

```
cat > /etc/my.cnf.d/openstack.cnf << EOF
[mysqld]
bind-address = 192.168.100.10
default-storage-engine = innodb
innodb_file_per_table = on
max_connections = 4096
collation-server = utf8_general_ci
character-set-server = utf8
EOF
```

- "cat > 文件名 << EOF" 格式的命令作用是创建文件，之后输入的任何内容都在该文件中，输入完成后以 EOF 结尾，代表结束。

重点是● expect 命令主要用于完成系统交互功能，由于 MySQL 在进行安全配置时，需要进行一些交互操作，在使用脚本时为了减少人的参与，所以在脚本中使用了一些 expect 命令。

- -c 参数表示执行 -c 参数之后的内容。
- spawn 命令是 expect 的初始命令，它用于启动一个进程，之后所有的 expect 操作都在这个进程中进行。spawn 命令后紧跟着是需要执行的 MySQL 安全配置的命令。
- expect \"Enter current password for root（enter for none）: \" 命令的作用是在返回的信息中查找并匹配双引号中的字符串，这里出现的 "\" 是转译符号无实际意义。后面所有与之结构类似的命令都是相同的作用。
- send 命令的作用是输出双引号中的内容，这里的 "\r" 表示回车键（Enter 键），"y\r"

表示输出"y"和"回车"。

- expect eof 命令表示上述命令结束，退出 expect 命令进程。

2. 脚本的执行

刚写好的脚本文件，在核实过内容后，是不能直接使用的。要想使用脚本，首先需要赋予脚本执行权限，命令如下。

```
[root@controller ~]# chmod a+x iaas-install-mysql.sh
```

或者执行

```
[root@controller ~]# chmod 777 iaas-install-mysql.sh
```

上述两种命令都可以实现赋予脚本执行权限的效果，这里的 iaas-install-mysql.sh"是示范用的脚本名称，读者在实际使用时可自行命名脚本名称。

说明：本次实验所使用的脚本均放在 root 用户的家目录里。

脚本有了执行权限后，就可以执行了。这里介绍两种执行脚本的方法。

方法 1：在脚本前加上脚本所在的路径，可以使用相对路径，也可以使用绝对路径，这取决于读者个人习惯，命令如下。

```
[root@controller ~]# ./iaas-install-mysql.sh
```

"."表示当前目录，脚本放在家目录中，而用户此时也在家目录中，所以为方便，这里使用的是当前目录。

方法 2：现将脚本所在路径加入环境变量 PATH 中，然后就可以像"ls"命令一样直接在命令中输入使用，命令如下。

```
[root@controller ~]# echo $PATH
/usr/local/sbin:/usr/local/bin:/sbin:/bin:/usr/sbin:/usr/bin:/root/bin
[root@controller ~]# PATH=${PATH}:/root/
[root@controller ~]# echo $PATH
/usr/local/sbin:/usr/local/bin:/sbin:/bin:/usr/sbin:/usr/bin:/root/bin:/root/
[root@controller ~]# iaas-install-mysql.sh
```

上述命令对比了将脚本所在目录加入环境变量 PATH 前后，环境变量 PATH 中的内容，最后可以直接在命令中像其他命令一样直接输入脚本名称。

习题 MySQL 数据库的安装及其配置

说明：直接在命令中改变的环境变量只对当前 Shell 和其子 Shell 有效，当重启或另外打开新的 Shell 时，当前修改的环境变量无效。然而可以通过修改 /root/.bash_profile 配置文件中 PATH 的内容，达到环境变量 PATH 永久有效的效果，注意修改 /root/.bash_profile 后需要通知系统内容变化，最直接的方法是重新打开新的 Shell。

第 6 章　Keystone 的安装及其配置

本章导读：
　　本章介绍 Keystone 组件的基本功能，作为云平台的认证服务模块 Keystone 对整个云平台的正常运行起着至关重要的作用，在实训项目 3 中重点介绍如何通过命令行以及服务组件的参数配置安装 Keystone 服务，同时解读 Keystone 在安装和使用过程中可能会出现的问题。在以上基础上，最后还将进一步介绍 Keystone 服务安装脚本的制作和使用。

电子资源：
　　电子教案　Keystone 的安装及其配置
　　PPT　Keystone 的安装及其配置
　　习题　Keystone 的安装及其配置

第 6 章　Keystone 的安装及其配置

6.1 Keystone 功能详解

电子教案 Keystone 功能详解

PPT Keystone 功能详解

1. Keystone 简介

Keystone 是 OpenStack 认证服务的项目名称，也是整个云平台的入口。如果把 OpenStack 云平台当作一个"屋子"，那么 Keystone 就是屋子的"大门"，想要拿到屋子里的任何一样东西，都必须通过这个大门。其实，在 OpenStack 中，每个服务都有自己的一扇"门"，要开启这扇门必须有对应的钥匙，而在 OpenStack 中 Keystone 也为这些具体的服务的小"门"提供了钥匙管理的功能。

Keystone 服务通过 OpenStack 的应用程序编程接口（API）提供令牌、策略以及目录功能。与其他 OpenStack 项目一样，Keystone 表示一个抽象的层，并不实际提供任何用户管理功能，而是会提供插件接口，以便组织可以利用其当前的身份验证服务，或者从市场上的各种身份管理系统（如 LDAP、AD）中进行选择，目前 Keystone 项目提供两种接口 WSGI 和 REST/HTTP。

2. Keystone 中的基本概念

Keystone 的组件包含了一系列的概念，这些概念之间又相互紧密关联，以下是 Keystone 中所涉及的概念。

● Keystone 身份认证：身份认证是确定"你是谁"的过程，Keystone 需要确认打开"大门"（所有传入的功能调用请求）的主人身份（发出请求的用户）。确认过程需要核对钥匙是否匹配（凭证测试）。这些钥匙可以是仅主人能够得到的数据信息，如密码、口令、密钥，或者是一些经过特殊处理的信息，如硬件令牌，也可以是一些生物信息，如指纹识别、视网膜识别等。

● Keystone 令牌（Token）：Keystone 在确认了"主人"身份后，会给其发放一张"邀请函"，即令牌，每张令牌都包含了用户的操作范围，用户只能在令牌列出的资源内对云平台进行操作，对于用户令牌内未标识的资源，用户将无法访问。令牌也有一定的时效性，超时的令牌也无法访问云平台资源。如果需要删除特定用户对特定资源的访问权限，可以对用户进行删除令牌操作，即使用户通过 Keystone 的身份认证，在没有令牌的环境下，也无法使用云平台资源。

● Keystone 用户（User）：用户是 Keystone 身份管理系统核心对象，是使用云平台的个人、系统或其他服务的 ID 表示，访问云平台的所有对象都必须具有用户标识。

● Keystone 项目（Project）：如果说用户是"房子"中的"人"，那么项目就是房子里的家族。用户通常会被分配给相对应的项目容器。项目容器中会存放一系列具有相同权限的用户，一般来说，项目标识一个客户、账户或一系列组织单位。一个用户可以隶属于多个项目，并具有这些项目的权限。

● Keystone 角色（Role）：当用户经历过"你是谁"的过程后，紧接着就需要确定"你能做什么"，即身份级别的认证。Keystone 定义了一个角色的概念，用于封装一组特定的权限。角色的权限信息会以令牌的形式发放给项目，在访问资源服务时，由资源服务将用户所属角色组（项目）与所请求的资源操作组进行匹配，继而决定是否运行其访问。在实训项目 3 中会看到整个用户、项目、角色相关联的操作过程。

- Keystone 服务（Service）：这里的服务可以理解为服务对象，OpenStack 中每个组件都是 Keystone 的服务对象，如 Nova、Glance、Swift、Heat、Ceilometer 等。Nova 提供云计算的服务，Glance 提供镜像管理服务，Swift 提供对象存储服务，Heat 提供资源编排服务，Ceilometer 则是提供告警计费服务，Cinder 提供块存储服务。

- Keystone 端点（Endpoint）：Endpoint 翻译为"端点"，也称为服务端点。Service 服务显得太抽象笼统，Endpoint 则具体化 Service。可以理解它是一个服务暴露出来的访问点，如果需要访问一个服务，则必须知道其 Endpoint，而 Endpoint 一般为 URL，知道了服务的 URL，就可以访问它。Endpoint 的 URL 具有 public、private 和 admin 这 3 种权限。public URL 可以被全局访问，private URL 只能被局域网访问，admin URL 被从常规的访问中分离，同样在后面的实训项目 3 中会看到上述端点的配置过程，通常来说端点的是基于 HTTP 的一个 URL。

- Keystone 服务目录：Keystone 除了提供认证服务外，还提供了一个附加服务，用于端点（Endpoint）发现的服务目录。服务目录包含了云平台可用服务的清单以及其 API 端点，端点即一个可供访问的网络地址（一般以 URL 的形式）以及端口号，用户可以选择其中一项或多项进行注册，在 Keystone 服务目录中注册的服务就可以直接到该服务所对应的 Keystone 端点，对服务资源进行操作。

3. Keystone 内部的安全机制

Keystone 服务的安全机制保障了整个云平台的安全与灵活，如图 6.1.1 所示是用户通过 Keystone "大门"进入 OpenStack "屋子"的过程，即 Keystone 内部安全机制的运行流程，可以帮助理解 Keystone 这扇"大门"是如何运作的。

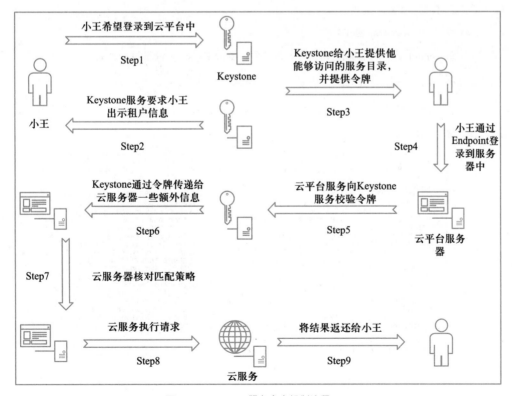

图 6.1.1　Keystone 服务安全机制流程

用户 Peter 想要进入云平台时，Keystone 会要求其声明所属项目，并提供邀请函（令牌），Keystone 通过匹配目录服务以及权限组告诉 Peter 所能访问的所有服务，并将 Peter 的请求转交给各项服务。云平台服务收到用户请求后将会执行请求，最后将结果返回给用户。

6.2 实训项目 3　Keystone 的手动安装与配置

电子教案　Keystone 的手动安装与配置

PPT　Keystone 的手动安装与配置

1. 实训前提环境

成功完成实训项目 2 中所有内容后开始本实训，或者从已完成实训项目 2 的镜像开始，继续完成本实训内容。

2. 实训涉及节点

Controller。

3. 实训目标

① 完成 Keystone 基本组件的安装。
② 完成 Keystone 数据库的创建以及授权。
③ 完成 Keystone 主配置文件的修改。
④ 完成 Keystone 安全与认证配置。
⑤ 完成 Keystone 用户、项目、角色以及服务和端点的创建。
⑥ 完成 Keystone 环境变量脚本的创建。

微课 4　Keystone 的手动安装与配置

4. 实训步骤及其详解

步骤 1：Keystone 基本组件的安装。

在 controller 节点上执行 yum 源安装命令，安装 Keystone 依赖包，命令如下。

```
[root@controller ~]# yum -y install openstack-keystone httpd mod_wsgi
```

执行上述安装命令成功后，可以看到成功标志，所有 Keystone 依赖包都安装完成，如图 6.2.1 所示。

图 6.2.1　Keystone 依赖包安装完成反馈结果

步骤 2：创建 Keystone 数据库并授权。

首先，登录 MySQL 数据库，命令如下。

```
[root@controller ~]# mysql -uroot -p000000
```

登录后，创建 Keystone 数据库，命令如下。

```
mysql>CREATE DATABASE keystone;
```

看到提示 Query OK，1 row affected（0.00 sec），表明数据库创建成功。

接着创建 MySQL 的 Keystone 用户，并赋予其 Keystone 数据库的操作权限，命令如下。

```
mysql> GRANT ALL PRIVILEGES ON keystone.* TO 'keystone'@'localhost' IDENTIFIED BY '000000';
mysql> GRANT ALL PRIVILEGES ON keystone.* TO 'keystone'@'%' IDENTIFIED BY '000000';
mysql> exit
```

上述 SQL 语句中，第 1 个 Keystone 为表名，第 2 个 Keystone 为 MySQL 的用户名，"*"代表该库中所有表，"%"代表所有主机，localhost 代表本机。通用的语法为：GRANT ALL PRIVILEGES ON 数据库名.表名 TO ' 用户名 '@' 主机 ' IDENTIFIED BY ' 密码 ';

执行上述 SQL 语句后，可以尝试使用 Keystone 用户、密码为 000000 登录 MySQL 数据库，登录命令如下。

```
[root@controller ~]# mysql -ukeystone -p000000
```

本书随后的实训项目中，均以上述方式创建数据库并进行数据库的授权，故后续实训项目中不再赘述。

步骤 3：修改配置 Keystone 文件。

```
[root@controller ~]# crudini --set /etc/keystone/keystone.conf database connection mysql+pymysql: //keystone: 000000@controller/keystone
```

上述命令的功能是在 /etc/keystone/keystone.conf 文件的 [database] 段落中添加 "connection = mysql+pymysql（数据库类型）: //keystone（登录数据库的用户名）: 000000（用户密码）@controller（数据库主机名）/keystone（数据库）"配置。用户通过 vi 命令直接修改 /etc/keystone/keystone.conf 文件，也可以达到相同效果。

```
[root@controller ~]# crudini --set /etc/keystone/keystone.conf token provider fernet
```

接着，需要同步数据库，为认证服务创建数据库表，命令如下。

```
[root@controller ~]# su -s /bin/sh -c "keystone-manage db_sync" keystone
```

可以通过执行下面一条语句，查看是否同步并创建成功，命令如下。

```
[root@controller ~]# mysql -u root -p000000 -e "use keystone; show tables; "
```

结果如图 6.2.2 所示。

步骤 4：Keystone 安全与认证配置。

默认情况下，Keystone 使用 fernet 令牌，创建签名密钥和证书，并且限制对生成的数据的访问，命令如下。

```
+-------------------------------+
| Tables_in_keystone            |
+-------------------------------+
| access_rule                   |
| access_token                  |
| application_credential        |
| application_credential_access_rule |
| application_credential_role   |
| assignment                    |
| config_register               |
| consumer                      |
| credential                    |
| endpoint                      |
| endpoint_group                |
| federated_user                |
| federation_protocol           |
| group                         |
| id_mapping                    |
| identity_provider             |
| idp_remote_ids                |
| implied_role                  |
| limit                         |
| local_user                    |
```

图 6.2.2　查看 Keystone 数据库同步情况

```
[root@controller ~]# keystone-manage fernet_setup --keystone-user keystone --keystone-group keystone
[root@controller ~]# keystone-manage credential_setup --keystone-user keystone --keystone-group keystone
```

步骤 5：创建端点（endpoint）。

创建一个端点时，需要分别为 public API、internal API 和 admin API 指定 URL。可以使用 IP 地址，也可以使用设定好的主机名即用户定义的 controller，命令如下。

```
[root@controller ~]# keystone-manage bootstrap --bootstrap-password 000000 --bootstrap-admin-url http://controller:5000/v3/ --bootstrap-internal-url http://controller:5000/v3/ --bootstrap-public-url http://controller:5000/v3/ --bootstrap-region-id RegionOne
```

步骤 6：启动认证服务。

```
[root@controller ~]# sed -i "s/#ServerName www.example.com:80/ServerName controller/g" /etc/httpd/conf/httpd.conf
[root@controller ~]# ln -s /usr/share/keystone/wsgi-keystone.conf /etc/httpd/conf.d/
[root@controller ~]# systemctl start httpd.service
```

接着将服务加入开机自启，命令如下。

```
[root@controller ~]# systemctl enable httpd.service
```

上述命令执行后，通过命令 systemctl 可以查看，看到 enabled 即表示成功加入开机自启，结果如下。

```
[root@controller ~]# systemctl is-enabled httpd
enabled
```

步骤 7：创建用户、项目和角色。

在安装认证服务之后，需要设置用户、项目以及角色来进行认证。

通常，需要指定一个用户名和密码通过认证服务进行身份验证。默认情况下，Keystone 已经创建好了 default 域、admin 用户、admin 项目以及 admin 角色，由于已经拥有 admin 相关的用户、项目和角色，下面将直接导入 admin 相关的环境变量来自定义其他所需要的域、项目以及用户，命令如下。

```
[root@controller ~]# export OS_USERNAME=admin
[root@controller ~]# export OS_PASSWORD=000000
[root@controller ~]# export OS_PROJECT_NAME=admin
[root@controller ~]# export OS_USER_DOMAIN_NAME=default
[root@controller ~]# export OS_PROJECT_DOMAIN_NAME=default
[root@controller ~]# export OS_AUTH_URL=http: //controller: 5000/v3
[root@controller ~]# export OS_IDENTITY_API_VERSION=3
```

用户可以通过 env 命令查看环境变量，来验证上述语句创建的环境变量是否生效，命令如下。

```
[root@controller ~]# env
```

现在将自定义 demo 域、service 项目、demo 项目、demo 用户，并授予 demo 用户权限，命令如下。

```
[root@controller~]# openstack domain create --description "My Domain" demo
```

结果如图 6.2.3 所示。

```
+-------------+----------------------------------+
| Field       | Value                            |
+-------------+----------------------------------+
| description | My Domain                        |
| enabled     | True                             |
| id          | 7acea2aa58a74c24906f87727e0131ce |
| name        | demo                             |
| tags        | []                               |
+-------------+----------------------------------+
```

图 6.2.3　创建域的反馈结果

上述命令用于创建域 demo，并为其添加描述即"--description My Domain"，这里创建了一个 admin 用户。

```
[root@controller~]# openstack project create --domain default --description "My Project" demo
```

结果如图 6.2.4 所示。

```
+-------------+----------------------------------+
| Field       | Value                            |
+-------------+----------------------------------+
| description | My Project                       |
| domain_id   | default                          |
| enabled     | True                             |
| id          | aa2becf12fcb4070ae7c3df5d235bb3a |
| is_domain   | False                            |
| name        | demo                             |
| parent_id   | default                          |
| tags        | []                               |
+-------------+----------------------------------+
```

图 6.2.4　创建项目的反馈结果

上述命令用于创建项目，这里创建了一个项目，名为 demo 并为其添加描述。

```
[root@controller~]# openstack user create --domain default --password 000000 demo
```

结果如图 6.2.5 所示。

上述命令用于创建用户，这里创建了一个 demo 用户。本书实训中所有的密码都为 000000，在实际生产环境中，应该按实际需要设定密码。

```
[root@controller~]# openstack role create user
[root@controller~]# openstack role add --project demo --user demo user
```

```
+--------------------+----------------------------------+
| Field              | Value                            |
+--------------------+----------------------------------+
| domain_id          | default                          |
| enabled            | True                             |
| id                 | 7b00af271df34935af0fdfc95992e68d |
| name               | demo                             |
| options            | {}                               |
| password_expires_at| None                             |
+--------------------+----------------------------------+
```

图 6.2.5　创建用户的反馈结果

上述命令可以实现用户、角色和组合的连接，即将用户添加到组合中，并赋予项目相应角色即权限，这里将 admin 用户添加到 admin 项目中并为 admin 用户绑定 admin 角色。同时，需要注意的是，执行绑定角色赋予权限的这条命令是没有任何输出的。

在 OpenStack 中，几乎所有的服务（包括 Keystone 服务）要想正常运行，都必须首先向 Keystone 服务器注册。每一个服务需要向 Keystone 注册用户信息、服务（service）和端点（endpoint）信息。为方便管理，这里创建一个 service 项目，将本书所有的服务都创建在 service 项目下，命令如下。

```
[root@controller~]# openstack project create --domain default --description "Service Project" service
```

结果如图 6.2.6 所示。

```
+-------------+----------------------------------+
| Field       | Value                            |
+-------------+----------------------------------+
| description | Service Project                  |
| domain_id   | default                          |
| enabled     | True                             |
| id          | 40d4894c641d4381bbc19ee24cd464b4 |
| is_domain   | False                            |
| name        | service                          |
| parent_id   | default                          |
| tags        | []                               |
+-------------+----------------------------------+
```

图 6.2.6　创建 service 项目的反馈结果

步骤 8：验证认证服务 Keystone 的安装。

接着，用管理凭证和管理终端建立一个 admin-openrc.sh 文件，命令如下。

```
[root@controller ~]# vi admin-openrc.sh
```

在 admin-openrc.sh 文件中写入以下内容。

```
export OS_USERNAME=admin
export OS_PASSWORD=000000
export OS_PROJECT_NAME=admin
export OS_USER_DOMAIN_NAME=default
export OS_PROJECT_DOMAIN_NAME=default
export OS_AUTH_URL=http://controller:5000/v3
export OS_IDENTITY_API_VERSION=3
```

然后，在环境变量中导入这个文件来读取，使环境变量生效，命令如下。

```
[root@controller ~]# source admin-openrc.sh
```

可以通过 Keystone 命令验证 admin-openrc.sh 文件是否被正确配置，命令如下。

```
[root@controller ~]# openstack token issue
```

上述命令会返回一个令牌和指定项目的 ID，这能验证已经正确配置的环境变量。

接下来需要验证管理员账户已经被授权执行管理命令的权限。可以通过查看 openstack user list 命令输出中的 name 是否匹配 openstack role list 命令中的 user name，以及管理角色是否对用户和相关组合罗列出来，这能验证用户账户是否拥有绑定角色相匹配的 admin 权限。命令如下。结果如图 6.2.7 和图 6.2.8 所示。

```
[root@controller ~]# openstack user list
```

```
+----------------------------------+-------+
| ID                               | Name  |
+----------------------------------+-------+
| 182ff4ebbe6d479e94efb4b4de489cc2 | admin |
| 7b00af271df34935af0fdfc95992e68d | demo  |
+----------------------------------+-------+
```

图 6.2.7　查看用户列表的反馈结果

```
[root@controller ~]# openstack role list --user admin --project admin
```

```
+----------------------------------+-------+---------+-------+
| ID                               | Name  | Project | User  |
+----------------------------------+-------+---------+-------+
| 0d8a96a47a6947c98a4937eee1bea040 | admin | admin   | admin |
+----------------------------------+-------+---------+-------+
```

图 6.2.8　查看权限的反馈结果

上述命令中，只要通过命令或环境变量定义的认证信息和认证服务端点，就可以从任意一台机器上运行所有的 OpenStack 命令。

步骤 9：常见错误及调试排错。

在执行 Keystone 相关命令时，经常提示认证错误信息，例如，查看用户列表命令如下。

```
[root@controller ~]# openstack user list
```

执行命令后出现了如下的错误提示。

```
The request you have made requires authentication. (HTTP 401)
```

一般出现这样的情况都是环境变量导入错误，在与 Keystone 进行认证时没有通过。

首先查看环境变量文件，命令如下。

```
[root@controller ~]# vi admin-openrc.sh
```

发现文件中的环境变量如下，其中第 2 行的用户密码不是创建用户时所对应的密码。

```
export OS_USERNAME=admin
export OS_PASSWORD=123456
export OS_PROJECT_NAME=admin
export OS_USER_DOMAIN_NAME=default
export OS_PROJECT_DOMAIN_NAME=default
export OS_AUTH_URL=http://controller:5000/v3
export OS_IDENTITY_API_VERSION=3
```

将其改为

```
export OS_PASSWORD=000000
```

保存后重新导入该环境变量文件，命令如下。

```
[root@controller ~]# source admin-openrc.sh
```

重新查看用户列表，如图 6.2.9 所示，发现用户显示正常。

```
+----------------------------------+---------+---------+---------------------+
|                id                |  name   | enabled |        email        |
+----------------------------------+---------+---------+---------------------+
| bc23d4f89c314521a030f620bf548a7e |  admin  |  True   |     ADMIN_EMAIL     |
| 757a77a7c7ee4815943aeab6184b2973 |  cinder |  True   | cinder@example.com  |
| 37d2da298f1b435a8acb29118780ba3a |  demo   |  True   |                     |
| fcdbedbd601e4785907e44f584aa7a44 |  glance |  True   | glance@example.com  |
| 1161e71d1c2243ef981c4cc133e32807 | neutron |  True   | neutron@example.com |
| 1094736d89f643b6800579b2b2981d12 |  nova   |  True   |  nova@example.com   |
| e40548f479854f8497381f17dc71b9c3 |  user1  |  True   |                     |
+----------------------------------+---------+---------+---------------------+
```

图 6.2.9 更正环境变量文件后 Keystone 命令正常执行

6.3 Keystone 安装脚本及其解读

电子教案 Keystone 安装脚本及其解读

PPT Keystone 安装脚本及其解读

在介绍了如何手动安装 Keystone 服务后，接下来将介绍另一种安装 Keystone 服务的方法，即利用已经写好的脚本自动安装 Keystone 服务。

1. 需要安装的脚本内容

```bash
#!/bin/bash
# Keystone 基本组件的安装
yum -y install openstack-keystone httpd mod_wsgi
# Keystone 数据库的创建以及授权
mysql -uroot -p000000 -e "create database IF NOT EXISTS keystone ; "
mysql -uroot -p000000 -e "GRANT ALL PRIVILEGES ON keystone.* TO 'keystone'@'localhost' IDENTIFIED BY '000000'; "
mysql -uroot -p000000 -e "GRANT ALL PRIVILEGES ON keystone.* TO 'keystone'@'%' IDENTIFIED BY '000000'; "
# Keystone 主配置文件的修改
crudini --set /etc/keystone/keystone.conf database connection mysql+pymysql://keystone:000000@controller/keystone
crudini --set /etc/keystone/keystone.conf token provider fernet
su -s /bin/sh -c "keystone-manage db_sync" keystone
# Keystone 安全与认证配置
keystone-manage fernet_setup --keystone-user keystone --keystone-group keystone
keystone-manage credential_setup --keystone-user keystone --keystone-group keystone
keystone-manage bootstrap --bootstrap-password 000000 \
  --bootstrap-admin-url http://controller:5000/v3/ \
  --bootstrap-internal-url http://controller:5000/v3/ \
  --bootstrap-public-url http://controller:5000/v3/ \
  --bootstrap-region-id RegionOne
sed -i "s/#ServerName www.example.com:80/ServerName controller/g" /etc/httpd/conf/httpd.conf
ln -s /usr/share/keystone/wsgi-keystone.conf /etc/httpd/conf.d/
systemctl enable httpd.service
systemctl start httpd.service
```

```
# 导入临时环境变量
export OS_USERNAME=admin
export OS_PASSWORD=000000
export OS_PROJECT_NAME=admin
export OS_USER_DOMAIN_NAME=default
export OS_PROJECT_DOMAIN_NAME=default
export OS_AUTH_URL=http: //controller: 5000/v3
export OS_IDENTITY_API_VERSION=3

# Keystone 用户、项目、角色以及服务和端点的创建
openstack domain create --description "My Domain" demo
openstack project create --domain default --description "My Project" demo
openstack user create --domain default --password 000000 demo
openstack role create user
openstack role add --project demo --user demo user
openstack project create --domain default --description "Service Project" service

# Keystone 环境变量脚本的创建
cat > /etc/keystone/admin-openrc.sh << EOF
export OS_PROJECT_DOMAIN_NAME=Default
export OS_USER_DOMAIN_NAME=Default
export OS_PROJECT_NAME=admin
export OS_USERNAME=admin
export OS_PASSWORD=000000
export OS_AUTH_URL=http: //controller: 5000/v3
export OS_IDENTITY_API_VERSION=3
export OS_IMAGE_API_VERSION=2
EOF
```

上面内容是写好的 Keystone 服务脚本，从中可以看到，在 Keystone 服务中，脚本安装与利用命令手动安装的内容基本一致。

这里，重点介绍脚本最后出现的"cat > FILENAME <<-EOF ……EOF"形式的命令。

在 Shell 脚本中，通常将 EOF 与 <<（输入重定向符号）结合使用，表示后续的输入作为子命令或子 Shell 的输入，直到再次在独立的一行遇到 EOF 为止，再返回主 Shell。这里的 EOF 只是一个分界符，也可以用其他任何字符串代替，只要是后面内容段中没有出现的字符串都可以，EOF 只是一个启示和结束的标志。这里需要说明的是，"<<-"与"<<"都是重定向操作符，两者的区别是，在使用 cat<<EOF 时，输入完成后，需要在新的一行输入 EOF 作为结束分界符，此时作为结束符的 EOF 必须顶行写，前面不能用制表符或者空格，而"<<-"就可以解决这个问题，即使 EOF 前有制表符或者空格也不会出错。通常使用命令 cat，来实现一些多行的屏幕输入或者创建临时文件。

2. 脚本的执行

刚写好的脚本文件，在核实过内容后，是不能直接使用的。要想使用脚本，首先需要

赋予脚本执行权限，命令如下。

```
[root@controller ~]# chmod a+x iaas-install-keystone.sh
```

或者执行

```
[root@controller ~]# chmod 777 iaas-install-keystone.sh
```

上述两种命令都可以实现赋予脚本执行权限的效果，这里的"iaas-install-keystone.sh"是示范用的脚本名称，读者在实际使用时可自行命名脚本名称。

说明：本次实训所使用的脚本均放在 root 用户的家目录里。

脚本有了执行权限后，就可以执行了。这里介绍两种执行脚本的方法。

方法 1：在脚本前加上脚本所在的路径，可以使用相对路径，也可以使用绝对路径，这取决于读者个人习惯，命令如下。

```
[root@controller ~]# ./iaas-install-keystone.sh
```

"."表示当前目录，用户的脚本放在家目录中，而用户此时也在家目录中，所以为方便，这里使用的是当前目录。

方法 2：现将脚本所在路径加入环境变量 PATH 中，然后就可以像 ls 命令一样直接在命令中输入使用，命令如下。

```
[root@controller ~]# echo $PATH
/usr/local/sbin:/usr/local/bin:/sbin:/bin:/usr/sbin:/usr/bin:/root/bin
[root@controller ~]# PATH=${PATH}:/root/
[root@controller ~]# echo $PATH
/usr/local/sbin:/usr/local/bin:/sbin:/bin:/usr/sbin:/usr/bin:/root/bin:/root/
[root@controller ~]# iaas-install-keystone.sh
```

上述命令对比了将脚本所在目录加入环境变量 PATH 前后，环境变量 PATH 中的内容，最后可以直接在命令中像其他命令一样直接输入脚本名称。

习题 Keystone 的安装及其配置

说明：直接在命令中改变的环境变量只对当前 Shell 和其子 Shell 有效，当重启或另外打开新的 Shell 时，当前修改的环境变量无效。也可以通过修改 /root/.bash_profile 配置文件中 PATH 的内容，达到环境变量 PATH 永久有效的效果，注意修改 /root/.bash_profile 后需要通知系统内容变化，最直接的方法是重新打开新的 Shell。

第 7 章　Glance 的安装及其配置

本章导读：

　　本章介绍 Glance 组件的基本功能，作为云平台的镜像服务模块，Glance 组件用于存储和管理云平台中的实例镜像文件。在实训项目 4 中介绍如何通过命令行以及服务组件的参数配置安装 Glance 服务，同时解读 Glance 在安装和使用过程中可能会出现的问题。在以上基础上，最后还将介绍 Glance 服务安装脚本的制作和使用。

电子资源：

　　电子教案　Glance 的安装及其配置
　　PPT　Glance 的安装及其配置
　　习题　Glance 的安装及其配置

第 7 章　Glance 的安装及其配置

电子教案　Glance 功能简介

PPT　Glance 功能简介

7.1　Glance 功能简介

Glance 是 OpenStack 的镜像服务，它提供了虚拟镜像的查询、注册和传输等服务。值得注意的是，Glance 本身并不实现对镜像的存储功能，它只是一个代理，充当了镜像存储服务与 OpenStack 的其他组件（特别是 Nova）之间的纽带。Glance 共支持两种镜像存储机制：简单文件系统和 Swift 服务存储镜像机制。

简单文件系统是指将镜像保存在 Glance 节点的文件系统中，这种机制相对比较简单，但也存在明显的不足。例如，没有备份机制，当文件系统损坏时，将导致所有的镜像不可用。

Swift 是 OpenStack 中用于管理对象存储的组件，Swift 服务存储镜像机制是指将镜像以对象的形式保存在 Swift 对象存储服务器中。由于 Swift 具有非常健壮的备份还原机制，因此可以降低因为文件系统损坏而造成的镜像不可用的风险。

Glance 服务支持多种格式的虚拟磁盘镜像，其中包括 ami、ari、aki、vhd、vmdk、raw、qcow2、vdi、iso 等。另外，也可以不把 Glance 当作镜像服务，而简单地把它当作一个对象存储代理服务，可以通过 Glance 存储任何其他格式的文件。

（1）Glance 主要由 Glance-API、Glance-Registry、Database 以及 Image Store 组成。

① Glance-API 用于接受镜像 API 需要的镜像发现、检索和存储。

② Glance-Registry 用于存储、处理和对镜像元数据检索。元数据包括大小和类型。需要注意的是，registry（注册）服务是一个内部的私有服务，仅为 Glance 服务本身所使用，不显示给用户。

③ Database 用于存储镜像元数据。用户可以根据自己的喜好选择数据库，大多数部署使用 MySQL 或 SQlite 数据库。

④ Image Store 作为镜像文件的存储库，Glance 除了常用的简单文件系统和 Swift 服务存储镜像机制外，还支持 RADOS 块设备、HTTP 和亚马逊 S3 系统，如图 7.1.1 所示。

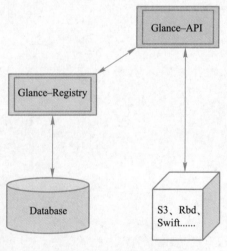

图 7.1.1　Glance 服务组成

（2）像所有的 OpenStack 项目一样，Glance 遵循以下思想

① 基于组件的架构，便于快速增加新特性。

② 高可用性，支持大负荷。

③ 容错性，独立的进程避免串行错误。

④ 开放标准，对社区驱动的 API 提供参考实现。

（3）Openstack 中 Glance 镜像的 4 种状态

① queued：标识该镜像 ID 已经被保留，但是镜像还未上传；

② saving：标识镜像正在被上传；

③ active：标识镜像在 Glance 中完全可用；

④ killed：标识镜像上传过程中出错，镜像完全不可用。

（4）在 Openstack 中，对于 KVM 应用到的 Image 格式

主要是 raw 和 qcow2 两种。

① raw 格式简单，容易转换为其他的格式。需要文件系统的支持才能支持 sparse file，性能相对较高。

② qcow2 是动态的，相对于 raw 来说，有下列的好处。

- 文件系统不支持 sparse file，文件大小也很小。
- Copy on write。
- Snapshot。
- 压缩。
- 加密。

7.2 实训项目 4　Glance 的手动安装与配置

1. 实训前提环境

成功完成实训项目 3 中所有内容后开始本实训内容，或者从已完成实训项目 3 的镜像开始，继续完成本实训内容。

电子教案　Glance 的手动安装与配置

2. 实训涉及节点

Controller。

3. 实训目标

① 完成 Glance 基本组件的安装。

② 完成 Glance 数据库的创建以及授权。

③ 完成 Glance 用户、服务及端点的创建。

④ 完成 Glance 主配置文件的修改。

⑤ 完成镜像的上传和验证。

PPT　Glance 的手动安装与配置

4. 实训步骤及其详解

步骤 1：安装 Glance 基本组件。

在 controller 节点上执行 yum 源安装命令安装 Glance 依赖包，命令如下。

```
[root@controller ~]# yum -y install openstack-glance
```

微课 5　Glance 的手动安装与配置

执行上述安装命令成功后，可以看到成功标志，所有 Glance 依赖包都安装完成，如图 7.2.1 所示。

```
Verifying    : 2:texlive-dvipng-bin-svn26509.0-43.20130427_r30134.el7.x86_64
Verifying    : lcms2-2.6-3.el7.x86_64
Verifying    : python-matplotlib-data-2.0.0-3.el7.noarch
Verifying    : mesa-libglapi-18.3.4-5.el7.x86_64
Verifying    : libfontenc-1.1.3-3.el7.x86_64
Verifying    : python-boto-2.34.0-4.el7.noarch
Installed:
  openstack-glance.noarch 1:18.0.0-1.el7
```

图 7.2.1　Glance 依赖包安装完成的反馈结果

步骤 2：创建 Glance 数据库并授权。

首先，用 root 用户登录 MySQL 数据库，命令如下。

`[root@controller ~]# mysql -uroot -p000000`

登录后，先创建 Glance 数据库，命令如下。

`mysql> CREATE DATABASE glance;`

看到提示 Query OK，1 row affected（0.00 sec），表明数据库创建成功。

接着，创建 MySQL 的 Glance 用户，并赋予其 Glance 数据库的操作权限，使得本地及远程都能访问，密码为 000000，命令如下。

```
mysql> GRANT ALL PRIVILEGES ON glance.* TO 'glance'@'localhost' IDENTIFIED BY '000000';
mysql> GRANT ALL PRIVILEGES ON glance.* TO 'glance'@'%' IDENTIFIED BY '000000';
mysql> exit
```

上述 SQL 语句与创建 Keystone 数据库的语句一致，可参考实训项目 3 中步骤 2 "创建 Keystone 数据库并授权"部分对创建数据库时 SQL 语句的说明，这里不做赘述。

然后，需要为 Glance 服务创建数据库表，创建数据库表之前，需要先修改 Glance 配置文件中用于数据库连接的内容，命令如下。

```
[root@controller ~]# crudini --set /etc/glance/glance-api.conf database connection mysql+pymysql://glance:000000@controller/glance
[root@controller ~]# crudini --set /etc/glance/glance-registry.conf database connection mysql+pymysql://glance:000000@controller/glance
```

上述命令的功能是分别在 /etc/glance/glance-api.conf 和 /etc/glance/glance-registry.conf 两个文件的 [database] 段落中添加 "connection = mysql+pymysql（数据库类型）://glance（登录数据库的用户名）:000000（用户密码）@controller（数据库主机名）/glance（数据库）" 配置。通过 vi/vim 命令直接修改 /etc/glance 下的两个配置文件，也可以达到相同效果。

接着，需要同步数据库，为 Glance 服务创建数据库表，命令如下。

`[root@controller ~]# su -s /bin/sh -c "glance-manage db_sync" glance`

可以通过执行下面一条语句，查看是否同步并创建成功，命令如下。

`[root@controller ~]# mysql -u root -p000000 -e "use glance; show tables;"`

结果如图 7.2.2 所示。

步骤 3：注册 Glance 服务至 Keystone 服务器。

```
Tables_in_glance
alembic_version
image_locations
image_members
image_properties
image_tags
images
metadef_namespace_resource_types
metadef_namespaces
metadef_objects
metadef_properties
metadef_resource_types
metadef_tags
migrate_version
task_info
tasks
```

图 7.2.2　查看数据库同步的反馈结果

在 OpenStack 中，几乎所有的服务（包括 Keystone 服务）要想正常运行，都必须首先向 Keystone 服务器注册。每一个服务需要向 Keystone 注册用户信息、服务（service）和端点（endpoint）信息。

首先，进行 Glance 用户信息的注册。注册用户信息的目的是为了认证用户身份的。当一个用户向 Glance 服务器发送请求时，Glance 服务器首先要认证该用户是否合法。此时，Glance 会使用已注册的 Glance 用户向 Keystone 服务器发送认证请求。

本书所有的服务都创建在 service 项目下。因此，Glance 用户必须在 service 项目下具有 admin 权限。

添加 Glance 用户，并为用户设置密码，在本书实训过程中所有的密码统一使用 000000，命令如下。

```
[root@controller ~]# openstack user create --domain default --password 000000 glance
```

结果如图 7.2.3 所示。

```
| Field               | Value                            |
| domain_id           | default                          |
| enabled             | True                             |
| id                  | 6be548ba485340c9acd42e7cff5ac458 |
| name                | glance                           |
| options             | {}                               |
| password_expires_at | None                             |
```

图 7.2.3　创建 Glance 用户反馈结果

再为 Glance 用户在 service 项目下分配 admin 权限，命令如下。

```
[root@controller ~]# openstack role add --project service --user glance admin
```

上述命令用法在实训项目 3 步骤 7 中已经提到，这里不再赘述。值得注意的是，执行赋予 Glance 用户 admin 权限的命令是没有任何输出的。

接下来，注册 Glance 服务（service）和端点（endpoint）信息，注册服务和端点的信息目的是为了确保用户能够顺利访问 Glance 服务，并执行一系列与 Glance 服务有关的操作。

注册 Glance 服务信息，命令如下。

```
[root@controller ~]# openstack service create --name glance --description "OpenStack Image" image
```

结果如图 7.2.4 所示。

```
+-------------+---------------------------------+
| Field       | Value                           |
+-------------+---------------------------------+
| description | OpenStack Image                 |
| enabled     | True                            |
| id          | 36cb55a8b66044b297a4d8f589c5760b|
| name        | glance                          |
| type        | image                           |
+-------------+---------------------------------+
```

图 7.2.4 Glance 服务创建反馈结果

注册 Glance 端点信息，命令如下，结果如图 7.2.5~ 图 7.2.7 所示。

```
[root@controller ~]# openstack endpoint create --region RegionOne $(openstack service list | awk '/ image / {print $4}') public http://controller:9292
```

```
+--------------+----------------------------------+
| Field        | Value                            |
+--------------+----------------------------------+
| enabled      | True                             |
| id           | 1a448745459e4a05930f78b00fc66adb |
| interface    | public                           |
| region       | RegionOne                        |
| region_id    | RegionOne                        |
| service_id   | 36cb55a8b66044b297a4d8f589c5760b |
| service_name | glance                           |
| service_type | image                            |
| url          | http://controller:9292           |
+--------------+----------------------------------+
```

图 7.2.5 Glance 的 public endpoint 创建反馈结果

```
[root@controller ~]# openstack endpoint create --region RegionOne $(openstack service list | awk '/ image / {print $4}') internal http://controller:9292
```

```
+--------------+----------------------------------+
| Field        | Value                            |
+--------------+----------------------------------+
| enabled      | True                             |
| id           | 2ac3ab6b19e14bddb2e100c34fdc9fc8 |
| interface    | internal                         |
| region       | RegionOne                        |
| region_id    | RegionOne                        |
| service_id   | 36cb55a8b66044b297a4d8f589c5760b |
| service_name | glance                           |
| service_type | image                            |
| url          | http://controller:9292           |
+--------------+----------------------------------+
```

图 7.2.6 Glance 的 internal endpoint 创建反馈结果

```
[root@controller ~]# openstack endpoint create --region RegionOne $(openstack service list | awk '/ image / {print $4}') admin http://controller:9292
```

```
+--------------+----------------------------------+
| Field        | Value                            |
+--------------+----------------------------------+
| enabled      | True                             |
| id           | cbe7f12f46ce435da70e0f6911d32e77 |
| interface    | admin                            |
| region       | RegionOne                        |
| region_id    | RegionOne                        |
| service_id   | 36cb55a8b66044b297a4d8f589c5760b |
| service_name | glance                           |
| service_type | image                            |
| url          | http://controller:9292           |
+--------------+----------------------------------+
```

图 7.2.7 Glance 的 admin endpoint 创建反馈结果

需要注意的是，Glance 服务端口号为 9292，同时各服务组件间的 URL 版本不尽相同，操作时应该格外注意。

步骤 4：修改 Glance 配置文件。

在步骤 3 进行了 Glance 用户信息、服务和端点信息的注册，为了能让 Glance 服务顺利通过 Keystone 的认证，还需要在两个配置文件 /etc/glance/glance-api.conf 以及 /etc/glance/

glance-registry.conf 中记录 Keystone 的认证信息，即在配置文件的 [keystone_authtoken] 字段中修改 Keystone 的相关身份认证信息，同时在 [paste_deploy] 字段中添加 Keystone 支持。有了这些认证信息，Glance 在向外提供服务的时候，能够顺利通过 Keystone 的认证。

首先修改 /etc/glance/glance-api.conf 配置文件，命令如下。

```
[root@controller ~]# crudini --set /etc/glance/glance-api.conf keystone_authtoken www_authenticate_uri http://controller:5000
[root@controller ~]# crudini --set /etc/glance/glance-api.conf keystone_authtoken auth_url http://controller:5000
[root@controller ~]# crudini --set /etc/glance/glance-api.conf keystone_authtoken memcached_servers controller:11211
[root@controller ~]# crudini --set /etc/glance/glance-api.conf keystone_authtoken auth_type password
[root@controller ~]# crudini --set /etc/glance/glance-api.conf keystone_authtoken project_domain_name default
[root@controller ~]# crudini --set /etc/glance/glance-api.conf keystone_authtoken user_domain_name default
[root@controller ~]# crudini --set /etc/glance/glance-api.conf keystone_authtoken project_name service
[root@controller ~]# crudini --set /etc/glance/glance-api.conf keystone_authtoken username glance
[root@controller ~]# crudini --set /etc/glance/glance-api.conf keystone_authtoken password 000000
[root@controller ~]# crudini --set /etc/glance/glance-api.conf paste_deploy flavor keystone
[root@controller ~]# crudini --set /etc/glance/glance-api.conf glance_store stores file,http
[root@controller ~]# crudini --set /etc/glance/glance-api.conf glance_store default_store file
[root@controller ~]# crudini --set /etc/glance/glance-api.conf glance_store filesystem_store_datadir /var/lib/glance/images/
```

也可以通过 vi/vim 直接编辑配置文件 /etc/glance/glance-api.conf，在相应的地方做上述内容的修改，以达到相同效果。修改完成后使用 grep 命令过滤 /etc/glance/glance-api.conf 文件中的内容，来检查已完成的配置。命令和结果如下所示。

```
[root@controller ~]# grep -Ev "^#|^$" /etc/glance/glance-api.conf
[DEFAULT]
[cinder]
[cors]
[database]
connection = mysql+pymysql://glance:000000@controller/glance
[glance_store]
stores = file,http
default_store = file
filesystem_store_datadir = /var/lib/glance/images/
```

```
[image_format]
[keystone_authtoken]
www_authenticate_uri = http: //controller: 5000
auth_url = http: //controller: 5000
memcached_servers = controller: 11211
auth_type = password
project_domain_name = default
user_domain_name = default
project_name = service
username = glance
password = 000000
[paste_deploy]
flavor = keystone
```

以下是 Keystone 认证信息中每条语句的作用。

- www_authenticate_uri 输入 Keystone 服务的 URI，即位置信息。对于外部用户 Keystone 服务定义了一个端口号为 5000 的 auth_url，这里输入的是 auth_url。

- memcached_servers 输入 Memcached 服务所在主机地址以及端口号。

- auth_type 输入 Keystone 认证所使用的类型，通常使用 password 类型。

- project_domain_name 输入项目所在的域名。

- user_domain_name 输入用户所在的域名。

- project_name 输入 Glance 服务所属的项目名，OpenStack 所有的组件均使用 service 项目；

- username 输入 Glance 服务在 Keystone 中注册时的用户名。

- password 输入 Glance 服务在 Keystone 中注册时所使用的密码，这里为 000000，在实际 OpenStack 生产环境的部署过程中该密码值必须与步骤 3 中添加 Glance 用户时所使用的密码一致。

接下来需要修改 /etc/glance/glance-registry.conf 配置文件，需要修改的内容与 /etc/glance/glance-api.conf 配置文件相同，命令如下。

```
[root@controller ~]# crudini --set /etc/glance/glance-registry.conf database connection mysql+pymysql: //glance: 000000@controller/glance
[root@controller ~]# crudini --set /etc/glance/glance-registry.conf keystone_authtoken www_authenticate_uri http: //controller: 5000
[root@controller ~]# crudini --set /etc/glance/glance-registry.conf keystone_authtoken auth_url http: //controller: 5000
[root@controller ~]# crudini --set /etc/glance/glance-registry.conf keystone_authtoken memcached_servers controller: 11211
[root@controller ~]# crudini --set /etc/glance/glance-registry.conf keystone_authtoken auth_type password
[root@controller ~]# crudini --set /etc/glance/glance-registry.conf keystone_authtoken project_domain_name default
[root@controller ~]# crudini --set /etc/glance/glance-registry.conf keystone_authtoken user_domain_name default
```

```
[root@controller ~]# crudini --set /etc/glance/glance-registry.conf keystone_authtoken project_name service
[root@controller ~]# crudini --set /etc/glance/glance-registry.conf keystone_authtoken username glance
[root@controller ~]# crudini --set /etc/glance/glance-registry.conf keystone_authtoken password 000000
[root@controller ~]# crudini --set /etc/glance/glance-registry.conf paste_deploy flavor keystone
```

同样，也可以通过 vi/vim 命令直接编辑配置文件 /etc/glance/glance-registry.conf，在相应的地方做上述内容的修改，以达到相同效果。修改完成后使用 grep 命令过滤 /etc/glance/glance-registry.conf 文件中的内容，检查已完成的配置。命令和结果如下所示。

```
[root@controller ~]# grep -Ev "^#|^$" /etc/glance/glance-registry.conf
[DEFAULT]
[database]
connection = mysql+pymysql://glance:000000@controller/glance
[keystone_authtoken]
www_authenticate_uri = http://controller:5000
auth_url = http://controller:5000
memcached_servers = controller:11211
auth_type = password
project_domain_name = default
user_domain_name = default
project_name = service
username = glance
password = 000000
[paste_deploy]
flavor = keystone
```

步骤 5：启动服务并加入开机自启。

为了让更新后的配置生效，需要重启 Glance 服务，命令如下。

```
[root@controller ~]# systemctl restart openstack-glance-api.service openstack-glance-registry.service
```

接着将两个服务加入开机自启，命令如下。

```
[root@controller ~]# systemctl enable openstack-glance-api.service openstack-glance-registry.service
```

上述命令执行后，可以通过 systemctl 命令查看，看到 enabled 即表示成功加入开机自启，结果如下。

```
[root@controller ~]# systemctl is-enabled openstack-glance-api.service openstack-glance-registry.service
enabled
enabled
```

步骤 6：上传镜像。

在 4.3 节"基本环境配置"的步骤 3 中已经通过 mount 命令将包含 OpenStack 可用镜像的 iso 文件挂载到 /opt/iaas 目录下，镜像存放在 /opt/iaas/images 目录下。OpenStack 支持多种镜像格式，具体支持哪些镜像格式将在下面内容中介绍。本书中的实训使用 qcow2 格式的镜像，可以通过 file 命令来验证镜像格式，命令和结果如下。

```
[root@controller ~]# file /opt/iaas/images/CentOS_7.1_x86_64.qcow2
/opt/iaas/images/CentOS_7.1_x86_64.qcow2: QEMU QCOW Image (v2), 8589934592 bytes
```

上传镜像的命令如下，结果如图 7.2.8 所示。

```
[root@controller ~]# glance image-create --name "centos7.1" --disk-format qcow2 --container-format bare --progress < /opt/iaas/images/CentOS_7.1_x86_64.qcow2
```

```
[================================>] 100%
+------------------+----------------------------------------------------------+
| Property         | Value                                                    |
+------------------+----------------------------------------------------------+
| checksum         | f5cdfc202cccc837def1244e0df75ddc                         |
| container_format | bare                                                     |
| created_at       | 2019-11-19T08:56:58Z                                     |
| disk_format      | qcow2                                                    |
| id               | 0f17819b-c2f3-41cd-a140-f20695523413                     |
| min_disk         | 0                                                        |
| min_ram          | 0                                                        |
| name             | centos7.1                                                |
| os_hash_algo     | sha512                                                   |
| os_hash_value    | f8b497e11afb0acc35087e68a71cdc8907f6e9e906be8a40         |
|                  | 42b983f7b818a55658ef7329f96cd3ea67860a687d39fe89         |
| os_hidden        | False                                                    |
| owner            | 887d1386bcc14be9bcf177550bc5a515                         |
| protected        | False                                                    |
| size             | 412221440                                                |
| status           | active                                                   |
| tags             | []                                                       |
| updated_at       | 2019-11-19T08:57:01Z                                     |
| virtual_size     | Not available                                            |
| visibility       | shared                                                   |
+------------------+----------------------------------------------------------+
```

图 7.2.8　上传镜像成功的反馈结果

上述命令中，--name 用来指定镜像名，centos7.1 仅为参考名，可自定义。

--disk-format 用来指定镜像格式，有效的镜像格式包括 ami、ari、aki、vhd、vmdk、raw、qcow2、vdi 和 iso。

--container-format 用来指定容器的格式，有效的格式包括 bare、ovf、aki、ari 和 ami。指定 bare 来表明镜像文件不是包含虚拟机元数据文件的格式。目前来说，这个字段是必需的，尽管它实际上并没有被任何 Openstack 服务使用，对系统也没有什么影响，但指定容器格式为 bare 总是安全的。

--progress 用来显示上传进度，可以省略。

< 后面的内容是要上传的具体镜像文件名。

其他具体参数可以查看 Glance 的帮助。

步骤 7：验证镜像服务。

上传镜像后，可以通过命令来验证是否成功上传，命令如下，结果如图 7.2.9 所示。

```
+--------------------------------------+-----------+
| ID                                   | Name      |
+--------------------------------------+-----------+
| 0f17819b-c2f3-41cd-a140-f20695523413 | centos7.1 |
+--------------------------------------+-----------+
```

图 7.2.9　查看镜像上传结果反馈信息

```
[root@controller ~]# glance image-list
```

这样通过上面的 7 步操作，就完成了对 Glance 服务的安装及配置，并且成功地上传了一个可用镜像，该镜像在后续实训环节中会使用到。

5. 常见错误及调试排错

错误 1：环境变量问题。

执行命令：

```
[root@controller ~]# glance image-list
```

返回如下结果：

```
You must provide a username via either --os-username or env[OS_USERNAME]
```

结果提示需要提供用户名或者设置环境变量，这种问题是由于没有环境变量导致的。

解决办法：

设置环境变量。source 之前编辑的环境变量脚本文件即可。source 后得到的环境变量只对当前 Shell 和子 Shell 有效，当关掉或重新打开一个 Shell 时，source 得到的环境变量就失效了，需要重新编辑 source 之前的环境变量脚本。

错误 2：配置文件错误。

配置完 Glance 服务后，需要通过上传一个镜像来验证服务是否正常运行，执行上传镜像命令时，出现如下错误，如图 7.2.10 所示：

```
[root@controller ~]# glance image-create --name "centos7.1" --disk-format qcow2 --container-format bare --progress < /opt/iaas/images/CentOS_7.1_x86_64.qcow2
```

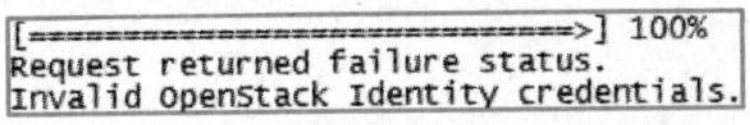

```
[================================>] 100%
Request returned failure status.
Invalid OpenStack Identity credentials.
```

图 7.2.10　上传镜像错误反馈信息

分析排错：

从提示信息"Invalid OpenStack Identity credentials"中文翻译为"无效的 OpenStack 认证证书"，可以初步猜测 Keystone 服务可能存在问题，接下来通过以下命令查看 Httpd 服务状态，以及 Keystone 服务所依赖的 MySQL 服务的状态，命令如下，如图 7.2.11 和图 7.2.12 所示。

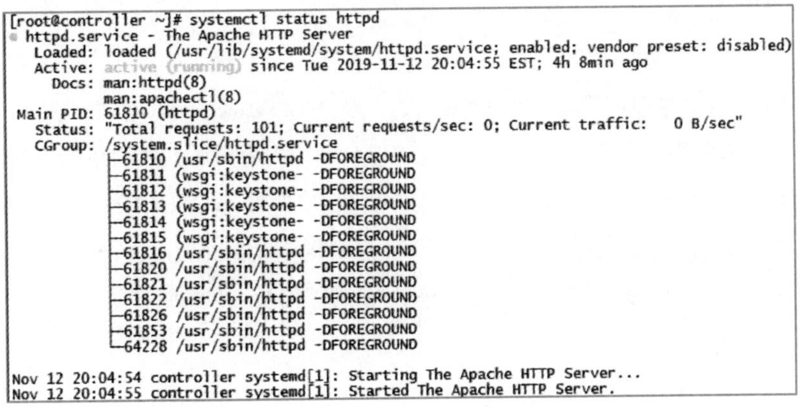

```
[root@controller ~]# systemctl status httpd
 httpd.service - The Apache HTTP Server
   Loaded: loaded (/usr/lib/systemd/system/httpd.service; enabled; vendor preset: disabled)
   Active: active (running) since Tue 2019-11-12 20:04:55 EST; 4h 8min ago
     Docs: man:httpd(8)
           man:apachectl(8)
 Main PID: 61810 (httpd)
   Status: "Total requests: 101; Current requests/sec: 0; Current traffic:   0 B/sec"
   CGroup: /system.slice/httpd.service
           ├─61810 /usr/sbin/httpd -DFOREGROUND
           ├─61811 (wsgi:keystone- -DFOREGROUND
           ├─61812 (wsgi:keystone- -DFOREGROUND
           ├─61813 (wsgi:keystone- -DFOREGROUND
           ├─61814 (wsgi:keystone- -DFOREGROUND
           ├─61815 (wsgi:keystone- -DFOREGROUND
           ├─61816 /usr/sbin/httpd -DFOREGROUND
           ├─61820 /usr/sbin/httpd -DFOREGROUND
           ├─61821 /usr/sbin/httpd -DFOREGROUND
           ├─61822 /usr/sbin/httpd -DFOREGROUND
           ├─61826 /usr/sbin/httpd -DFOREGROUND
           ├─61853 /usr/sbin/httpd -DFOREGROUND
           └─64228 /usr/sbin/httpd -DFOREGROUND

Nov 12 20:04:54 controller systemd[1]: Starting The Apache HTTP Server...
Nov 12 20:04:55 controller systemd[1]: Started The Apache HTTP Server.
```

图 7.2.11　查询 Httpd 状态信息

```
[root@controller ~]# systemctl status httpd
[root@controller ~]# systemctl status mariadb
```

```
[root@controller ~]# systemctl status mariadb
● mariadb.service - MariaDB 10.3 database server
   Loaded: loaded (/usr/lib/systemd/system/mariadb.service; en
   Active: active (running) since Tue 2019-11-12 07:47:12 EST;
     Docs: man:mysqld(8)
           https://mariadb.com/kb/en/library/systemd/
 Main PID: 60113 (mysqld)
   Status: "Taking your SQL requests now..."
   CGroup: /system.slice/mariadb.service
           └─60113 /usr/libexec/mysqld --basedir=/usr

Nov 12 07:47:12 controller mysql-prepare-db-dir[60010]: Please
Nov 12 07:47:12 controller mysql-prepare-db-dir[60010]: The la
Nov 12 07:47:12 controller mysql-prepare-db-dir[60010]: You ca
Nov 12 07:47:12 controller mysql-prepare-db-dir[60010]: http:/
Nov 12 07:47:12 controller mysql-prepare-db-dir[60010]: Consid
Nov 12 07:47:12 controller mysql-prepare-db-dir[60010]: https:
Nov 12 07:47:12 controller mysqld[60113]: 2019-11-12  7:47:12
Nov 12 07:47:12 controller mysqld[60113]: 2019-11-12  7:47:12
Nov 12 07:47:12 controller mysqld[60113]: 2019-11-12  7:47:12
Nov 12 07:47:12 controller systemd[1]: Started MariaDB 10.3 da
```

图 7.2.12　查询 MySQL 状态信息

反馈的结果是两个服务都在运行，没问题。

接下来通过如下命令查看 Keystone 的日志文件，命令如下。

```
[root@controller ~]# tail -f /var/log/keystone/keystone.log
```

在日志文件中发现如下信息，如图 7.2.13 所示。

```
WARNING keystone.common.wsgi [-] Could not find role, admin.
WARNING keystone.common.wsgi [-] Could not find project, admin.
WARNING keystone.common.wsgi [-] Could not find user, admin.
WARNING keystone.common.wsgi [-] Could not find role, _member_.
WARNING keystone.common.wsgi [-] Could not find project, admin.
WARNING keystone.common.wsgi [-] Could not find project, admin.
WARNING keystone.common.wsgi [-] Could not find user, admin.
WARNING keystone.common.wsgi [-] Could not find user, glance.
WARNING keystone.common.wsgi [-] Could not find role, admin.
WARNING keystone.common.wsgi [-] Could not find project, service.
```

图 7.2.13　Keystone 日志文件反馈信息

从图 7.2.13 中可以看到，Keystone 服务没有找到 Glance 服务的内容，进一步猜测是 Glance 在连接认证的地方有错误，再查看一下 glance-api 的日志，部分内容如图 7.2.14 所示。

```
[root@controller ~]# tail -f /var/log/glance/api.log
2015-10-03 13:38:24.171 2068 WARNING keystoneclient.middleware.auth_token [-] Retrying on HTTP
connection exception: [Errno -2] Name or service not known
2015-10-03 13:38:24.672 2068 INFO urllib3.connectionpool [-] Starting new HTTP connection (1):
controllr
2015-10-03 13:38:24.673 2068 WARNING keystoneclient.middleware.auth_token [-] Retrying on HTTP
connection exception: [Errno -2] Name or service not known
2015-10-03 13:38:25.674 2068 INFO urllib3.connectionpool [-] Starting new HTTP connection (1):
controllr
2015-10-03 13:38:25.674 2068 WARNING keystoneclient.middleware.auth_token [-] Retrying on HTTP
connection exception: [Errno -2] Name or service not known
2015-10-03 13:38:27.677 2068 INFO urllib3.connectionpool [-] Starting new HTTP connection (1):
controllr
```

图 7.2.14　glance-api 日志文件反馈信息

从上述日志文件中，基本可以判断错误的原因，进入 glance-api 配置文件找到 Keystone 认证部分，某个值拼写错误，这与日志文件中显示的内容一致。修改错误信息，重启 openstack-glance-api 服务，再次上传镜像，成功解决。命令如下，如图 7.2.15 所示。

```
[root@controller ~]# glance image-create --name "centos" --disk-format qcow2
--container-format bare --progress < /opt/iaas/images/CentOS_7.1_x86_64.qcow2
```

```
[===========================>] 100%
+------------------+--------------------------------------+
| Property         | value                                |
+------------------+--------------------------------------+
| checksum         | dfbd01ddbb81c9e8254de236e5e83b0f     |
| container_format | bare                                 |
| created_at       | 2015-10-03T17:48:48                  |
| deleted          | False                                |
| deleted_at       | None                                 |
| disk_format      | qcow2                                |
| id               | c187b59e-b4f1-4eba-b043-50ea1dcc865f |
| is_public        | True                                 |
| min_disk         | 0                                    |
| min_ram          | 0                                    |
| name             | centos                               |
| owner            | dfd90e1e4de54f0a82b8bf365e711929     |
| protected        | False                                |
| size             | 305397760                            |
| status           | active                               |
| updated_at       | 2015-10-03T17:48:51                  |
| virtual_size     | None                                 |
+------------------+--------------------------------------+
```

图 7.2.15　镜像上传成功

7.3　Glance 安装脚本及其解读

前面已经介绍了如何手动安装 Glance 服务，接下来介绍另一种安装 Glance 服务的方法，即利用已经写好的脚本自动安装 Glance 服务。

1. 需要安装的脚本内容

```
#!/bin/bash
source /etc/keystone/admin-openrc.sh
# 安装 Glance 服务软件包
yum -y install openstack-glance
# 数据库配置
mysql -uroot -p000000 -e "create database IF NOT EXISTS glance ; "
mysql -uroot -p000000 -e "GRANT ALL PRIVILEGES ON glance.* TO 'glance'@'localhost' IDENTIFIED BY '000000' ; "
mysql -uroot -p000000 -e "GRANT ALL PRIVILEGES ON glance.* TO 'glance'@'%' IDENTIFIED BY '000000' ; "
# Keystone 认证
openstack user create --domain default --password 000000 glance
openstack role add --project service --user glance admin
openstack service create --name glance --description "OpenStack Image" image
openstack endpoint create --region RegionOne image public http://controller:9292
openstack endpoint create --region RegionOne image internal http://controller:9292
openstack endpoint create --region RegionOne image admin http://controller:9292
# API 主配置文件设置
crudini --set /etc/glance/glance-api.conf database connection mysql+pymysql://glance:000000@controller/glance
crudini --set /etc/glance/glance-api.conf keystone_authtoken www_authenticate_uri http://controller:5000
crudini --set /etc/glance/glance-api.conf keystone_authtoken auth_url http://controller:5000
crudini --set /etc/glance/glance-api.conf keystone_authtoken memcached_servers controller:11211
```

```
    crudini --set /etc/glance/glance-api.conf keystone_authtoken auth_type password
    crudini --set /etc/glance/glance-api.conf keystone_authtoken project_domain_name
default
    crudini --set /etc/glance/glance-api.conf keystone_authtoken user_domain_name default
    crudini --set /etc/glance/glance-api.conf keystone_authtoken project_name service
    crudini --set /etc/glance/glance-api.conf keystone_authtoken username glance
    crudini --set /etc/glance/glance-api.conf keystone_authtoken password 000000
    crudini --set /etc/glance/glance-api.conf paste_deploy flavor keystone
    crudini --set /etc/glance/glance-api.conf glance_store stores file,http
    crudini --set /etc/glance/glance-api.conf glance_store default_store file
    crudini --set /etc/glance/glance-api.conf glance_store filesystem_store_datadir /var/
lib/glance/images/
    # Registry 主配置文件设置
    crudini --set /etc/glance/glance-registry.conf database connection mysql+pymysql://
glance:000000@controller/glance
    crudini --set /etc/glance/glance-registry.conf keystone_authtoken www_authenticate_
uri http://controller:5000
    crudini --set /etc/glance/glance-registry.conf keystone_authtoken auth_url http://
controller:5000
    crudini --set /etc/glance/glance-registry.conf keystone_authtoken memcached_servers
controller:11211
    crudini --set /etc/glance/glance-registry.conf keystone_authtoken auth_type password
    crudini --set /etc/glance/glance-registry.conf keystone_authtoken project_domain_
name default
    crudini --set /etc/glance/glance-registry.conf keystone_authtoken user_domain_name
default
    crudini --set /etc/glance/glance-registry.conf keystone_authtoken project_name
service
    crudini --set /etc/glance/glance-registry.conf keystone_authtoken username glance
    crudini --set /etc/glance/glance-registry.conf keystone_authtoken password 000000
    crudini --set /etc/glance/glance-registry.conf paste_deploy flavor keystone
    # 同步数据库
    su -s /bin/sh -c "glance-manage db_sync" glance
    # 落定服务
    systemctl enable openstack-glance-api.service openstack-glance-registry.service
    systemctl restart openstack-glance-api.service openstack-glance-registry.service
```

上面内容是 Glance 服务安装脚本的全部内容。从中可以看到，在 Glance 服务中，脚本安装与利用命令手动安装的内容基本一致。

2. 脚本的执行

刚写好的脚本文件，在核实过内容后，是不能直接使用的。要想使用脚本，首先需要赋予脚本执行权限，命令如下。

```
[root@controller ~]# chmod a+x iaas-install-glance.sh
```

或者执行

```
[root@controller ~]# chmod 777 iaas-install-glance.sh
```

上述两种命令都可以实现赋予脚本执行权限的效果，这里的"iaas-install-glance.sh"是示范用的脚本名称，读者在实际使用时可自行命名脚本名称。

说明：本次实验所使用的脚本均放在 root 用户的家目录里。

脚本有了执行权限后就可以执行了。这里介绍两种执行脚本的方法。

方法 1：在脚本前加上脚本所在的路径，可以使用相对路径，也可以使用绝对路径，这取决于读者个人习惯，命令如下。

```
[root@controller ~]# ./iaas-install-glance.sh
```

"."表示当前目录，用户的脚本放在家目录中，而用户此时也在家目录中，所以为方便，这里使用的是当前目录。

方法 2：现将脚本所在路径加入环境变量 PATH 中，然后就可以像 ls 命令一样直接在命令中输入使用了，命令如下。

```
[root@controller ~]# echo $PATH
/usr/local/sbin:/usr/local/bin:/sbin:/bin:/usr/sbin:/usr/bin:/root/bin
[root@controller ~]# PATH=${PATH}:/root/
[root@controller ~]# echo $PATH
/usr/local/sbin:/usr/local/bin:/sbin:/bin:/usr/sbin:/usr/bin:/root/bin:/root/
[root@controller ~]# iaas-install-glance.sh
```

上述命令中对比了将脚本所在目录加入环境变量 PATH 前后，环境变量 PATH 中的内容，最后可以直接在命令中像其他命令一样直接输入脚本名称了。

说明：直接在命令中改变的环境变量只对当前 Shell 和其子 Shell 有效，当重启或另外打开新的 Shell 时，当前修改的环境变量无效。可以通过修改 /root/.bash_profile 配置文件中 PATH 的内容，达到环境变量 PATH 永久有效的效果，注意修改 /root/.bash_profile 后需要通知系统内容变化，最直接的方法是重新打开新的 Shell。

习题 Glance 的安装及其配置

相对于手动安装来说，脚本安装的好处是简单省事，可以在安装之前写好脚本，安装时直接执行即可，而且脚本可以方便以后多次利用。

第 8 章　Placement 的安装及其配置

本章导读:

　　本章介绍 placement 组件的基本功能,作为云平台的镜像服务模块,Placement 组件用于便捷地使用、管理、监控整个 OpenStack 的系统资源。在实训项目 5 中介绍如何通过命令行以及服务组件的参数配置安装 Placement 服务,同时解读 Placement 在安装和使用过程中可能会出现的问题。在以上基础上,最后还将介绍 Placement 服务安装脚本的制作和使用。

电子资源:

　　电子教案　Placement 的安装及其配置
　　PPT　Placement 的安装及其配置
　　习题　Placement 的安装及其配置

第 8 章　Placement 的安装及其配置

8.1　Placement 功能简介

电子教案　Placement 功能简介

PPT　Placement 功能简介

从用户的角度出发，作为使用共享存储解决方案的用户，会希望 Nova 和 Horizon 能够正确报告共享存储磁盘资源的总量和使用量信息。作为高级的 Neutron 用户，预期会使用外部的第三方路由网络功能，希望 Nova 能够掌握和使用特定的网络端口与特定的子网池相关联，确保虚拟机能够在该子网池上启动。作为高级的 Cinder 用户，希望当用户在 nova boot 命令中指定了 cinder volume-id 后 Nova 能够知道哪一些计算节点与 Request Volume 所在的 Cinder 存储池相关联。

所以，OpenStack 除了要处理计算节点 CPU、内存、PCI 设备、本地磁盘等内部资源外，还经常需要纳管有如 SDS、NFS 提供的存储服务，SDN 提供的网络服务等外部资源。

但在以往，Nova 只能处理由计算节点提供的资源。Nova Resource Tracker 假定所有资源均来自计算节点，因此在周期性上报资源状况时，Resource Tracker 只会单纯对计算节点清单进行资源总量和使用量的加和统计。显然，这无法满足上述复杂的生产需求，也违背了 OpenStack 一向赖以自豪的开放性原则。而且随着 OpenStack 的定义被社区进一步升级为"一个开源基础设施集成引擎"，意味 OpenStack 的资源系统将会由更多外部资源类型构成。

当资源类型和提供者变得多样时，自然就需要一种高度抽象且简单统一的管理方法，让用户和代码能够便捷地使用、管理、监控整个 OpenStack 的系统资源，这就是 Placement（布局）。

最早在 Newton 版本被引入到 openstack/nova repo，以 API 的形式进行孵化，所以也经常被称为 Placement API。它参与到 nova-scheduler 选择目标主机的调度流程中，负责跟踪记录 Resource Provider 的 Inventory 和 Usage，并使用不同的 Resource Classes 来划分资源类型，使用不同的 Resource Traits 来标记资源特征。在 Ocata 版本中，Placement API 是一个可选项，建议用户启用并替代 CpuFilter、CoreFilter 和 DiskFilter。到了 Pike 版本，则强制要求启动 Placement API 服务，否则 nova-compute service 无法正常运行。

Placement 中的基本概念如下。

① Resource Provider：资源提供者，实际提供资源的实体，如 Compute Node、Storage Pool、IP Pool 等。

② Resource Class：资源种类，即资源的类型，Placement 为 Compute Node 缺省了下列几种类型，同时支持 Custom Resource Classes。

③ Inventory：资源清单，资源提供者所拥有的资源清单，如 Compute Node 拥有的 vCPU、Disk、RAM 等 Inventories。

④ Provider Aggregate：资源聚合，类似 HostAggregate 的概念，是一种聚合类型。

⑤ Traits：资源特征，不同资源提供者可能会具有不同的资源特征。Traits 作为资源提供者特征的描述，它不能够被消费，但在某些 Workflow 或者会非常有用。例如，标识可用的 Disk 具有 SSD 特征，有助于 Scheduler 灵活匹配 Launch Instance 的请求。

⑥ Resource Allocations：资源分配状况，包含了 Resource Class、Resource Provider 以及 Consumer 的映射关系。记录消费者使用了该类型资源的数量。

8.2 实训项目 5 Placement 的手动安装与配置

电子教案 Placement 的手动安装与配置

PPT Placement 的手动安装与配置

1. 实训前提环境

成功完成实训项目 4 中所有内容后开始本实训，或者从已完成实训项目 4 的镜像开始，继续完成本实训内容。

2. 实训涉及节点

Controller。

3. 实训目标

① 完成 Placement 基本组件的安装。
② 完成 Placement 数据库的创建以及授权。
③ 完成 Placement 用户、服务及端点的创建。
④ 完成 Placement 主配置文件的修改。

微课 6 Placement 的手动安装与配置

4. 实训步骤及其详解

步骤 1：安装 Placement 基本组件。

在 controller 节点上执行 yum 源安装命令安装 Placement 依赖包，命令如下。

```
[root@controller ~]# yum -y install openstack-placement-api
```

执行上述安装命令成功后，可以看到成功标志，所有 Glance 依赖包都安装完成，如图 8.2.1 所示。

```
Installed:
  openstack-placement-api.noarch 0:1.1.0-1.el7

Dependency Installed:
  openstack-placement-common.noarch 0:1.1.0-1.el7
  python2-os-resource-classes.noarch 0:0.3.0-1.el7
  python2-placement.noarch 0:1.1.0-1.el7

Complete!
```

图 8.2.1 安装 Placement 结果反馈

步骤 2：创建 Placement 数据库并授权。

首先，用 root 用户登录 MySQL 数据库，命令如下。

```
[root@controller ~]# mysql -uroot -p000000
```

登录后，先创建 Placement 数据库，命令如下。\

```
mysql> CREATE DATABASE placement;
```

若提示 Query OK，1 row affected（0.00 sec），表明数据库创建成功。

接着创建 MySQL 的 Placement 用户，并赋予其 Placement 数据库的操作权限，使得本地及远程都能访问，密码为 000000，命令如下。

```
mysql> GRANT ALL PRIVILEGES ON placement.* TO placement@'localhost' IDENTIFIED BY '000000';
```

```
mysql> GRANT ALL PRIVILEGES ON placement.* TO placement@'%' IDENTIFIED BY '000000';
mysql> exit
```

上述 SQL 语句与创建 Keystone 数据库的语句一致，请参考实训项目 3 中步骤 2 "创建 Keystone 数据库并授权" 部分对创建数据库时 SQL 语句的说明，在此不做赘述。

然后，需要为 placement 服务创建数据库表，创建数据库表之前，需要先修改 placement 配置文件中用于数据库连接的内容，命令如下。

```
[root@controller ~]# crudini --set /etc/placement/placement.conf placement_database
connection mysql+pymysql: //placement: 000000@controller/placement
```

上述命令的功能是在 /etc/placement/placement.conf 文件的 [database] 段落中添加 "connection = mysql+pymysql（数据库类型）://placement（登录数据库的用户名）:000000（用户密码）@controller（数据库主机名）/placement（数据库）" 配置。通过 vi/vim 命令直接修改 /etc/placement/placement.conf 配置文件，也可以达到相同效果。

接着，需要同步数据库，为 Placement 服务创建数据库表，命令如下。

```
[root@controller ~]# su -s /bin/sh -c "placement-manage db sync" placement
```

可以通过执行下面一条语句，查看是否同步并创建成功，命令如下，结果如图 8.2.2 所示。

```
[root@controller ~]# mysql -u root -p000000 -e "use placement; show tables;"
```

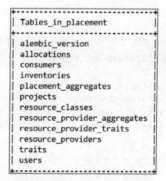

图 8.2.2 Placement 数据库中表信息

步骤 3：注册 Glance 服务至 Keystone 服务器。

在 OpenStack 中，几乎所有的服务（包括 Keystone 服务）要想正常运行，都必须首先向 Keystone 服务器注册。每一个服务需要向 Keystone 注册以下两个方面的信息，即用户信息、服务（service）和端点（endpoint）信息。

首先，进行 Placement 用户信息的注册。注册用户信息的目的是为了认证用户身份的。当一个用户向 Placement 服务器发送请求时，Placement 服务器首先要认证该用户是否合法。此时，Placement 会使用已注册的 Placement 用户向 Keystone 服务器发送认证请求。

本书所有的服务都创建在 service 项目下。因此，Placement 用户必须在 service 项目下具有 admin 权限。

添加 Placement 用户，并为用户设置密码，在本书实训过程中所有的密码统一使用

000000，命令如下，结果如图 8.2.3 所示。

```
[root@controller ~]# openstack user create --domain default --password 000000 placement
```

```
+---------------------+----------------------------------+
| Field               | Value                            |
+---------------------+----------------------------------+
| domain_id           | default                          |
| enabled             | True                             |
| id                  | 9fb240337ae24b3095d9ff226a3af345 |
| name                | placement                        |
| options             | {}                               |
| password_expires_at | None                             |
+---------------------+----------------------------------+
```

图 8.2.3　创建 Placement 用户反馈结果

再为 Glance 用户在 service 项目下分配 admin 权限，命令如下。

```
[root@controller ~]# openstack role add --project service --user placement admin
```

上述命令用法在实训项目 3 步骤 7 中已经提到，在此不再赘述。注意执行赋予 Glance 用户 admin 权限的命令是没有任何输出的。

接下来，注册 Placement 服务（service）和端点（endpoint）信息，注册服务和端点的信息目的是为了确保用户能够顺利访问 Glance 服务，并执行一系列与 Glance 服务有关的操作。

注册 Glance 服务信息，命令如下，结果如图 8.2.4 所示。

```
[root@controller ~]# openstack service create --name placement --description "Placement API" placement
```

```
+-------------+----------------------------------+
| Field       | Value                            |
+-------------+----------------------------------+
| description | Placement API                    |
| enabled     | True                             |
| id          | 7f3c4bb8988a448c9775607c5acc5349 |
| name        | placement                        |
| type        | placement                        |
+-------------+----------------------------------+
```

图 8.2.4　Placement 服务创建反馈结果

注册 Glance 端点信息，命令如下，结果如图 8.2.5 所示。

```
[root@controller ~]# openstack endpoint create --region RegionOne placement public http://controller:8778
[root@controller ~]# openstack endpoint create --region RegionOne placement internal http://controller:8778
[root@controller ~]# openstack endpoint create --region RegionOne placement admin http://controller:8778
```

```
+-----------+------------------------+
| Interface | URL                    |
+-----------+------------------------+
| admin     | http://controller:8778 |
| internal  | http://controller:8778 |
| public    | http://controller:8778 |
+-----------+------------------------+
```

图 8.2.5　Placement endpoint

需要注意，Placement 服务端口号为 8778，同时各服务组件间的 url 版本不尽相同，在操作时应该格外注意。

步骤 4：修改 Placement 配置文件。

在步骤 3 进行了 Placement 用户信息、服务和端点信息的注册，为了能让 Placement 服务顺利通过 Keystone 的认证，还需要在配置文件 /etc/placement/placement.conf 中记录 Keystone 的认证信息，即在配置文件的 [keystone_authtoken] 字段中修改 Keystone 的相关身份认证信息，同时在 [paste_deploy] 字段中添加 Keystone 支持。有了这些认证信息，Placement 在向外提供服务的时候，能够顺利通过 Keystone 的认证。

首先修改 /etc/placement/placement.conf 配置文件，命令如下。

```
[root@controller ~]# crudini --set /etc/placement/placement.conf api auth_strategy keystone
[root@controller ~]# crudini --set /etc/placement/placement.conf keystone_authtoken auth_url http://controller:5000/v3
[root@controller ~]# crudini --set /etc/placement/placement.conf keystone_authtoken memcached_servers controller:11211
[root@controller ~]# crudini --set /etc/placement/placement.conf keystone_authtoken auth_type password
[root@controller ~]# crudini --set /etc/placement/placement.conf keystone_authtoken project_domain_name default
[root@controller ~]# crudini --set /etc/placement/placement.conf keystone_authtoken user_domain_name default
[root@controller ~]# crudini --set /etc/placement/placement.conf keystone_authtoken project_name service
[root@controller ~]# crudini --set /etc/placement/placement.conf keystone_authtoken username placement
[root@controller ~]# crudini --set /etc/placement/placement.conf keystone_authtoken password 000000
```

当然也可以通过 vi/vim 直接编辑配置文件 /etc/placement/placement.conf，在相应的地方做上述内容的修改，以达到相同的效果。修改完成后使用 grep 命令过滤 /etc/glance/glance-api.conf 文件中的内容，来检查已完成的配置。命令和结果如下所示。

```
[root@controller ~]# grep -Ev "^#|^$" /etc/placement/placement.conf
[DEFAULT]
[api]
auth_strategy = keystone
[keystone_authtoken]
auth_url = http://controller:5000/v3
memcached_servers = controller:11211
auth_type = password
project_domain_name = default
user_domain_name = default
project_name = service
username = placement
password = 000000
```

```
[placement]
[placement_database]
connection = mysql+pymysql://placement:000000@controller/placement
```

以下是 Keystone 认证信息中每条语句的作用。

对于外部用户 Keystone 服务定义了一个端口号为 5000 的 auth_url，这里填写的是 auth_url；

memcached_servers 输入 Memcached 服务所在主机地址以及端口号；

auth_type 输入 Keystone 认证所使用的类型，通常我们使用 password 类型；

project_domain_name 输入项目所在的域名；

user_domain_name 输入用户所在的域名；

project_name 输入 Placement 服务所属的项目名，OpenStack 所有的组件均使用 service 项目；

username 输入 Placement 服务在 Keystone 中注册时的用户名；

password 输入 Placement 服务在 Keystone 中注册时所使用的密码，这里为 000000，在实际 OpenStack 生产环境的部署过程中该密码值必须与步骤 3 中添加 Placement 用户时所使用的密码一致。

步骤 5：在 httpd 服务中开启 Placement 服务的权限，命令如下。

```
[root@controller ~]# vi /etc/httpd/conf.d/00-placement-api.conf
```

添加如下内容：

```
<Directory /usr/bin>
   <IfVersion >= 2.4>
      Require all granted
   </IfVersion>
   <IfVersion < 2.4>
      Order allow,deny
      Allow from all
   </IfVersion>
</Directory>
```

为了让更新的配置生效，需要重启 Httpd 服务，命令如下。

```
[root@controller ~]# systemctl restart httpd
```

步骤 6：执行状态检查以确保一切正常，如图 8.2.6 所示，命令如下。

```
[root@controller ~]# placement-status upgrade check
```

```
+----------------------------------+
| Upgrade Check Results            |
+----------------------------------+
| Check: Missing Root Provider IDs |
| Result: Success                  |
| Details: None                    |
+----------------------------------+
| Check: Incomplete Consumers      |
| Result: Success                  |
| Details: None                    |
+----------------------------------+
```

图 8.2.6　检查服务状态

8.3 Placement 安装脚本及其解读

在介绍了手工安装 Placement 服务后，接下来介绍另一种安装 Placement 服务的方法，即利用已经写好的脚本自动安装 Placement 服务。

1. 需要安装的脚本内容

```bash
#!/bin/bash
source /etc/keystone/admin-openrc.sh
# 数据库配置
mysql -uroot -p000000 -e "create database IF NOT EXISTS placement ; "
mysql -uroot -p000000 -e "GRANT ALL PRIVILEGES ON placement.* TO 'placement' @ 'localhost' IDENTIFIED BY '000000' ; "
mysql -uroot -p000000 -e "GRANT ALL PRIVILEGES ON placement.* TO 'placement' @'%' IDENTIFIED BY '000000' ; "
# Keystone 认证
openstack user create --domain default --password 000000 placement
openstack role add --project service --user placement admin
openstack service create --name placement --description "Placement API" placement
openstack endpoint create --region RegionOne placement public http://controller:8778
openstack endpoint create --region RegionOne placement internal http://controller:8778
openstack endpoint create --region RegionOne placement admin http://controller:8778
# 安装 Placement 软件包
yum -y install openstack-placement-api
# 主配置文件设置
crudini --set /etc/placement/placement.conf placement_database connection mysql+pymysql://placement:000000@controller/placement
crudini --set /etc/placement/placement.conf api auth_strategy keystone
crudini --set /etc/placement/placement.conf keystone_authtoken auth_url http://controller:5000/v3
crudini --set /etc/placement/placement.conf keystone_authtoken memcached_servers controller:11211
crudini --set /etc/placement/placement.conf keystone_authtoken auth_type password
crudini --set /etc/placement/placement.conf keystone_authtoken project_domain_name default
crudini --set /etc/placement/placement.conf keystone_authtoken user_domain_name default
crudini --set /etc/placement/placement.conf keystone_authtoken project_name service
crudini --set /etc/placement/placement.conf keystone_authtoken username placement
crudini --set /etc/placement/placement.conf keystone_authtoken password 000000
# 添加 Httpd 认证
echo "
<Directory /usr/bin>
<IfVersion >= 2.4>
```

```
Require all granted
</IfVersion>
<IfVersion < 2.4>
Order allow,deny
Allow from all
</IfVersion>
</Directory>
" >> /etc/httpd/conf.d/00-placement-api.conf
# 同步数据库
su -s /bin/sh -c "placement-manage db sync" placement
systemctl restart httpd
```

以上是 Placement 服务安装脚本的全部内容，从内容上看到，在 Placement 服务中，脚本安装与利用命令手动安装的内容基本一致，在前面已经详细地向读者讲解过安装过程中各部分命令的作用。

电子教案　Placement 安装及其配置

PPT　Placement 安装及其配置

2. 脚本的执行

刚写好的脚本文件，在核实过内容后，是不能直接使用的。要想使用脚本，首先需要赋予脚本执行权限，命令如下。

```
[root@controller ~]# chmod a+x iaas-install-glance.sh
```

或者执行

```
[root@controller ~]# chmod 777 iaas-install-glance.sh
```

上述两种命令都可以实现赋予脚本执行权限的效果，这里的"iaas-install-placement.sh"是示范用的脚本名称，读者在实际使用时可自行命名脚本名称。

说明：本次实验所使用的脚本均放在 root 用户的家目录里。

脚本有了执行权限后就可以执行了。这里介绍两种执行脚本的方法。

方法 1：在脚本前加上脚本所在的路径，可以使用相对路径，也可以使用绝对路径，这取决于读者个人习惯，命令如下。

```
[root@controller ~]# ./iaas-install-placement.sh
```

"."表示当前目录，脚本放在家目录中，而用户此时也在家目录中，所以为方便，这里使用的是当前目录。

方法 2：现将脚本所在路径加入环境变量 PATH 中，然后就可以像 ls 命令一样直接在命令中输入使用了，命令如下。

```
[root@controller ~]# echo $PATH
/usr/local/sbin:/usr/local/bin:/sbin:/bin:/usr/sbin:/usr/bin:/root/bin
[root@controller ~]# PATH=${PATH}:/root/
[root@controller ~]# echo $PATH
/usr/local/sbin:/usr/local/bin:/sbin:/bin:/usr/sbin:/usr/bin:/root/bin:/root/
[root@controller ~]# iaas-install-placement.sh
```

习题 Placement 的安装及其配置

上述命令对比了将脚本所在目录加入环境变量 PATH 前后、环境变量 PATH 中的内容，最后可以直接在命令中像其他命令一样直接输入脚本名称了。

说明：直接在命令中改变的环境变量只对当前 shell 和其子 shell 有效，当重启或另外打开新的 shell 时，当前修改的环境变量无效。可以通过修改 /root/.bash_profile 配置文件中 PATH 的内容，达到环境变量 PATH 永久有效的效果，注意，修改 /root/.bash_profile 后需要通知系统内容变化，最直接的方法是重新打开新的 shell。

相对于手动安装来说，脚本安装的好处是简单省事，读者可以在安装之前写好脚本，安装时直接执行即可，且脚本可以方便以后多次利用。

第 9 章 Nova 的安装及其配置

本章导读：

本章介绍 Nova 组件的基本功能。作为云平台的计算服务提供模块，Nova 组件用于调度并管理整个云平台的计算资源。在本章的实训项目 6 中介绍如何通过命令行以及服务组件的参数配置安装 Nova 服务，同时解读 Nova 在安装和使用过程中可能出现的问题。在此基础上，本章最后还将介绍 Nova 服务安装脚本的制作和使用。

电子资源：

    电子教案　Nova 的安装及其配置
    PPT　Nova 的安装及其配置
    习题　Nova 的安装及其配置

9.1 Nova 功能简介

电子教案　Nova 功能简介

PPT　Nova 功能简介

计算服务是云计算结构控制器，这是原生 OpenStack 云平台的主要部分。用它来主持和管理云计算系统。主要模块用 Python 实现。Nova 可以说是 OpenStack 中最核心的组件。OpenStack 的其他组件，归根结底都是为 Nova 组件服务的。Nova 组件如此重要，也注定它是 OpenStack 中最为复杂的组件。Nova 服务由多个子服务构成，这些子服务通过 RPC 实现通信。因此，服务之间具有很松的耦合性。以下是 Nova 组件中各个子服务的功能。

Nova API：一个 HTTP 服务，用于接收和处理客户端发送的 HTTP 请求；Nova API 负责接收和响应终端用户有关虚拟机和云硬盘的请求，提供了 OpenStack API、亚马逊 EC2API 以及管理员控制 API。Nova API 是整个 Nova 的入口，它接收用户请求，将指令发送至消息队列，由相应的服务执行相关的指令消息。

Nova Cert：用于管理证书，为了兼容 AWS(Amazon Web Service)，提供了一整套基础设施和应用程序服务，使得几乎所有的应用程序都能在云上运行）。

Nova Compute：Nova 组件中最核心的服务，实现虚拟机管理的功能，并实现在计算节点上创建、启动、暂停、关闭和删除虚拟机、虚拟机在不同节点间（在线或离线）迁移、虚拟机安全控制、管理虚拟机磁盘镜像以及快照等功能。

Nova Conductor：OpenStack 中的一个 RPC 服务，主要提供数据库查询功能。在 G 版本之前的 OpenStack 中，Nova Compute 子服务里定义了许多数据库查询的方法。但是，由于 Nova Compute 子服务需要在每个计算节点上启动，一旦某个计算节点被攻击，那么攻击者将获得数据库的完全访问权限。有了 Nova Conductor 子服务后，便可在 Nova Conductor 中实现数据库访问权限的控制。同时，应该避免 Nova Conductor 与 Nova Compute 部署在同一个计算节点，所以本书中将 Nova Conductor 部署在 Controller 节点上，而 Nova Compute 部署在 Compute 节点上。

Nova Scheduler：Nova 的调度子服务。当客户端向 Nova 服务器发起创建虚拟机请求时，Nova Scheduler 子服务决定虚拟机创建在哪个计算节点上。

Nova Consoleauth 与 Nova Vncproxy：Nova 控制台子服务。其功能是实现客户端通过代理服务器远程访问虚拟机实例的控制台界面。

Nova 中的各个子组件之间具有很松的耦合性，这意味着，就算其中有些服务不启动，Nova 服务也能正常工作。在 Nova 中，要实现基本的虚拟机管理功能，至少需要启动 Nova API、Nova Conductor、Nova Compute 和 Nova Scheduler 服务。其中，Nova Compute 服务需要在每个计算节点中启动。而 Nova API、Nova Conductor 和 Nova Scheduler 服务只需要在 Controller 节点启动即可。

当然，在各子组件之间传输还需要 Queue。Queue 也就是消息队列，它就像网络上的一个 hub，nova 各个组件之间的通信几乎都是靠它进行的，当前的 stein 是用 RabbitMQ 实现的，它和 database 一起在各个守护进程之间传递消息。

下面以创建虚拟机为例，介绍 Nova 不同子服务之间的调用关系。启动一个新的 instance 涉及很多 OpenStack 组件的共同协作。

① 通过调用 nova-api 创建虚拟机接口，nova-api 对参数进行解析以及初步合法性校验，检查 quota（配额）。

② 在数据库中创建 Instance 实例，并发送 rpc 给 nova-conductor。

③ 接下来需要调用具体的物理机实现虚拟机部署，在这里就会涉及调度 nova-scheduler,nova-conductor 发送 rpc 给 nova-scheduler（选择一个物理机来创建虚拟机称之为 schedule 的过程）。

④ nova-scheduler 将选出的 host 发送 rpc 给 nova-compute，nova-compute 接收到请求后，通过 resource tracker 将创建虚拟机所需要的资源声明占用。

⑤ nova-compute 调用 neutron api 配置 network 虚拟机处于 networking 的状态，neutron 接收到消息根据私网资源池，结合 DHCP，实现 IP 分配和 IP 地址绑定。

⑥ 最后调用底层虚拟化 Hypervisor 技术，部署虚拟机。

1. 配额 (quota) 管理

配额管理就是控制用户资源的数量。在 OpenStack 里，管理员为每一个工程 (project) 分配的资源都是有一定限制的，如图 9.1.1 所示。这些资源包括实例（instance）、CPU、内存、存储空间等，不能让一个工程无限制地使用资源，配额管理针对的单位是工程 (project)。

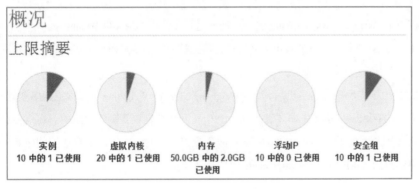

图 9.1.1 OpenStack 配额概况

管理员给这个工程的资源配额是最多创建 10 个实例，最多使用 20 个 VCPU，最多使用 50 GB 的内存，等等，只要达到某一个资源的使用上限，就会出现异常，这就是配额管理。

可以在命令行中通过 nova 命令查看系统配额，命令如下，结果如图 9.1.2 所示。

```
[root@controller ~]# nova quota-show
```

Quota	Limit
instances	10
cores	20
ram	51200
metadata_items	128
key_pairs	100
server_groups	10
server_group_members	10

图 9.1.2 查看配额默认值的反馈结果

由于还没有对配额进行修改，所以通过上面的命令看到的是系统默认的配额值。在 OpenStack 运维章节，会深入介绍如何更改 quota 的默认值，这里只做简单介绍。

对于创建 instance 的过程，schedule 是一个非常重要的环节，功能实现的过程中涉及一些具体显现的方法函数，暂且忽略，不做详解，只对 schedule 的过程进行简单介绍。

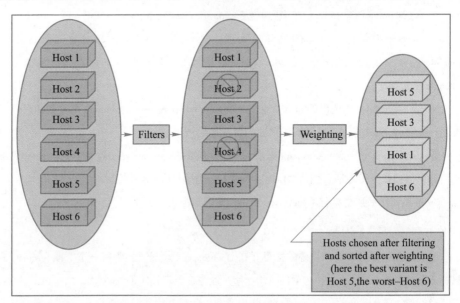

图 9.1.3　系统筛选的过程图

图 9.1.3 所示是 nova-scheduler 的一个经典过程图。先 Filters 再 Weighting。

① 对 Host 进行 Filtering。Filter 主要是利用主机的状态信息对所有可管理的主机进行筛选，选出匹配的主机（Host）。

② 选出了满足条件的主机后，进行 Weighting 的操作。Weighting 可以根据很多变量操作，一般来说内存和硬盘是最先需要满足的，CPU 和网络则是次要需要考虑的。在众多经过筛选满足基本条件的主机中，nova-scheduler 服务调用一些方法，找出最合适的创建实例的主机，将信息通过 RPC 传给 nova-compute 以供后面使用。

2. Libvirt

现阶段有很多种虚拟化技术。而对于 OpenStack 来说，Nova 通过独立的软件管理模块实现 XenServer、Hyper-V 和 VMWare ESX 的调用与管理，同时对于其他的 Hypervisor，如 KVM、LXC、QEMU、UML 和 Xen 则是通过 Libvirt 标准接口统一实现，其中 KVM 是 Nova-Compute 中 Libvirt 默认调用的底层虚拟化平台。Libvirt 是一种实现虚拟化平台能力交互的工具集，它为所支持的 Hypervisor 提供了一种通用的 API 接口套件，上层管理平台（如 Nova）通过 Libvirt 来实现对虚拟机的生命周期管理。Libvirt 当前支持以下底层虚拟化平台：

KVM：Linux 平台仿真器。

QEMU：面向各种架构的平台仿真器。

Xen：面向 IA-32、IA-64 和 PowerPC970 架构的虚拟机监控程序。

LXC：用于操作系统虚拟化的 Linux（轻量级）容器。

OpenVZ：基于 Linux 内核的操作系统级虚拟化。

User Mode Linux：面向各种架构的 Linux 平台仿真器。

VirtualBox：x86 虚拟化虚拟机监控程序。

ESX、GSX：VMW 爱热企业级虚拟化平台。

VMWare Workstation、VMWare Player：VMWare 用户级虚拟化平台。

Hyper-V：Microsoft 虚拟化平台。

另外，Libvirt 支持以 Bridging、NAT、VEPA 和 VN-LINK 方式构建虚拟网络，以及支持基于不同制式的 IDE/SCSI/USB disks、FibreChannel、LVM、iSCSI、NFS 和 Filesystems 存储。因此，Libvirt 在功能性、兼容性以及管理等方面的优势十分明显。

Libvirt 对底层虚拟化平台的调用与管理有本地管理与远程管理两种方式。

图 9.1.4 和图 9.1.5 中显示了 Libvirt 所用术语对照。Libvirt 将物理主机称作节点，将来宾操作系统称作域。这里需要注意的是，Libvirt（及其应用程序）在宿主 Linux 操作系统（域 0）中运行。

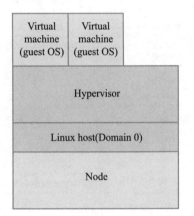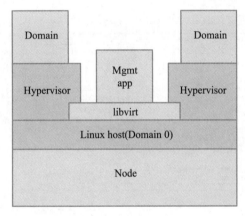

图 9.1.4　Libvirt 的本地管理

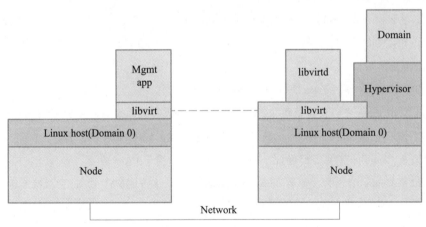

图 9.1.5　Libvirt 的远程管理

如图 9.1.4 所示，管理应用程序和域位于同一节点上，管理应用程序通过 Libvirt 工作，以控制本地域。该种方式在安全性、可靠性以及可扩展性方面存在一定弊端。

当管理应用程序和域位于不同节点上时，便产生了另一种控制方式，需要远程通信，如图 9.1.5 所示。该模式使用一种运行于远程节点，名为 libvirtd 的特殊守护进程。当在新节点上安装 Libvirt 时该程序会自动启动，且可自动确定本地虚拟机监控程序并为其安装驱动程序。该管理应用程序通过一种通用协议从本地 Libvirt 连接到远程 libvirtd。

因此，相比而言基于 Libvirt 的远程控制模式可以较好地解决本地管理模式所遇到的问题。

Nova 组件中，会调用 nova VNC Proxy 子服务来登录控制台。以下是 nova VNC Proxy 的基本工作原理，如图 9.1.6 所示。

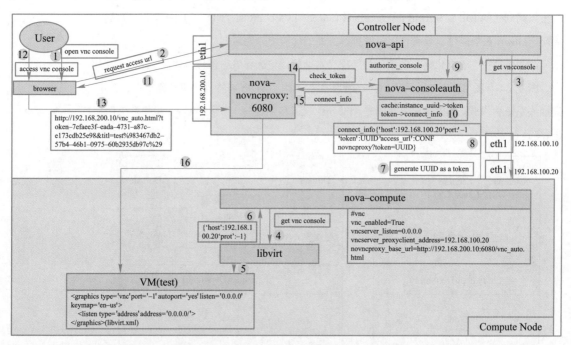

图 9.1.6　nova vnc proxy 基本工作流程图

（1）VNC Proxy 的功能

① 将公网 (public network) 和私网 (private network) 隔离。

② VNC Client 运行在公网上，VNC Server 运行在私网上，VNC Proxy 作为中间的桥梁将二者连接起来。

③ VNC Proxy 通过 Token 对 VNC Client 进行验证。

④ VNC Proxy 不仅使得私网的访问更加安全，而且将具体的 VNC Server 实现分离，可以支持不同 Hypervisor 的 VNC Server 但不影响用户体验。VNC Proxy 的部署。

⑤ 在 Controller 节点上部署 nova-consoleauth 进程，用于 Token 验证。

⑥ 在 Controller 节点上部署 nova-novncproxy 服务，用户的 VNC Client 会直接连接这个服务。

⑦ Controller 节点一般有两张网卡，连接到两个网络，一张用于外部访问，称为 public network，或者 API network，其 IP 地址是外网 IP，如图 9.1.6 所示中 192.168.200.10，另外一张网卡用于 openstack 各个模块之间的通信，称为 management network，一般是内网 IP，如图 9.1.6 所示中的 192.168.100.10。

⑧ 在 Compute 节点上部署 nova-compute，在 nova.conf 文件中对 [vnc] 组进行下面的配置。

```
enabled = true
server_listen = 0.0.0.0
VNC Server 的监听地址
server_proxyclient_address = 192.168.100.20
```

nova vnc proxy 是通过内网 IP 来访问 vnc server 的，所以 nova-compute 会告知 vnc proxy 用这个 IP 来连接。

novncproxy_base_url = http://192.168.100.10:6080/vnc_auto.html

这个 url 是返回给客户的 url，因而里面的 IP 是外网 IP。

（2）VNC Proxy 的运行过程大致分为以下几个步骤

① 一个用户试图从浏览器里面打开连接到虚拟机的 VNC Client。

② 浏览器向 nova-api 发送请求，要求返回访问 vnc 的 url。

③ nova-api 调用 nova-compute 的 get vnc console 方法，要求返回连接 VNC 的信息。

④ nova-compute 调用 libvirt 的 get vnc console 函数。

⑤ libvirt 会通过解析虚拟机运行的 /etc/libvirt/qemu/instance-00000001.xml 文件来获得 VNC Server 的信息（注意，instance 后跟的数字指的是虚拟机的序号，'00000001' 是指用户创建的第一个虚拟机）。

⑥ libvirt 将 host、port 等信息以 JSON 格式返回给 nova-compute。

⑦ nova-compute 会随机生成一个 UUID 作为 Token。

⑧ nova-compute 将 libvirt 返回的信息以及配置文件中的信息综合成 connect_info 返回给 nova-api。

⑨ nova-api 会调用 nova-consoleauth 的 authorize_console 函数。

⑩ nova-consoleauth 会将 instance->token, token->connect_info 的信息 cache 起来。

⑪ nova-api 将 connect_info 中的 access url 信息返回给浏览器：http://192.168.100.10/vnc_auto.html?token=7efaee3f-eada-4731-a87c-e173cdb25e98&titl=test%983467db2-57b4-46b1-0975-60b2935db97c%29。

⑫ 浏览器会试图打开上面的链接。

⑬ 链接会将请求发送给 nova-novncproxy。

⑭ nova-novncproxy 调用 nova-consoleauth 的 check_token 函数。

⑮ nova-consoleauth 验证了这个 token，将这个 instance 对应的 connect_info 返回给 nova-novncproxy。

⑯ nova-novncproxy 通过 connect_info 中的 host、port 等信息，连接 compute 节点上的 VNC Server，从而开始了 proxy 的工作。

9.2 实训项目 6　Nova 的手动安装与配置

1. 实训前提环境

成功完成实训项目 5 中所有内容后开始本实验，或者从已完成实训项目 5 的镜像开始，继续完成本实训内容。

2. 实训涉及节点

controller 和 compute

3. 实训目标

① 完成 Nova 基本组件的安装。

② 完成 Nova 数据库的创建以及授权。

③ 完成 Nova 用户、服务及端点的创建。

④ 完成 Nova 主配置文件的修改。

⑤ 完成与 vnc 服务的连接。

⑥ 完成 nova 使用 placement 服务的配置。

⑦ 完成 Nova 服务验证。

4. 实训步骤及其详解

微课 7 Nova 的手动安装与配置（1）

Nova 是一个服务的集合，可以启动虚拟机实例。可以配置这些服务运行在单独或相同的节点。本书中，将服务分别部署在两个节点，即 controller 和 compute 节点上，让虚拟机实例的服务运行在 compute 节点上，其他的服务部署在 controller 节点上。下面分别介绍这两个节点上部署的内容。

先在 controller 节点配置要运行的服务。

步骤 1：安装 Nova 基本组件。

通过 yum 命令安装 controller 节点 Nova 服务所需要的依赖包，命令如下。

```
[root@controller ~]# yum -y install openstack-nova-api openstack-nova-conductor openstack-nova-novncproxy openstack-nova-scheduler openstack-nova-console
```

执行上述安装命令成功后，可以看到安装成功标志，所有 Nova 依赖包都安装完成，如图 9.2.1 所示：

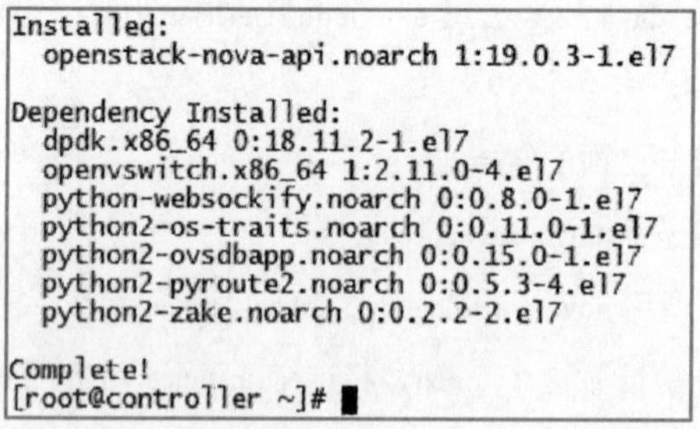

图 9.2.1 Nova 依赖包安装完成的反馈结果

步骤 2：创建 Nova 数据库并添加授权。

首先，用 root 用户登录 MySQL 数据库，命令如下：

```
[root@controller ~]# mysql -uroot -p000000
```

登录后，首先创建 Nova 数据库，命令如下：

```
mysql> CREATE DATABASE nova_api;
mysql> CREATE DATABASE nova;
mysql> CREATE DATABASE nova_cell0;
```

看到提示 Query OK, 1 row affected (0.00 sec)，表明数据库创建成功。

接着，创建 MySQL 的 Nova 用户，并赋予其 Nova 数据库的操作权限，使得本地及远程都能访问，密码为 000000，命令如下：

```
mysql> GRANT ALL PRIVILEGES ON nova_api.* TO 'nova'@'localhost' IDENTIFIED BY'000000';
mysql> GRANT ALL PRIVILEGES ON nova_api.* TO 'nova'@'%' IDENTIFIED BY '000000';
mysql> GRANT ALL PRIVILEGES ON nova.* TO 'nova'@'localhost' IDENTIFIED BY '000000';
mysql> GRANT ALL PRIVILEGES ON nova.* TO 'nova'@'%' IDENTIFIED BY '000000';
mysql> GRANT ALL PRIVILEGES ON nova_cell0.* TO 'nova'@'localhost' IDENTIFIED BY '000000';
mysql> GRANT ALL PRIVILEGES ON nova_cell0.* TO 'nova'@'%' IDENTIFIED BY '000000';
mysql> exit
```

上述 SQL 语句与创建 Keystone 数据库的语句一致，请参考实训项目 3 步骤 2 "创建 Keystone 数据库并授权"部分创建数据库时 SQL 语句的说明，在此不再赘述。

然后，需要为 Nova 服务创建数据库表。与其他 OpenStack 组件一样，在创建数据库表之前，需要先修改 Nova 配置文件中用于数据库连接的内容，命令如下：

```
[root@controller ~]# crudini --set /etc/nova/nova.conf api_database connection mysql+pymysql://nova:000000@controller/nova_api
[root@controller ~]# crudini --set /etc/nova/nova.conf database connection mysql+pymysql://nova:000000@controller/nova
```

同样，也可以通过 vi/vim 编辑器编辑 /etc/nova/nova.conf 文件中 [database] 字段与数据库连接的内容，以达到相同效果，编辑时须注意语法格式。

需要同步数据库，为 Nova 服务创建数据库表，命令如下：

```
[root@controller ~]# su -s /bin/sh -c "nova-manage api_db sync" nova
[root@controller ~]# su -s /bin/sh -c "nova-manage cell_v2 map_cell0" nova
[root@controller ~]# su -s /bin/sh -c "nova-manage cell_v2 create_cell --name=cell1 --verbose" nova
[root@controller ~]# su -s /bin/sh -c "nova-manage db sync" nova
[root@controller ~]# su -s /bin/sh -c "nova-manage cell_v2 list_cells" nova
```

与前面几个组件一样，也可以通过一条 SQL 命令来查看同步后创建的表，命令如下，结果如图 9.2.2 所示。

```
[root@controller ~]# mysql -u root -p000000 -e "use nova;show tables;"
```

```
+-----------------------+
| Tables_in_nova        |
+-----------------------+
| agent_builds          |
| aggregate_hosts       |
| aggregate_metadata    |
| aggregates            |
| allocations           |
| block_device_mapping  |
| bw_usage_cache        |
| cells                 |
| certificates          |
| compute_nodes         |
```

图 9.2.2　查看数据库同步的反馈结果

步骤 3：注册 Nova 服务至 Keystone 服务器。

在实训项目 4 步骤 3 中介绍过，在 Openstack 中，几乎所有的服务（包括 Keystone 服务）

要想正常运行，都必须首先向 Keystone 服务器注册。每一个服务需要向 Keystone 注册以下两方面的信息，即用户信息、服务 (service) 和端点 (endpoint) 信息。

首先，进行 Nova 用户信息的注册。注册用户信息的目的是认证用户身份。当一个用户向 Nova 服务器发送请求时，Nova 服务器首先要认证该用户是否合法。此时，Nova 会使用已注册的 Nova 用户向 Keystone 服务器发送认证请求。

本书所有的服务都创建在 service 项目下。因此，Nova 用户必须在 service 项目下具有 admin 权限。

添加 Nova 用户，并为用户设置密码，本书中所有的密码统一使用 000000，命令如下，结果如图 9.2.3 所示。

```
[root@controller ~]# openstack user create --domain default --password 000000 nova
```

```
+---------------------+----------------------------------+
| Field               | Value                            |
+---------------------+----------------------------------+
| domain_id           | default                          |
| enabled             | True                             |
| id                  | aa8d1085ef1d49119f707e64642ed528 |
| name                | nova                             |
| options             | {}                               |
| password_expires_at | None                             |
+---------------------+----------------------------------+
```

图 9.2.3　创建 Nova 用户的反馈结果

再为 Nova 用户在 service 项目下分配 admin 权限，命令如下：

```
[root@controller ~]# openstack role add --project service --user nova admin
```

上述命令用法在实训项目 3 步骤 7 中已经提到，在此不再赘述。

接下来，注册 Nova 服务 (service) 和端点 (endpoint) 信息，注册服务和端点信息目的是确保用户能够顺利访问 Nova 服务，并执行一系列与 Nova 服务有关的操作。

注册 Nova 服务和端点信息，命令如下，结果如图 9.2.4~ 图 9.2.7 所示。

```
[root@controller ~]# openstack service create --name nova --description "OpenStack Compute" compute
```

```
+-------------+----------------------------------+
| Field       | Value                            |
+-------------+----------------------------------+
| description | OpenStack Compute                |
| enabled     | True                             |
| id          | acdeb8e189d444198bf04528f611ebc0 |
| name        | nova                             |
| type        | compute                          |
+-------------+----------------------------------+
```

图 9.2.4　创建 Nova 服务的反馈结果

```
[root@controller ~]# openstack endpoint create --region RegionOne $(openstack service list | awk '/ compute / {print $4}') public http://controller:8774/v2.1
```

```
+--------------+----------------------------------+
| Field        | Value                            |
+--------------+----------------------------------+
| enabled      | True                             |
| id           | 50be31ba7db74ccf883b2eadd420865a |
| interface    | public                           |
| region       | RegionOne                        |
| region_id    | RegionOne                        |
| service_id   | acdeb8e189d444198bf04528f611ebc0 |
| service_name | nova                             |
| service_type | compute                          |
| url          | http://controller:8774/v2.1      |
+--------------+----------------------------------+
```

图 9.2.5　创建 Nova 的 public endpoint 的反馈结果

9.2 实训项目 6 Nova 的手动安装与配置

```
[root@controller ~]# openstack endpoint create --region RegionOne $(openstack service list | awk '/ compute / {print $4}') internal http://controller:8774/v2.1
```

```
| Field        | Value                            |
| enabled      | True                             |
| id           | 9a2bad15159a43ecba3387e1fcf52799 |
| interface    | internal                         |
| region       | RegionOne                        |
| region_id    | RegionOne                        |
| service_id   | acdeb8e189d444198bf04528f611ebc0 |
| service_name | nova                             |
| service_type | compute                          |
| url          | http://controller:8774/v2.1      |
```

图 9.2.6 创建 Nova 的 internal endpoint 的反馈结果

```
[root@controller ~]# openstack endpoint create --region RegionOne $(openstack service list | awk '/ compute / {print $4}') admin http://controller:8774/v2.1
```

```
| Field        | Value                            |
| enabled      | True                             |
| id           | 36904c2fe02342f4aa48851ea8394ebe |
| interface    | admin                            |
| region       | RegionOne                        |
| region_id    | RegionOne                        |
| service_id   | acdeb8e189d444198bf04528f611ebc0 |
| service_name | nova                             |
| service_type | compute                          |
| url          | http://controller:8774/v2.1      |
```

图 9.2.7 创建 Nova 的 admin endpoint 的反馈结果

需要注意的是，Nova 的服务端口号为 8774，同时注意 Nova 服务端点的语法结构与 Keystone 和 Glance 的服务端点不一样。"v2.1" 即版本 2.1。

步骤 4：配置 Nova 使用消息队列服务。

Nova 中的每个组件都会连接消息队列服务器，一个组件可能是一个消息发送者，也可能是一个消息接收者。本书实训过程中所使用的消息队列为 RabbitMQ，当然 Openstack 也支持其他的消息队列服务。需要修改配置文件中的消息队列部分的内容，命令如下：

```
[root@controller ~]# crudini --set /etc/nova/nova.conf DEFAULT transport_url rabbit://openstack:000000@controller
```

因为 RabbitMQ 服务部署在 controller 节点，所以命令最后的值为 "controller"。

步骤 5：修改 Nova 的配置文件。

首先，在 [default] 组下，需增加一些配置，命令如下：

```
[root@controller ~]# crudini --set /etc/nova/nova.conf DEFAULT enabled_apis osapi_compute,metadata
[root@controller ~]# crudini --set /etc/nova/nova.conf DEFAULT my_ip 192.168.100.10
[root@controller ~]# crudini --set /etc/nova/nova.conf DEFAULT use_neutron true
[root@controller ~]# crudini --set /etc/nova/nova.conf DEFAULT firewall_driver nova.virt.firewall.NoopFirewallDriver
```

同实训项目 4 步骤 4 的 Glance 配置一样，在上一步骤中进行了 Nova 用户信息、服务和端点信息的注册，为了能让 Nova 服务顺利通过 Keystone 的认证，需要在配置文件 /etc/nova/nova.conf 中记录 Keystone 的认证信息，即在配置文件的 [keystone_authtoken] 字段中修改 Keystone 的相关身份认证信息，同时在 [DEFAULT] 字段中添加 Keystone 支持。有了这些认证信息，Nova 在向外提供服务的时候，能够顺利通过 Keystone 的认证。

```
[root@controller ~]# crudini --set /etc/nova/nova.conf api auth_strategy keystone
[root@controller ~]# crudini --set /etc/nova/nova.conf keystone_authtoken auth_url http://controller:5000/v3
[root@controller ~]# crudini --set /etc/nova/nova.conf keystone_authtoken memcached_servers controller:11211
[root@controller ~]# crudini --set /etc/nova/nova.conf keystone_authtoken auth_type password
[root@controller ~]# crudini --set /etc/nova/nova.conf keystone_authtoken project_domain_name default
[root@controller ~]# crudini --set /etc/nova/nova.conf keystone_authtoken user_domain_name default
[root@controller ~]# crudini --set /etc/nova/nova.conf keystone_authtoken project_name service
[root@controller ~]# crudini --set /etc/nova/nova.conf keystone_authtoken username nova
[root@controller ~]# crudini --set /etc/nova/nova.conf keystone_authtoken password 000000
```

对于上述命令，也可以通过 vi/vim 直接编辑配置文件 /etc/nova/nova.conf，在相应的地方做上述内容的修改，以达到相同的效果。修改完成后使用 grep 命令过滤 /etc/nova/nova.conf 文件中的内容，来检查已完成的配置。命令和结果如下：

```
[root@controller ~]# grep -Ev "^#|^$" /etc/nova/nova.conf
[DEFAULT]
transport_url = rabbit://openstack:000000@controller
enabled_apis = osapi_compute,metadata
my_ip = 192.168.100.10
use_neutron = true
firewall_driver = nova.virt.firewall.NoopFirewallDriver
[api]
auth_strategy = keystone
[api_database]
connection = mysql+pymysql://nova:000000@controller/nova_api
[database]
connection = mysql+pymysql://nova:000000@controller/nova
[keystone_authtoken]
auth_url = http://controller:5000/v3
memcached_servers = controller:11211
auth_type = password
project_domain_name = default
user_domain_name = default
project_name = service
username = nova
password = 000000
```

步骤 6：更改配置文件，连接 controller 节点 vnc 服务。

```
[root@controller ~]# crudini --set /etc/nova/nova.conf vnc enabled true
[root@controller ~]# crudini --set /etc/nova/nova.conf vnc server_listen 192.168.100.10
[root@controller ~]# crudini --set /etc/nova/nova.conf vnc server_proxyclient_address 192.168.100.10
```

要修改 [vnc] 字段 enabled、server_listen、server_proxyclient_address 这 3 个值，需要连接 controller 节点，一般使用 controller 节点的管理地址。

步骤 7：配置 glance 服务的 api 地址以及锁路径。

```
[root@controller ~]# crudini --set /etc/nova/nova.conf glance api_servers http://controller:9292
[root@controller ~]# crudini --set /etc/nova/nova.conf oslo_concurrency lock_path /var/lib/nova/tmp
```

步骤 8：配置 nova 使用 placement 服务。

```
[root@controller ~]# crudini --set /etc/nova/nova.conf placement region_name RegionOne
[root@controller ~]# crudini --set /etc/nova/nova.conf placement project_domain_name default
[root@controller ~]# crudini --set /etc/nova/nova.conf placement project_name service
[root@controller ~]# crudini --set /etc/nova/nova.conf placement auth_type password
[root@controller ~]# crudini --set /etc/nova/nova.conf placement user_domain_name default
[root@controller ~]# crudini --set /etc/nova/nova.conf placement auth_url http://controller:5000/v3
[root@controller ~]# crudini --set /etc/nova/nova.conf placement username placement
[root@controller ~]# crudini --set /etc/nova/nova.conf placement password 000000
```

步骤 9：启动服务并将其加入开机自启。

为了让更新后的配置生效，需要重启 Nova 服务，在这之前再次清空 iptables 规则，命令如下：

```
[root@controller ~]# iptables -F
[root@controller ~]# iptables -X
[root@controller ~]# iptables -Z
[root@controller ~]# /usr/libexec/iptables/iptables.init save
[root@controller ~]# systemctl restart openstack-nova-api.service
[root@controller ~]# systemctl restart openstack-nova-consoleauth.service
[root@controller ~]# systemctl restart openstack-nova-scheduler.service
[root@controller ~]# systemctl restart openstack-nova-conductor.service
[root@controller ~]# systemctl restart openstack-nova-novncproxy.service
[root@controller ~]# systemctl enable openstack-nova-api.service
[root@controller ~]# systemctl enable openstack-nova-consoleauth.service
[root@controller ~]# systemctl enable openstack-nova-scheduler.service
[root@controller ~]# systemctl enable openstack-nova-conductor.service
[root@controller ~]# systemctl enable openstack-nova-novncproxy.service
```

在 controller 节点配置完 Nova 服务后，因本次是使用双节点模式，所以还需要配置启动虚拟机实例的 compute 节点。

步骤 10：安装 Nova 计算服务软件包。

```
[root@compute ~]# yum -y install openstack-nova-compute
```

执行上述安装命令成功后，可以看到成功标志，所有 Nova 依赖包都安装完成，如图 9.2.8 所示。

微课 8　Nova 的手动安装与配置（2）

```
Installed:
  openstack-nova-compute.noarch 1:19.0.3-1.el7
Dependency Installed:
  OpenIPMI.x86_64 0:2.0.27-1.el7
  adobe-mappings-cmap.noarch 0:20171205-3.el7
  atk.x86_64 0:2.28.1-1.el7
  augeas-libs.x86_64 0:1.4.0-9.el7
  blosc.x86_64 0:1.11.1-3.el7
  cairo.x86_64 0:1.15.12-4.el7
  cups-libs.x86_64 1:1.6.3-40.el7
  dejavu-fonts-common.noarch 0:2.33-6.el7
  device-mapper-multipath.x86_64 0:0.4.9-127.el7
```

图 9.2.8　compute 节点安装 Nova 软件包反馈结果

步骤 11：修改 Nova 主配置文件。

```
[root@compute ~]# crudini --set /etc/nova/nova.conf api auth_strategy keystone
[root@compute ~]# crudini --set /etc/nova/nova.conf keystone_authtoken auth_url http://controller:5000/v3
[root@compute ~]# crudini --set /etc/nova/nova.conf keystone_authtoken memcached_servers controller:11211
[root@compute ~]# crudini --set /etc/nova/nova.conf keystone_authtoken auth_type password
[root@compute ~]# crudini --set /etc/nova/nova.conf keystone_authtoken project_domain_name default
[root@compute ~]# crudini --set /etc/nova/nova.conf keystone_authtoken user_domain_name default
[root@compute ~]# crudini --set /etc/nova/nova.conf keystone_authtoken project_name service
[root@compute ~]# crudini --set /etc/nova/nova.conf keystone_authtoken username nova
[root@compute ~]# crudini --set /etc/nova/nova.conf keystone_authtoken password 000000
```

上述命令与 controller 节点的配置相同，参照 controller 节点的说明。接下来需要修改 [default] 组中的配置，命令如下：

```
[root@compute ~]# crudini --set /etc/nova/nova.conf DEFAULT enabled_apis osapi_compute,metadata
[root@compute ~]# crudini --set /etc/nova/nova.conf DEFAULT my_ip 192.168.100.20
[root@compute ~]# crudini --set /etc/nova/nova.conf DEFAULT use_neutron true
[root@compute ~]# crudini --set /etc/nova/nova.conf DEFAULT firewall_driver nova.virt.firewall.NoopFirewallDriver
```

步骤 12：配置消息代理服务。

```
[root@compute ~]# crudini --set /etc/nova/nova.conf DEFAULT transport_url rabbit://openstack:000000@controller
```

步骤 13：配置控制台对实例访问。

```
[root@compute ~]# crudini --set /etc/nova/nova.conf vnc enabled true
[root@compute ~]# crudini --set /etc/nova/nova.conf vnc vncserver_listen 0.0.0.0
[root@compute ~]# crudini --set /etc/nova/nova.conf vnc vncserver_proxyclient_address 192.168.100.20
[root@compute ~]# crudini --set /etc/nova/nova.conf vnc novncproxy_base_url http://192.168.100.10:6080/vnc_auto.html
```

compute 节点的 server_proxyclient_address 的 IP 地址为 compute 节点的管理网段 IP 地址，同时将 cserver_listen 的值改为"0.0.0.0"，即为监听所有的主机，此外 compute 节点还需要加上 enabled 和 novncproxy_base_url 两个变量值。

需要注意的是，novncproxy_base_url 字段中，也可以把 url 中的 IP 地址换成 controller 节点的主机名，如果写为主机名，则需要在使用浏览器访问云平台 Dashboard 界面的主机中添加对应 controller 主机名的域名解析，这样在 Dashboard 界面访问实例控制台时才不会有问题。

步骤 14：指定运行镜像服务的主机。

```
[root@compute ~]# crudini --set /etc/nova/nova.conf glance api_servers http://controller:9292
```

因 Glance 服务部署在 controller 节点，因此，需要告诉系统镜像服务的位置即指定镜像服务主机为 controller。

```
[root@compute ~]# crudini --set /etc/nova/nova.conf oslo_concurrency lock_path /var/lib/nova/tmp
```

步骤 15：在 Nova 中配置 placement 服务。

```
[root@compute ~]# crudini --set /etc/nova/nova.conf placement region_name RegionOne
[root@compute ~]# crudini --set /etc/nova/nova.conf placement project_domain_name default
[root@compute ~]# crudini --set /etc/nova/nova.conf placement project_name service
[root@compute ~]# crudini --set /etc/nova/nova.conf placement auth_type password
[root@compute ~]# crudini --set /etc/nova/nova.conf placement user_domain_name default
[root@compute ~]# crudini --set /etc/nova/nova.conf placement auth_url http://controller:5000/v3
[root@compute ~]# crudini --set /etc/nova/nova.conf placement username placement
[root@compute ~]# crudini --set /etc/nova/nova.conf placement password 000000
```

步骤 16：检查系统处理器是否支持虚拟机的硬件加速。

执行命令：

```
[root@compute ~]# egrep -c '(vmx|svm)' /proc/cpuinfo
```

如果该命令返回一个 1 或更大的值，说明系统支持硬件加速。

如果该指令返回一个 0 值，说明系统不支持硬件加速，必须配置 Libvirt 取代 KVM 来使用 QEMU。

但是一般情况下，在 CentOS7.3 及以上版本的操作系统中，都需要使用 QEMU 而不是 KVM，否则会报如图 9.2.9 所示的错误。

图 9.2.9　云主机启动报错

由于使用的操作系统版本较新，所以需使用 QEMU 替代 KVM，则对文件修改的命令如下：

```
[root@compute ~]# crudini --set /etc/nova/nova.conf libvirt virt_type qemu
[root@compute ~]# crudini --set /etc/nova/nova.conf libvirt inject_key True
```

步骤 17：启动服务并将其加入开机自启，在这之前再次清空 iptables 规则，命令如下。

```
[root@compute ~]# iptables -F
[root@compute ~]# iptables -X
[root@compute ~]# iptables -Z
[root@compute ~]# /usr/libexec/iptables/iptables.init save
[root@compute ~]# systemctl restart libvirtd
[root@compute ~]# systemctl restart openstack-nova-compute
[root@compute ~]# systemctl enable libvirtd
[root@compute ~]# systemctl enable openstack-nova-compute
```

在安装的最后，需要将计算节点添加到单元数据库中，在控制节点执行的命令如下。

```
[root@controller ~]# su -s /bin/sh -c "nova-manage cell_v2 discover_hosts --verbose" nova
```

步骤 18：验证 Nova 服务。

完成上述的所有配置，controller 节点的安装就完成了，可以通过 Nova 命令进行验证，命令如下。

```
[root@controller ~]# nova list
```

当返回如下结果时，表明成功配置了 controller 节点的 Nova 服务，结果如图 9.2.10 所示。

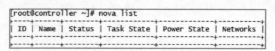

图 9.2.10　查看主机列表的反馈结果

5. 常见错误及调试排错

错误 1：endpoint 错误。

在 controller 节点部署完 Nova 服务后，通过 nova list 命令进行验证服务是否处于正常状态。

```
[root@controller ~]# nova list
ERROR: 'servers'
```

系统提示"servers（服务）"错误。当遇到一个错误却没有解决方案时，首先看一下 nova 的日志，进入存放 nova 日志的目录 /var/log/nova/，Nova 服务的每一个子服务都有自己的日志，先看 nova-api 的日志，命令如下：

```
[root@controller ~]# tail -f /var/log/nova/api.log
```

在日志中看到如图 9.2.11 所示内容。

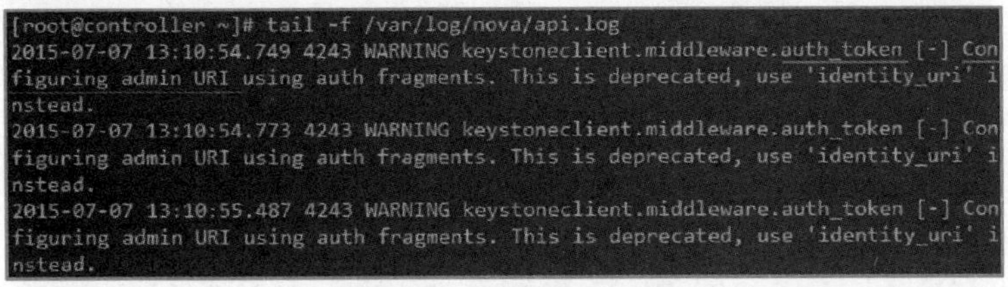

图 9.2.11　nova api 日志输出

从日志中可以初步猜测应该是 URI 有问题，接着将 nova 子服务的日志逐一看完，并没有看到有帮助的信息，因为 nova-api 日志中有提到 Keystone 认证信息。接下来，再看一下 Keystone 的日志，命令如下：

```
[root@controller nova]# tail -f /var/log/keystone/keystone.log
```

在日志中发现如图 9.2.12 所示信息：

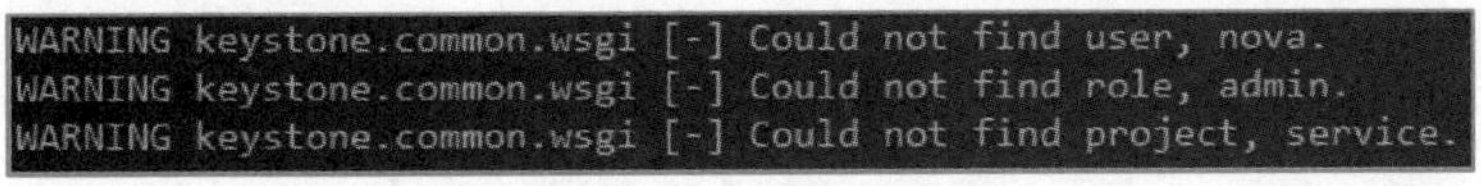

图 9.2.12　Keystone 日志输出

结合 nova-api 的日志内容，既然与 URI 有关，基本可以判断一般由两种原因引起，一种可能是配置文件 etc/nova/nova.conf 中与 Keystone 认证有关的内容出现拼写错误；另一种可能是在注册 endpoint（端点）时出现错误。

锁定原因后，逐一进行排查，经检查配置文件没有错误，那么可以肯定是 endpoint（端点）错误，通过 keystone 命令查看已经创建的 endpoint（端点），命令如下：

```
[root@controller ~]# openstack endpoint list
```

仔细对照发现，endpoint（端点）内容拼写错误。删掉错误的 endpoint（端点），重新创建新的正确的 endpoint（端点），注意，为同一服务创建的多个 endpoint（端点）不会覆盖，所以为避免不必要的麻烦，先删除错误的内容，然后重新创建新的内容。创建好正确的 endpoint（端点），重启 Nova 子服务，再次验证，成功解决。

错误 2：两个节点间的时间没有同步。

有时候突然发现 nova-compute 挂掉，或者不正常了，通过 nova-manage 查看服务的状态是 XXX 而不是 :)。之前一直是正常的，然后莫名其妙地出错，这时需要检查是否是两个节点的时间没有同步。所以 Nova 环境中各节点时间同步很重要，一定要确保时间同步。

电子教案 Nova 安装脚本及其解读

PPT Nova 安装脚本及其解读

9.3 Nova 安装脚本及其解读

在介绍了手动安装 Nova 服务后，接下来介绍另一种安装 Nova 服务的方法，即利用已经写好的脚本自动安装 Nova 服务。

1. 需要安装的脚本内容

Nova 服务手动安装时，需要分别在 controller 和 compute 节点配置，所以脚本也需要分别在两个节点完成各自的配置内容。

controller 节点的脚本内容如下：

```
#!/bin/bash
source /etc/keystone/admin-openrc.sh
# 安装 Nova 需要的软件包
yum -y install openstack-nova-api openstack-nova-conductor openstack-nova-novncproxy openstack-nova-scheduler openstack-nova-console
# 数据库配置
mysql -uroot -p000000 -e "create database IF NOT EXISTS nova_api ;"
mysql -uroot -p000000 -e "create database IF NOT EXISTS nova ;"
mysql -uroot -p000000 -e "create database IF NOT EXISTS nova_cell0 ;"
mysql -uroot -p000000 -e "GRANT ALL PRIVILEGES ON nova_api.* TO 'nova'@'localhost' IDENTIFIED BY '000000' ;"
mysql -uroot -p000000 -e "GRANT ALL PRIVILEGES ON nova_api.* TO 'nova'@'%' IDENTIFIED BY '000000' ;"
mysql -uroot -p000000 -e "GRANT ALL PRIVILEGES ON nova.* TO 'nova'@'localhost' IDENTIFIED BY '000000' ;"
mysql -uroot -p000000 -e "GRANT ALL PRIVILEGES ON nova.* TO 'nova'@'%' IDENTIFIED BY '000000' ;"
mysql -uroot -p000000 -e "GRANT ALL PRIVILEGES ON nova_cell0.* TO 'nova'@'localhost' IDENTIFIED BY '000000' ;"
mysql -uroot -p000000 -e "GRANT ALL PRIVILEGES ON nova_cell0.* TO 'nova'@'%' IDENTIFIED BY '000000' ;"
# Keystone 认证
openstack user create --domain default --password 000000 nova
openstack role add --project service --user nova admin
openstack service create --name nova --description "OpenStack Compute" compute
openstack endpoint create --region RegionOne compute public http://controller:8774/v2.1
openstack endpoint create --region RegionOne compute internal http://controller:8774/v2.1
openstack endpoint create --region RegionOne compute admin http://controller:8774/v2.1
# 主配置文件设置
crudini --set /etc/nova/nova.conf api_database connection mysql+pymysql://nova:000000@controller/nova_api
```

```
    crudini --set /etc/nova/nova.conf database connection mysql+pymysql://nova:000000@controller/nova
    crudini --set /etc/nova/nova.conf DEFAULT enabled_apis osapi_compute,metadata
    crudini --set /etc/nova/nova.conf DEFAULT transport_url rabbit://openstack:000000@controller
    crudini --set /etc/nova/nova.conf DEFAULT my_ip 192.168.100.10
    crudini --set /etc/nova/nova.conf DEFAULT use_neutron true
    crudini --set /etc/nova/nova.conf DEFAULT firewall_driver nova.virt.firewall.NoopFirewallDriver
    crudini --set /etc/nova/nova.conf api auth_strategy keystone
    crudini --set /etc/nova/nova.conf keystone_authtoken auth_url http://controller:5000/v3
    crudini --set /etc/nova/nova.conf keystone_authtoken memcached_servers controller:11211
    crudini --set /etc/nova/nova.conf keystone_authtoken auth_type password
    crudini --set /etc/nova/nova.conf keystone_authtoken project_domain_name default
    crudini --set /etc/nova/nova.conf keystone_authtoken user_domain_name default
    crudini --set /etc/nova/nova.conf keystone_authtoken project_name service
    crudini --set /etc/nova/nova.conf keystone_authtoken username nova
    crudini --set /etc/nova/nova.conf keystone_authtoken password 000000
    # VNC 设置
    crudini --set /etc/nova/nova.conf vnc enabled true
    crudini --set /etc/nova/nova.conf vnc server_listen 192.168.100.10
    crudini --set /etc/nova/nova.conf vnc server_proxyclient_address 192.168.100.10
    crudini --set /etc/nova/nova.conf glance api_servers http://controller:9292
    crudini --set /etc/nova/nova.conf oslo_concurrency lock_path /var/lib/nova/tmp
    # 配置 Placement 认证
    crudini --set /etc/nova/nova.conf placement region_name RegionOne
    crudini --set /etc/nova/nova.conf placement project_domain_name default
    crudini --set /etc/nova/nova.conf placement project_name service
    crudini --set /etc/nova/nova.conf placement auth_type password
    crudini --set /etc/nova/nova.conf placement user_domain_name default
    crudini --set /etc/nova/nova.conf placement auth_url http://controller:5000/v3
    crudini --set /etc/nova/nova.conf placement username placement
    crudini --set /etc/nova/nova.conf placement password 000000
    # 同步数据库
    su -s /bin/sh -c "nova-manage api_db sync" nova
    su -s /bin/sh -c "nova-manage cell_v2 map_cell0" nova
    su -s /bin/sh -c "nova-manage cell_v2 create_cell --name=cell1 --verbose" nova
    su -s /bin/sh -c "nova-manage db sync" nova
    su -s /bin/sh -c "nova-manage cell_v2 list_cells" nova
    iptables -F
    iptables -X
    iptables -Z
    /usr/libexec/iptables/iptables.init save
```

```bash
# 落定服务
systemctl enable openstack-nova-api.service \
  openstack-nova-consoleauth.service openstack-nova-scheduler.service \
  openstack-nova-conductor.service openstack-nova-novncproxy.service
systemctl start openstack-nova-api.service \
  openstack-nova-consoleauth.service openstack-nova-scheduler.service \
  openstack-nova-conductor.service openstack-nova-novncproxy.service
```

以上内容是写好的 controller 节点 Nova 服务的脚本。从上述内容上可以看出，除了命令执行的顺序会有所变化外，脚本安装的内容与利用命令手动安装的内容基本一致，前面几个小节已经详细讲解过安装过程中各部分命令的作用，这里不再赘述。

compute 节点的脚本内容如下：

```bash
#!/bin/bash
# 加载环境变量
source openrc.sh
# 安装 Nova 需要的软件包
yum -y install openstack-nova-compute
crudini --set /etc/nova/nova.conf DEFAULT enabled_apis osapi_compute,metadata
# 连接消息队列
crudini --set /etc/nova/nova.conf DEFAULT transport_url rabbit://openstack:000000@controller
# 配置 Keystone 认证
crudini --set /etc/nova/nova.conf api auth_strategy keystone
crudini --set /etc/nova/nova.conf keystone_authtoken auth_url http://controller:5000/v3
crudini --set /etc/nova/nova.conf keystone_authtoken memcached_servers controller:11211
crudini --set /etc/nova/nova.conf keystone_authtoken auth_type password
crudini --set /etc/nova/nova.conf keystone_authtoken project_domain_name default
crudini --set /etc/nova/nova.conf keystone_authtoken user_domain_name default
crudini --set /etc/nova/nova.conf keystone_authtoken project_name service
crudini --set /etc/nova/nova.conf keystone_authtoken username nova
crudini --set /etc/nova/nova.conf keystone_authtoken password 000000
crudini --set /etc/nova/nova.conf DEFAULT my_ip 192.168.100.20
crudini --set /etc/nova/nova.conf DEFAULT use_neutron true
crudini --set /etc/nova/nova.conf DEFAULT firewall_driver nova.virt.firewall.NoopFirewallDriver
# 配置 VNC 连接
crudini --set /etc/nova/nova.conf vnc enabled true
crudini --set /etc/nova/nova.conf vnc vncserver_listen 0.0.0.0
crudini --set /etc/nova/nova.conf vnc vncserver_proxyclient_address 192.168.100.20
crudini --set /etc/nova/nova.conf vnc novncproxy_base_url http://192.168.100.10:6080/vnc_auto.html
crudini --set /etc/nova/nova.conf glance api_servers http://controller:9292
crudini --set /etc/nova/nova.conf oslo_concurrency lock_path /var/lib/nova/tmp
```

```
# 配置连接 Placement 认证
crudini --set /etc/nova/nova.conf placement region_name RegionOne
crudini --set /etc/nova/nova.conf placement project_domain_name default
crudini --set /etc/nova/nova.conf placement project_name service
crudini --set /etc/nova/nova.conf placement auth_type password
crudini --set /etc/nova/nova.conf placement user_domain_name default
crudini --set /etc/nova/nova.conf placement auth_url http://controller:5000/v3
crudini --set /etc/nova/nova.conf placement username placement
crudini --set /etc/nova/nova.conf placement password 000000
crudini --set /etc/nova/nova.conf libvirt virt_type qemu
crudini --set /etc/nova/nova.conf libvirt inject_key True
iptables -F
iptables -X
iptables -Z
/usr/libexec/iptables/iptables.init save
# 落定服务
systemctl enable libvirtd.service openstack-nova-compute.service
systemctl start libvirtd.service openstack-nova-compute.service
```

以上是写好的 compute 节点 Nova 服务的脚本,同样从上述内容可以看出,除了命令执行的顺序会有所变化外,脚本安装的内容与利用命令手动安装的内容基本一致。

2. 脚本的执行

刚写好的脚本文件,在核实过内容后,是不能直接使用的。要想使用脚本,首先需要赋予脚本执行权限,这里以 controller 节点的脚本为例讲解,命令如下:

```
[root@controller ~]# chmod a+x iaas-install-nova-controller.sh
```

或者执行:

```
[root@controller ~]# chmod 777 iaas-install-nova-controller.sh
```

上述两种命令都可以实现赋予脚本执行权限的效果,这里的"iaas-install-nova-controller.sh"是示范用的脚本名称,读者在实际使用时可自行命名脚本名称(说明:本次实验所使用的脚本均放在 root 用户的家目录里)。

脚本有了执行权限后就可以执行了。这里介绍两种执行脚本的方法。

方法 1:在脚本前加上脚本所在的路径,可以使用相对路径,也可以使用绝对路径,这取决于读者个人习惯,命令如下:

```
[root@controller ~]# ./iaas-install-nova-controller.sh
```

"."表示当前目录,用户的脚本放在家目录中,而用户此时也在家目录中,所以为方便,这里使用的是当前目录。

方法 2:现将脚本所在路径加入环境变量 PATH 中,然后就可以像 ls 命令一样直接在命令中输入使用了,命令如下:

```
[root@controller ~]# echo $PATH
/usr/local/sbin:/usr/local/bin:/sbin:/bin:/usr/sbin:/usr/bin:/root/bin
```

```
[root@controller ~]# PATH=${PATH}:/root/
[root@controller ~]# echo $PATH
/usr/local/sbin:/usr/local/bin:/sbin:/bin:/usr/sbin:/usr/bin:/root/bin:/root/
[root@controller ~]# iaas-install-nova-controller.sh
```

上述命令对比了将脚本所在目录加入环境变量 PATH 前后，环境变量 PATH 中的内容，最后可以直接在命令中像其他命令一样直接输入脚本名称了。

说明：直接在命令中改变的环境变量只对当前 shell 和其子 shell 有效，当重启或另外打开新的 shell 时，当前修改的环境变量无效。然而可以通过修改 /root/.bash_profile 配置文件中 PATH 的内容，达到环境变量 PATH 永久有效的效果。注意，修改 /root/.bash_profile 后需要通知系统内容变化，最直接的方法是重新打开新的 shell。

第 10 章　Neutron 的安装及其配置

 本章导读：

　　本章介绍 Neutron 组件的基本功能。作为云平台的网络服务提供模块，Neutron 组件用于实现实例间以及实例对外的所有基于 IP 的网络访问功能，并嵌入了 SDN 的设计思想。在本章的实训项目 7 中介绍如何通过命令行以及服务组件的参数配置安装 Neutron 的依赖服务，主要讲解 Neutron 服务的组件配置与网络创建，同时解读 Neutron 在安装和使用过程中可能会出现的问题。在以上基础上，本章最后还将介绍 Neutron 服务安装脚本的制作和使用。

电子资源：
　　电子教案　Neutron 的安装及其配置
　　PPT　Neutron 的安装及其配置
　　习题　Neutron 的安装及其配置

第 10 章 Neutron 的安装及其配置

10.1 Neutron 功能简介

电子教案 Neutron 功能简介

PPT Neutron 功能简介

自 OpenStack 的 Austin 版本发布以来，软件定义网络（SDN）的设计就已经融入到 OpenStack 项目当中并作为 Nova 的一部分（nova-network）。

在 2012 年发布的 Folsom 版本中，网络服务被独立，Quantum 就此诞生。Quantum 的功能相较于之前的 Nova-network 而言有了很大的突破。例如，提供了单独的服务 API，可以通过 Keystone 调用网络服务；提供全新的虚拟网络模型，将 OSI 模型中二层网络概念和三层网络概念进行封装；提供插件式的网络组件，典型的有 OpenvSwitch 组件、Cisco 组件等，支持三层转发和多路由器功能。

由于 Quantum 的版权问题，在 OpenStack 的 Havana 版本中，Quantum 被命名为 Neutron，也就有了今天的 Openstack 核心项目 Neutron。

1. 软件定义网络（SDN）

Neutron 是提供虚拟网络服务的，但谈及 "虚拟网络"，需先了解 "软件定义网络(SDN)"。2006 年 SDN 在斯坦福大学 Clean Slate 课题中被提出，随后得到了飞速发展，思科、IBM、微软、等 IT 界知名厂商也发起了攻势。2013 年 4 月，思科和 IBM 联合微软、Big Switch、博科、思杰、戴尔、爱立信、富士通、英特尔、瞻博网络、NEC、惠普、红帽和 VMware 等发起成立了 Open Daylight，与 Linux 基金会合作，开发 SDN 控制器、南向 / 北向 API 等软件，旨在打破大厂商对网络硬件的垄断，驱动网络技术创新力，使网络管理更容易、更廉价。我国在 2012 年的 "863 计划"（国家高科技研究发展计划）的 "未来网络体系结构和创新环境" 中，SDN 也被加入作为主要研究方向。

传统网络将像交换、路由这些网络功能和像负载均衡、冗余设计、虚拟化设计这些业务功能与网络设备硬件紧耦合，网络功能被独立出来，划分为 "核心层" "汇聚层" "接入层" 3 部分。

在传统网络中，由物理网络设备（交换机、路由器等）控制网络功能，对于拓扑相对简单的企业网络而言，这样的传统网络似乎没有什么问题，而对于融合负载均衡、集群等融合业务的云计算网络而言，传统网络的紧耦合架构必将成为业务上线的瓶颈，业务功能复杂后传统网络的实现难度和成本都会上升，所以有了全新的解决方案：软件定义网络（SDN）。

SDN 的核心理念是将网络功能和业务处理功能与网络设备硬件解耦合，将原本独立的网络功能和业务功能结合，变成一个个抽象化的功能，再通过外置的控制器来控制这些抽象化的对象。SDN 网络架构分为应用、控制、转发三层分离的架构，如图 10.1.1 所示。

SDN 使得网络部署成为制约业务上线和云业务效率的瓶颈，网络将和计算、存储等资源一样，成为可自由调度的资源。

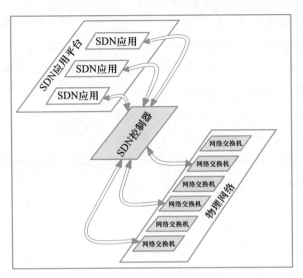

图 10.1.1　SDN 网络架构图

2. SDN 在 OpenStack 中的实现

随着云计算技术的不断发展，诞生出了很多出色的云管理工具，OpenStack 作为云管理工具的代表，集成了云管理所必需的组件，这些组件中就包括了 SDN 控制器 Neutron。Neutron 添加了一层虚拟的网络服务，能让用户（实际是项目）构建自己的虚拟网络。Neutron 公开了一组可扩展的 API，可以通过改组 API 创建和管理虚拟网络，实现方式如图 10.1.2 所示。

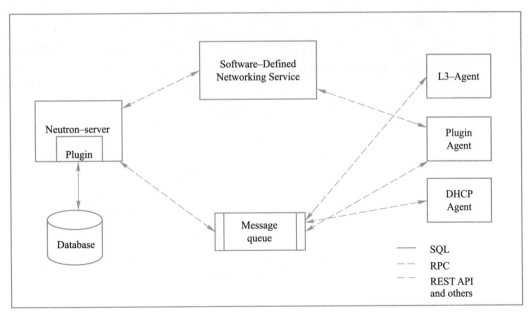

图 10.1.2　Neutron 可扩展 API 实现方式

在 Neutron 的架构图中，可以看到 Neutron 通过代理服务的方式去实现 SDN 的抽象功能，其中 Neutron Server 负责创建 Neutron 守护进程（Daemon），包括 Neutron Server 主进程（neutron-server）和 Neutron 各插件进程（neutron-*-plugin）。Neutron 的插件包括如

LinuxBridge、OpenvSwitch、Cisco、Cisco 等，本书实验采用开源的 OpenvSwitch 插件。

Neutron Server 主进程负责提供 API 接口，并对其他服务的 API 的调用请求传递给 Neutron 插件进行处理。Neutron 插件会通过访问数据库来查询状态、策略、配置等信息。在 Neutron 中，真正的数据包不由插件直接进行转发，而会通过相应的插件代理（Plugin Agent）服务来处理。

Neutron 的主要代理服务包括以下几个：

DHCP 代理（DHCP Agent）：DHCP 代理可以为相应的项目网络提供 DHCP 服务，虚拟机实例可以通过 DHCP 服务获取 IP 地址。

三层转发代理（L3 Agent）：为虚拟机实例提供三层路由转发功能，可以通过 L3 代理创建 GRE 网络。

Metadata 代理（Metadata Agent）：负责为相应项目创建虚拟机实例传递属性，主要用于传递虚拟机实例的 ssh 公钥。

负载均衡代理（LBaaS Agent）：提供负载均衡服务，通过调用编配（Heat）服务的模板，实现网络流量负载均衡。

Neutron 作为 OpenStack 原生的 SDN 控制器，可以支持多种网络类型，包括扁平化网络（FLAT）、三层网络（GRE）、虚拟局域网（VLAN）等。OpenStack 在 Havana 版本中，新增了 ML2（The Modular Layer2）插件。ML2 插件可以通过 Type Drivers 和 Mechanism Drivers 来让多种插件并存。原生的 ML2 插件支持 OpenStack 自家的 OpenvSwitch 插件，可以通过 ML2 插件配合 OpenvSwith 快速部署 FLAT 网络。

3. Open vSwitch 的功能简介

Open vSwitch（以下简称 OVS）作为 OpenStack 自家的插件，在 OpenStack 的 SDN 实现上起到了至关重要的作用，OVS 在云网络中的位置如图 10.1.3 所示。

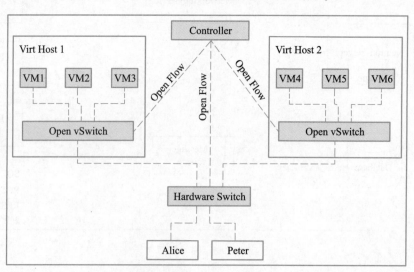

图 10.1.3　OVS 在云网络中的位置

OVS 在整个云网络中承担虚拟交换机（virtual switch）的角色，通过 Open Flow 协议与 Controller 节点统一管理，可以被 Controller 所管理的还有一些支持 Open Flow 的硬件交换机，它们与虚拟机实例一起组成了整个云计算网络。

OVS 通过 Open Flow 协议去构建自己的流控表（Flow Table），当数据包通过虚拟交换机时，OVS 会根据自己的 Flow Table 与数据包进行匹配，从而控制数据包的流向或改变数据包结构。通过一系列对数据包的操作，使 OVS 实现 SDN 的业务需求。

OVS 在根据业务需求对数据包进行修改时，会通过查询数据库获取业务和云网络中其他设备（包括物理设备和虚拟设备）的信息，从而改变自己的 Flow Table，如图 10.1.4 所示。

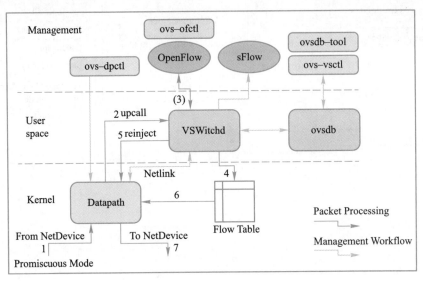

图 10.1.4　OpenvSwitch 架构图

4. OpenStack 网络划分

OpenStack 以节点的方式分布相应服务，对于 Neutron 来说也不例外，从 Havana 版本开始，OpenStack 就针对 Neutron 构建了三节点（Controller 节点、Compute 节点、Network 节点）云计算网络，如图 10.1.5 所示。本书中将节点合并为两个，将 Network 的功能集中到了 Controller 节点上。

该网络模型主要将 OpenStack 网络划分为以下 3 种：

① External Network（外部网络）：主要用作虚拟机实例访问外网，该网络还可以用作外网服务（如 ssh）连入内部虚拟机实例。

② Data Network（数据网络）：用于虚拟机实例之间的数据传递，包括虚拟机实例与虚拟路由器之间的数据传递都是依托于数据网络。

③ Management Network（管理网络）：负责处理 OpenStack 各个组件之间的交互、连接数据库，AMQP 通信数据都是在管理网络中传输。

三个网络互不干扰，相对隔离，这样的设计既出于安全考虑同样也方便了流量隔离，对于流量较小的管理网络，显然不需要与数据网络抢占资源，也不需要进行任何形式的流量控制，整个网络可以相对稳定地运转。

本书中，为了方便读者理解，实验采用双节点模式，将 Network 节点与 Compute 节点整合，但整体网络框架不变。

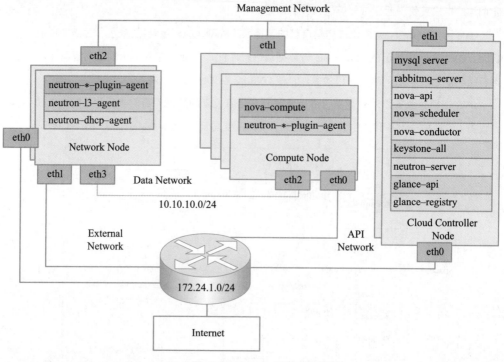

图 10.1.5 Neutron 三节点架构图

5. Neutron 的未来

Neutron（Quantum）的出现将 OpenStack 推向了一个前所未有的高度，OpenStack 终于能在 SDN 领域占有一席之地。在随后推出的多插件整合功能 ML2，也使 Neutron 得以商业化，支持 Open Flow 协议的 OpenvSwitch 插件更让人眼前一亮。

尽管作为 OpenStack 的 SDN 控制器，Neutron 确实让 OpenStack 跻身于 SDN 这个大家庭，但 Neutron 和真正商用 SDN 控制器还是有很大差距。OVS 插件虽然支持 Open Flow 协议，也提供了 Open Flow 协议中的南向接口（模块内部接口），但 Neutron 并没有通过标准接口调用 Open Flow，而是使用 OVS 的 CLI 的配置方式，对于北向接口（模块对外接口），Neutron 采用 Rest API 的标准技术实现，并没有直接调用标准接口。在集中化管理层面，Neutron 通过 Rest API 与 Horizon 进行交互，而 Horizon 并不是一个专用于管理 SDN 的软件，在真正交互时，Horizon 仅是通过 CLI 的方式去实现对网络资源的集中化管理，这也使得 Neutron 在集中化控制方面存在很大的瓶颈。

虽然 Neutron 不及真正的 SDN 控制器，但 Neutron 可以与真正的 SDN 控制器之间进行结合，将 OpenStak 云网络带入商用 SDN 的领域，相信在不久的未来，Neutron 会给人们一个意想不到的惊喜。

10.2 实训项目 7 Neutron 的手动安装与外部环境配置

1. 实训前提环境

成功完成实训项目 6 中所有内容后开始本实验，或者从已完成实训项目 6 的镜像开始，继续完成本实训内容。

10.2 实训项目 7 Neutron 的手动安装与外部环境配置

2. 实训涉及节点

controller 和 compute

电子教案 Neutron 的手动安装与外部环境配置

3. 实训目标

① 完成 Neutron 基本组件的安装。

② 完成 Linux 内核的修改。

③ 完成 Neutron 数据库的创建以及授权。

PPT Neutron 的手动安装与外部环境配置

④ 完成 Neutron 用户、项目、角色以及服务端点的创建。

⑤ 完成 Neutron 主配置文件认证部分的配置。

⑥ 完成虚拟网卡的创建。

4. 实训步骤及其详解

步骤 1：Neutron 基本组件的安装。

在 controller 节点上执行 yum 源安装命令安装 Neutron 相关依赖包，命令如下：

```
[root@controller ~]# yum -y install openstack-neutron openstack-neutron-ml2 ebtables openstack-neutron-openvswitch
```

执行上述安装命令成功后，可以看到安装成功标志，所有 controller 节点 Neutron 依赖包都安装完成，如图 10.2.1 所示。

微课 9 Neutron 的手动安装与外部环境配置

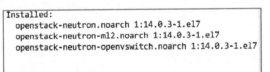

图 10.2.1 Neutron 软件包安装完成的反馈结果（1）

在 compute 节点上执行 yum 源安装命令安装 Neutron 相关依赖包，命令如下：

```
[root@compute ~]# yum -y install openstack-neutron openstack-neutron-ml2 openstack-neutron-openvswitch ebtables ipset
```

执行上述安装命令成功后，可以看到成功标志，所有 compute 节点 Neutron 依赖包都安装完成，如图 10.2.2 所示。

图 10.2.2 Neutron 软件包安装完成的反馈结果（2）

步骤 2：修改 Linux 内核参数。

为使 Neutron 完美实现 SDN 功能，需要通过修改内核文件开启内核 IP 转发功能，并关闭数据包加强过滤（rp_filter）功能。该功能在 controller 及 compute 节点都需实现，故在两个节点分别修改 /etc/sysctl.conf 内核文件，实现 SDN 功能，命令如下：

173

```
[root@controller ~]# modprobe br_netfilter
[root@controller ~]# vi /etc/sysctl.conf
```

添加：

```
net.bridge.bridge-nf-call-iptables=1
net.bridge.bridge-nf-call-ip6tables=1
net.ipv4.ip_forward=1
net.ipv4.conf.default.rp_filter=0
net.ipv4.conf.all.rp_filter=0
```

修改完成后需要使其生效，命令如下：

```
[root@controller ~]# sysctl -p
```

compute 节点修改方式相同，不再赘述。

步骤 3：创建 Neutron 数据库并授权。

相关操作与前面几个服务创建数据库类似，首先用 root 用户登录 MySQL 数据库，命令如下：

```
[root@controller ~]# mysql -uroot -p000000
```

接着创建 Neutron 数据库，命令如下：

```
mysql> CREATE DATABASE neutron;
```

看到提示 Query OK, 1 row affected (0.00 sec)，表明数据库创建成功。

下面，创建 MySQL 的 Neutron 用户，并赋予其 Neutron 数据库的操作权限，使得本地及远程都能访问，密码为 000000，命令如下：

```
mysql> GRANT ALL PRIVILEGES ON neutron.* TO'neutron'@'localhost' IDENTIFIED BY'000000';
mysql> GRANT ALL PRIVILEGES ON neutron.* TO 'neutron'@'%' IDENTIFIED BY '000000';
mysql> exit
```

上述 SQL 语句与创建 Keystone 数据库的语句一致，请参考实训项目 3 中步骤 2 "创建 Keystone 数据库并授权"部分对创建数据库时 SQL 语句的说明，在此不再赘述。

然后，对 controller 节点的 Neutron 主配置文件进行修改，使 Neutron 服务可以访问 MySQL 数据库，具体命令如下：

```
[root@controller ~]# crudini --set /etc/neutron/neutron.conf database connection mysql+pymysql://neutron:000000@controller/neutron
```

完成所有的配置后再同步 Neutron 数据库。

也可以通过 vi/vim 编辑器编辑 /etc/neutron/neutron.conf 文件中 [database] 字段与数据库连接的内容，以达到相同效果，编辑时请注意语法格式。

步骤 4：注册 Neutron 服务至 Keystone 服务器。

在 OpenStack 中，几乎所有的服务（包括 Keystone 服务）要想正常运行，都必须首先向 Keystone 服务器注册。每一个服务需要向 Keystone 注册两个方面的信息，即用户信息、服务 (service) 和端点 (endpoint) 信息。

首先，注册 Neutron 用户信息，注册用户信息的目的是为了认证用户身份的。现在添加 Neutron 用户，并为用户设置密码，在本书实训过程中所有的密码统一使用 000000，命令如下，结果如图 10.2.3 所示。

```
[root@controller ~]# openstack user create --domain default --password 000000 neutron
```

```
+---------------------+----------------------------------+
| Field               | Value                            |
+---------------------+----------------------------------+
| domain_id           | default                          |
| enabled             | True                             |
| id                  | 53288081a2ee401a89feb48f3afc87c8 |
| name                | neutron                          |
| options             | {}                               |
| password_expires_at | None                             |
+---------------------+----------------------------------+
```

图 10.2.3　创建 Neutron 用户的反馈结果

本书所有的服务都创建在 service 项目下。因此，Neutron 用户必须在 service 项目下具有 admin 权限。以下为 Neutron 用户在 service 项目下分配 admin 权限，命令如下：

```
[root@controller ~]# openstack role add --project service --user neutron admin
```

上述命令的用法在实训项目 3 步骤 7 中已经提到，在此不再赘述。

接下来，注册 Neutron 服务 (service) 和端点 (endpoint) 信息，注册服务和端点信息目的是确保用户能够顺利访问 Neutron 服务，并执行一系列与 Neutron 服务有关的操作，前面已经说过。现在进行 Neutron 服务和端点信息的注册，命令如下，结果如图 10.2.4 和图 10.2.5 所示。

```
[root@controller ~]# openstack service create --name neutron --description "OpenStack Networking" network
```

```
+-------------+----------------------------------+
| Field       | Value                            |
+-------------+----------------------------------+
| description | OpenStack Networking             |
| enabled     | True                             |
| id          | f55a6dbce5f04b3e977d7c97ec779245 |
| name        | neutron                          |
| type        | network                          |
+-------------+----------------------------------+
```

图 10.2.4　创建 Neutron 服务的反馈结果

```
[root@controller ~]# openstack endpoint create --region RegionOne network public http://controller:9696
[root@controller ~]# openstack endpoint create --region RegionOne network internal http://controller:9696
[root@controller ~]# openstack endpoint create --region RegionOne network admin http://controller:9696
```

```
+-----------+------------------------+
| Interface | URL                    |
+-----------+------------------------+
| admin     | http://controller:9696 |
| public    | http://controller:9696 |
| internal  | http://controller:9696 |
+-----------+------------------------+
```

图 10.2.5　创建 Neutron 端点的反馈结果

上述命令的用法在实训项目 3 步骤 8 中已经提到，在此不再赘述。需要注意的是，Neutron 的服务端口号为 9696。

步骤 5：配置 Neutron 使用消息队列服务。

消息队列仍然使用 RabbitMQ 队列，可以通过如下命令配置 controller 节点及 compute 节点的 Neutron 的消息队列。

controller 节点中执行如下命令：

```
[root@controller ~]# crudini --set /etc/neutron/neutron.conf DEFAULT transport_url rabbit://openstack:000000@controller
```

compute 节点中执行如下命令：

```
[root@compute ~]# crudini --set /etc/neutron/neutron.conf DEFAULT transport_url rabbit://openstack:000000@controller
```

也可以通过修改 /etc/neutron/neutron.conf 文件来实现同样的效果。

步骤6：配置 Neutron 的 Keystone 认证信息。

与 OpenStack 其他组件类似，Neutron 同样需要与 Keystone 之间进行交互，可以通过修改 controller 节点及 compute 的 Neutron 的主配置文件来实现，也可以通过如下命令来配置 Neuron 的 Keystone 认证信息。

controller 节点中执行如下命令。

```
[root@controller ~]# crudini --set /etc/neutron/neutron.conf DEFAULT auth_strategy keystone
[root@controller ~]# crudini --set /etc/neutron/neutron.conf keystone_authtoken www_authenticate_uri http://controller:5000
[root@controller ~]# crudini --set /etc/neutron/neutron.conf keystone_authtoken auth_url http://controller:5000
[root@controller ~]# crudini --set /etc/neutron/neutron.conf keystone_authtoken memcached_servers controller:11211
[root@controller ~]# crudini --set /etc/neutron/neutron.conf keystone_authtoken auth_type password
[root@controller ~]# crudini --set /etc/neutron/neutron.conf keystone_authtoken project_domain_name default
[root@controller ~]# crudini --set /etc/neutron/neutron.conf keystone_authtoken user_domain_name default
[root@controller ~]# crudini --set /etc/neutron/neutron.conf keystone_authtoken project_name service
[root@controller ~]# crudini --set /etc/neutron/neutron.conf keystone_authtoken username neutron
[root@controller ~]# crudini --set /etc/neutron/neutron.conf keystone_authtoken password 000000
```

compute 节点中执行如下命令。

```
[root@controller ~]# crudini --set /etc/neutron/neutron.conf DEFAULT auth_strategy keystone
[root@controller ~]# crudini --set /etc/neutron/neutron.conf keystone_authtoken www_authenticate_uri http://controller:5000
[root@controller ~]# crudini --set /etc/neutron/neutron.conf keystone_authtoken auth_url http://controller:5000
[root@controller ~]# crudini --set /etc/neutron/neutron.conf keystone_authtoken memcached_servers controller:11211
```

```
[root@controller ~]# crudini --set /etc/neutron/neutron.conf keystone_authtoken auth_type password
[root@controller ~]# crudini --set /etc/neutron/neutron.conf keystone_authtoken project_domain_name default
[root@controller ~]# crudini --set /etc/neutron/neutron.conf keystone_authtoken user_domain_name default
[root@controller ~]# crudini --set /etc/neutron/neutron.conf keystone_authtoken project_name service
[root@controller ~]# crudini --set /etc/neutron/neutron.conf keystone_authtoken username neutron
[root@controller ~]# crudini --set /etc/neutron/neutron.conf keystone_authtoken password 000000
```

Neutron 的 Keystone 认证信息包括默认认证类型、认证服务端点、认证端口、用户名、项目名、密码等。在完成步骤 6 后，可以通过在 controller 节点和 compute 节点分别执行 grep 命令过滤 /etc/neutron/neutron.conf 文件中的内容，来检查已完成的配置。

步骤 7：配置 lock path。

controller 节点中执行如下命令：

```
[root@controller ~]# crudini --set /etc/neutron/neutron.conf oslo_concurrency lock_path /var/lib/neutron/tmp
```

compute 节点中执行如下命令：

```
[root@compute ~]# crudini --set /etc/neutron/neutron.conf oslo_concurrency lock_path /var/lib/neutron/tmp
```

controller 节点中执行如下命令：

```
[root@controller ~]# grep -Ev "^#|^$" /etc/neutron/neutron.conf
```

执行后，可以看到 controller 节点 Neutron 主配置文件过滤后的内容，如下所示。

```
[DEFAULT]
transport_url = rabbit://openstack:000000@controller
auth_strategy = keystone
[cors]
[database]
connection = mysql+pymysql://neutron:000000@controller/neutron
[keystone_authtoken]
www_authenticate_uri = http://controller:5000
auth_url = http://controller:5000
memcached_servers = controller:11211
auth_type = password
project_domain_name = default
user_domain_name = default
project_name = service
username = neutron
password = 000000
```

```
[oslo_concurrency]
lock_path = /var/lib/neutron/tmp
[oslo_messaging_amqp]
[oslo_messaging_kafka]
[oslo_messaging_notifications]
[oslo_messaging_rabbit]
[oslo_middleware]
[oslo_policy]
[privsep]
[ssl]
```

compute 节点中执行如下命令。

```
[root@compute ~]# grep -Ev "^#|^$" /etc/neutron/neutron.conf
```

同样能看到过滤后的 compute 节点 Neutron 主配置文件的参数，如下所示。

```
[DEFAULT]
transport_url = rabbit://openstack:000000@controller
auth_strategy = keystone
[cors]
[database]
[keystone_authtoken]
www_authenticate_uri = http://controller:5000
auth_url = http://controller:5000
memcached_servers = controller:11211
auth_type = password
project_domain_name = default
user_domain_name = default
project_name = service
username = neutron
password = 000000
[oslo_concurrency]
lock_path = /var/lib/neutron/tmp
[oslo_messaging_amqp]
[oslo_messaging_kafka]
[oslo_messaging_notifications]
[oslo_messaging_rabbit]
[oslo_middleware]
[oslo_policy]
[privsep]
[ssl]
```

步骤 8：配置外网网卡。

Neutron 通过外网网卡与 Openstack 创建的虚拟机进行通信。本书中的实验为保证全网通信正常，在 controller 与 compute 节点都配置了外网网卡，可以根据实际情况在相应节点配置，以 controller 节点为例。首先进入 controller 节点 CentOS7.6 的网卡配置目录，命令如下：

```
[root@controller ~]# vi /etc/sysconfig/network-scripts/ifcfg-ens34
```

接下来，通过 vi 命令修改第 2 块网卡的配置文件，具体内容如下：

```
DEVICE=ens34
TYPE=Ethernet
BOOTPROTO=none
ONBOOT=yes
```

网卡文件修改完成后，将物理网卡与网桥绑定，本书实训中，第 2 块网卡地址在 192.168.200.0/24 网段，故与处于该网段的 ens34 网卡进行绑定，绑定前先重启网桥服务并将其加入开机自启动：

```
[root@controller ~] systemctl restart openvswitch
[root@controller ~] systemctl enable openvswitch
```

网桥服务成功重启后，可以执行以下绑定命令。

```
[root@controller ~] ovs-vsctl add-br br-int
[root@controller ~] ovs-vsctl add-br br-ex
[root@controller ~] ovs-vsctl add-port br-ex ens34
```

绑定完成后，关闭物理网卡的 GRO 功能，GRO 功能会将来自同一数据流的小流量数据包合并为一个大流量数据包交给内核协议栈，当虚拟网桥处理此类比 MTU（最大传输单元）值更大的包时，就会选择丢弃。为了保证虚拟网络之间不出现丢包及传输过慢的情况，需要关闭物理网卡的 GRO 功能。关闭后重启网络服务，命令如下：

```
[root@controller ~] ethtool -K ens34 gro off
[root@controller ~] systemctl restart network
```

controller 节点外网网卡创建完成后，配置 compute 节点的外网网卡，以及创建网桥，修改方法与 controller 节点类似，不再赘述。

修改成功后，可以通过命令来查看是否绑定成功，命令如下：

```
[root@controller ~]# ovs-vsctl show
```

若绑定成功，结果如图 10.2.6 所示。

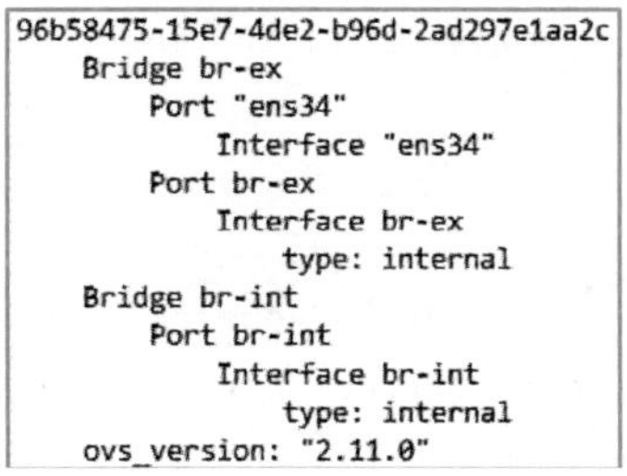

```
96b58475-15e7-4de2-b96d-2ad297e1aa2c
    Bridge br-ex
        Port "ens34"
            Interface "ens34"
        Port br-ex
            Interface br-ex
                type: internal
    Bridge br-int
        Port br-int
            Interface br-int
                type: internal
    ovs_version: "2.11.0"
```

图 10.2.6　查看虚拟网卡与物理网卡绑定情况的反馈结果

电子教案 Neutron 的主要服务组件配置与网络创建

PPT Neutron 的主要服务组件配置与网络创建

微课 10 Neutron 的主要服务组件配置与网络创建（1）

10.3 实训项目 8 Neutron 的主要服务组件配置与网络创建

1. 实训前提环境

成功完成实训项目 7 中所有内容后开始本实验，或者从已完成实训项目 7 的镜像开始，继续完成本实训内容。

2. 实训涉及节点

controller 和 compute

3. 实训目标

① 完成 Neutron 与 Nova 组件之间的交互配置。
② 完成 OpenvSwitch 的配置。
③ 完成 GRE 网络的配置。
④ 完成 Neutron Metadata 的配置。
⑤ 完成 Neutron DHCP 组件的配置。
⑥ 完成 GRE 网络的创建。

4. 实训步骤及其详解

步骤 1：在 Neutron 组件中配置 Nova 的认证信息。

Neutron 为了保证各虚拟机实例之间通信正常，需要在 Neutron 中配置 Nova 的认证信息，在 Neutron 组件运行时，会通过 Keystone 组件获取 Nova 的相关信息以获取虚拟机实例的状态信息，首先在 controller 节点的 Neutron 的主文件中添加 Nova 的相关信息，命令如下：

```
[root@controller ~]# crudini --set /etc/neutron/neutron.conf nova auth_url http://controller:5000
[root@controller ~]# crudini --set /etc/neutron/neutron.conf nova auth_type password
[root@controller ~]# crudini --set /etc/neutron/neutron.conf nova project_domain_name default
[root@controller ~]# crudini --set /etc/neutron/neutron.conf nova user_domain_name default
[root@controller ~]# crudini --set /etc/neutron/neutron.conf nova region_name RegionOne
[root@controller ~]# crudini --set /etc/neutron/neutron.conf nova project_name service
[root@controller ~]# crudini --set /etc/neutron/neutron.conf nova username nova
[root@controller ~]# crudini --set /etc/neutron/neutron.conf nova password 000000
```

接下来，对网络拓扑变化更新通告进行配置，当网络中虚拟节点发生变化时，Neutron 会自动通过从 Nova 中获取的信息来判断虚拟机实例状态，从而变更网络拓扑。同样需要对 controller 节点的 Neutron 主文件进行配置，在其中添加告知网络拓扑变化的相关信息，命令如下：

10.3 实训项目 8　Neutron 的主要服务组件配置与网络创建

```
[root@controller ~]# crudini --set /etc/neutron/neutron.conf DEFAULT notify_nova_on_port_status_changes true
[root@controller ~]# crudini --set /etc/neutron/neutron.conf DEFAULT notify_nova_on_port_data_changes true
```

步骤 2：在 Nova 组件中添加 Neutron 的相关信息。

为了使虚拟机实例可以获取 Neutron 提供的 IP 地址等信息，需要在 Nova 组件中添加 Neutron 的相关信息。Nova 通过 Keystone 组件获取 Neutron 的相关信息，compute 节点的 Nova 服务同样也需要获取 Neutron 的相关信息，故在两个节点的 nova 主配置文件中需同时配置 Nova 信息。

controller 节点命令如下：

```
[root@controller ~]# crudini --set /etc/nova/nova.conf neutron url http://controller:9696
[root@controller ~]# crudini --set /etc/nova/nova.conf neutron auth_url http://controller:5000
[root@controller ~]# crudini --set /etc/nova/nova.conf neutron auth_type password
[root@controller ~]# crudini --set /etc/nova/nova.conf neutron project_domain_name default
[root@controller ~]# crudini --set /etc/nova/nova.conf neutron user_domain_name default
[root@controller ~]# crudini --set /etc/nova/nova.conf neutron region_name RegionOne
[root@controller ~]# crudini --set /etc/nova/nova.conf neutron project_name service
[root@controller ~]# crudini --set /etc/nova/nova.conf neutron username neutron
[root@controller ~]# crudini --set /etc/nova/nova.conf neutron password 000000
```

compute 节点命令如下：

```
[root@compute ~]# crudini --set /etc/nova/nova.conf neutron url http://controller:9696
[root@compute ~]# crudini --set /etc/nova/nova.conf neutron auth_url http://controller:5000
[root@compute ~]# crudini --set /etc/nova/nova.conf neutron auth_type password
[root@compute ~]# crudini --set /etc/nova/nova.conf neutron project_domain_name default
[root@compute ~]# crudini --set /etc/nova/nova.conf neutron user_domain_name default
[root@compute ~]# crudini --set /etc/nova/nova.conf neutron region_name RegionOne
[root@compute ~]# crudini --set /etc/nova/nova.conf neutron project_name service
[root@compute ~]# crudini --set /etc/nova/nova.conf neutron username neutron
[root@compute ~]# crudini --set /etc/nova/nova.conf neutron password 000000
```

步骤 3：配置 OpenvSwitch。

Linux 默认的虚拟网络插件是 LinuxBrige，本书中采用 OpenvSwitch 替代 LinuxBrige，需

要在 Nova 的主配置文件中对默认插件信息及 OpenvSwitch 提供的防火墙、安全组等功能进行修改。

OpenvSwitch 配置在 controller 和 compute 节点的主配置文件都需要进行修改。

controller 节点命令如下：

```
[root@controller ~]# crudini --set /etc/nova/nova.conf DEFAULT auto_assign_floating_ip true
[root@controller ~]# crudini --set /etc/nova/nova.conf DEFAULT metadata_listen 0.0.0.0
[root@controller ~]# crudini --set /etc/nova/nova.conf DEFAULT metadata_listen_port 8775
[root@controller ~]# crudini --set /etc/nova/nova.conf DEFAULT compute_driver libvirt.LibvirtDriver
```

compute 节点命令如下：

```
[root@compute ~]# crudini --set /etc/nova/nova.conf DEFAULT linuxnet_interface_driver nova.network.linux_net.LinuxOVSInterfaceDriver
[root@compute ~]# crudini --set /etc/nova/nova.conf DEFAULT firewall_driver nova.virt.firewall.NoopFirewallDriver
[root@compute ~]# crudini --set /etc/nova/nova.conf DEFAULT use_neutron true
[root@compute ~]# crudini --set /etc/nova/nova.conf DEFAULT vif_plugging_is_fatal true
[root@compute ~]# crudini --set /etc/nova/nova.conf DEFAULT vif_plugging_timeout 300
```

配置完成后，可以通过以下命令过滤 Nova 主配置文件检查配置是否正确。

controller 节点命令如下：

```
[root@controller ~]# grep -Ev "^#|^$" /etc/neutron/neutron.conf
```

执行后可以看到过滤后的配置：

```
[DEFAULT]
transport_url = rabbit://openstack:000000@controller
auth_strategy = keystone
notify_nova_on_port_status_changes = true
notify_nova_on_port_data_changes = true
[cors]
[database]
connection = mysql+pymysql://neutron:000000@controller/neutron
[keystone_authtoken]
www_authenticate_uri = http://controller:5000
auth_url = http://controller:5000
memcached_servers = controller:11211
auth_type = password
project_domain_name = default
user_domain_name = default
project_name = service
```

```
username = neutron
password = 000000
[oslo_concurrency]
lock_path = /var/lib/neutron/tmp
[oslo_messaging_amqp]
[oslo_messaging_kafka]
[oslo_messaging_notifications]
[oslo_messaging_rabbit]
[oslo_middleware]
[oslo_policy]
[privsep]
[ssl]
[nova]
auth_url = http://controller:5000
auth_type = password
project_domain_name = default
user_domain_name = default
region_name = RegionOne
project_name = service
username = nova
password = 000000
```

compute 节点：

```
[root@compute ~]# grep -Ev "^#|^$" /etc/neutron/neutron.conf
[DEFAULT]
transport_url = rabbit://openstack:000000@controller
auth_strategy = keystone
[cors]
[database]
[keystone_authtoken]
www_authenticate_uri = http://controller:5000
auth_url = http://controller:5000
memcached_servers = controller:11211
auth_type = password
project_domain_name = default
user_domain_name = default
project_name = service
username = neutron
password = 000000
[oslo_concurrency]
lock_path = /var/lib/neutron/tmp
[oslo_messaging_amqp]
[oslo_messaging_kafka]
[oslo_messaging_notifications]
[oslo_messaging_rabbit]
[oslo_middleware]
```

```
[oslo_policy]
[privsep]
[ssl]
```

微课 11　Neutron 的主要服务组件配置与网络创建（2）

步骤4：通过 ML2 插件，搭建 GRE 网络。

OpenStack 从 Havana 版本开始。引入了 ML2 插件，ML2 插件可以作为一个框架同时支持不同的二层网络，对于 OpenvSwith 插件，使用 ML2 创建虚拟二层网络是一个理想的方案。

首先需要在 controller 节点和 compute 节点的 Neutron 的主配置文件中声明虚拟网络内核插件和服务插件，服务插件采用"router"。

controller 节点命令如下：

```
[root@controller ~]# crudini --set /etc/neutron/neutron.conf DEFAULT core_plugin ml2
[root@controller ~]# crudini --set /etc/neutron/neutron.conf DEFAULT service_plugins router
[root@controller ~]# crudini --set /etc/neutron/neutron.conf DEFAULT allow_overlapping_ips true
```

compute 节点命令如下：

```
[root@compute ~]# crudini --set /etc/neutron/neutron.conf DEFAULT core_plugin ml2
[root@compute ~]# crudini --set /etc/neutron/neutron.conf DEFAULT service_plugins router
[root@compute ~]# crudini --set /etc/neutron/neutron.conf DEFAULT allow_overlapping_ips true
```

声明完成后，对 Neutron 的 ML2 插件进行配置，首先配置 ML2 支持网络类型。

controller 节点命令如下：

```
[root@controller ~]# crudini --set /etc/neutron/plugins/ml2/ml2_conf.ini ml2 type_drivers flat,vlan,gre,vxlan,local
[root@controller ~]# crudini --set /etc/neutron/plugins/ml2/ml2_conf.ini ml2 tenant_network_types gre
[root@controller ~]# crudini --set /etc/neutron/plugins/ml2/ml2_conf.ini ml2_type_gre tunnel_id_ranges 1:1000
[root@controller ~]# crudini --set /etc/neutron/plugins/ml2/ml2_conf.ini ml2_type_flat flat_networks external
```

compute 节点命令如下：

```
[root@compute ~]# crudini --set /etc/neutron/plugins/ml2/ml2_conf.ini ml2 type_drivers flat,vlan,gre,vxlan,local
[root@compute ~]# crudini --set /etc/neutron/plugins/ml2/ml2_conf.ini ml2 tenant_network_types gre
[root@compute ~]# crudini --set /etc/neutron/plugins/ml2/ml2_conf.ini ml2_type_gre tunnel_id_ranges 1:1000
```

也能通过 vim 修改 /etc/neutron/plugins/ml2/ml2_conf.ini 文件来实现同样的效果。

接着为 ML2 插件配置 OpenvSwitch 插件以及防火墙、安全组。

controller 节点命令如下:

```
[root@controller ~]# crudini --set /etc/neutron/plugins/ml2/ml2_conf.ini securitygroup enable_ipset true
[root@controller ~]# crudini --set /etc/neutron/plugins/ml2/ml2_conf.ini ml2 mechanism_drivers openvswitch,l2population
[root@controller ~]# crudini --set /etc/neutron/plugins/ml2/ml2_conf.ini securitygroup firewall_driver iptables_hybrid
[root@controller ~]# crudini --set /etc/neutron/plugins/ml2/ml2_conf.ini securitygroup enable_security_group true
[root@controller ~]# crudini --set /etc/neutron/plugins/ml2/ml2_conf.ini ml2 extension_drivers port_security
```

compute 节点命令如下:

```
[root@compute ~]# crudini --set /etc/neutron/plugins/ml2/ml2_conf.ini securitygroup enable_ipset true
[root@compute ~]# crudini --set /etc/neutron/plugins/ml2/ml2_conf.ini ml2 mechanism_drivers openvswitch,l2population
[root@compute ~]# crudini --set /etc/neutron/plugins/ml2/ml2_conf.ini securitygroup firewall_driver iptables_hybrid
[root@compute ~]# crudini --set /etc/neutron/plugins/ml2/ml2_conf.ini securitygroup enable_security_group true
[root@compute ~]# crudini --set /etc/neutron/plugins/ml2/ml2_conf.ini ml2 extension_drivers port_security
```

最后,需要对 OpenvSwitch 服务的默认启动脚本进行修改,使其使用 ML2 插件代替传统的 OVS 插件。

首先,创建 ML2 配置文件的软连接。

controller 节点命令如下:

```
[root@controller ~]# ln -s /etc/neutron/plugins/ml2/ml2_conf.ini /etc/neutron/plugin.ini
```

compute 节点命令如下。

```
[root@compute ~]# ln -s /etc/neutron/plugins/ml2/ml2_conf.ini /etc/neutron/plugin.ini
```

接着,对原始的启动脚本进行备份。

controller 节点命令如下:

```
[root@controller ~]# cp /etc/neutron/plugins/ml2/openvswitch_agent.ini /etc/neutron/plugins/ml2/openvswitch_agent.ini.bak
```

compute 节点命令如下:

```
[root@compute ~]# cp /etc/neutron/plugins/ml2/openvswitch_agent.ini /etc/neutron/plugins/ml2/openvswitch_agent.ini.bak
```

最后,通过 crudini 命令对 OpenvSwitch 启动脚本进行修改。

controller 节点命令如下：

```
[root@controller ~]# crudini --set /etc/neutron/plugins/ml2/openvswitch_agent.ini agent l2_population true
[root@controller ~]# crudini --set /etc/neutron/plugins/ml2/openvswitch_agent.ini agent prevent_arp_spoofing true
[root@controller ~]# crudini --set /etc/neutron/plugins/ml2/openvswitch_agent.ini ovs integration_bridge br-int
[root@controller ~]# crudini --set /etc/neutron/plugins/ml2/openvswitch_agent.ini securitygroup firewall_driver iptables_hybrid
[root@controller ~]# crudini --set /etc/neutron/plugins/ml2/openvswitch_agent.ini ovs bridge_mappings external:br-ex
[root@controller ~]# crudini --set /etc/neutron/plugins/ml2/openvswitch_agent.ini agent tunnel_types gre
[root@controller ~]# crudini --set /etc/neutron/plugins/ml2/openvswitch_agent.ini ovs local_ip 192.168.100.10
[root@controller ~]# crudini --set /etc/neutron/plugins/ml2/openvswitch_agent.ini ovs enable_tunneling true
```

compute 节点命令如下：

```
[root@compute ~]# crudini --set /etc/neutron/plugins/ml2/openvswitch_agent.ini agent l2_population true
[root@compute ~]# crudini --set /etc/neutron/plugins/ml2/openvswitch_agent.ini agent prevent_arp_spoofing true
[root@compute ~]# crudini --set /etc/neutron/plugins/ml2/openvswitch_agent.ini ovs integration_bridge br-int
[root@compute ~]# crudini --set /etc/neutron/plugins/ml2/openvswitch_agent.ini securitygroup firewall_driver iptables_hybrid
[root@compute ~]# crudini --set /etc/neutron/plugins/ml2/openvswitch_agent.ini ovs bridge_mappings external:br-ex
[root@compute ~]# crudini --set /etc/neutron/plugins/ml2/openvswitch_agent.ini agent tunnel_types gre
[root@compute ~]# crudini --set /etc/neutron/plugins/ml2/openvswitch_agent.ini ovs local_ip 192.168.100.20
[root@compute ~]# crudini --set /etc/neutron/plugins/ml2/openvswitch_agent.ini ovs enable_tunneling true
```

步骤 5：配置 Neutron 的 DHCP 和三层路由服务。

DHCP 服务可以自动为虚拟机实例提供 IP 地址。本书中的网络节点实际为 controller 节点，故在 controller 节点启用 Neutron 的 DHCP 服务，在真实环境中，可以根据实际情况进行选择。

首先，需要声明 DHCP 服务对应的接口驱动，本书实验采用 OpenvSwitch，具体命令如下：

```
[root@controller ~]# crudini --set /etc/neutron/dhcp_agent.ini DEFAULT interface_driver neutron.agent.linux.interface.OVSInterfaceDriver
```

接着，配置 Dnsmasq 为默认的 DHCP 驱动，命令如下：

```
[root@controller ~]# crudini --set /etc/neutron/dhcp_agent.ini DEFAULT dhcp_driver neutron.agent.linux.dhcp.Dnsmasq
```

最后，打开 DHCP 插件的 isolated_metadata 功能，使虚拟机实例能够从名字空间找到 DHCP 服务，命令如下：

```
[root@controller ~]# crudini --set /etc/neutron/dhcp_agent.ini DEFAULT enable_isolated_metadata true
```

下面来配置三层路由服务。本书中的网络节点实际为 controller 节点，故在 controller 节点启用 Neutron 的三层路由服务，在真实环境中，可以根据实际情况进行选择。

```
[root@controller ~]# crudini --set /etc/neutron/l3_agent.ini DEFAULT interface_driver neutron.agent.linux.interface.OVSInterfaceDriver
[root@controller ~]# crudini --set /etc/neutron/l3_agent.ini DEFAULT ovs_integration_bridge br-int
[root@controller ~]# crudini --set /etc/neutron/l3_agent.ini DEFAULT external_network_bridge br-ex
```

步骤 6：配置 Neutron 的 Metadata 服务。

Metadata 服务的作用是向虚拟机实例传递公钥等信息，在实例需要通过公钥的方式连入时，Metadata 服务就显得尤为重要。

Metadata 服务在 controller 节点通过 Neutron 代理的方式实现，在 controller 节点将启用单独的 Metadata 服务。

接下来，为 controller 节点 Metadata 指定 Nova 服务的位置，命令如下：

```
[root@controller ~]# crudini --set /etc/neutron/metadata_agent.ini DEFAULT nova_metadata_host controller
```

将 Nova 的位置指定到 controller 节点，即 Metadata 将通过 controller 节点访问 Nova 服务，这里的 "controller" 可以根据实际情况改为 IP 地址。

最后，配置 Metadata 的预共享密钥及端口，命令如下：

```
[root@controller ~]# crudini --set /etc/neutron/metadata_agent.ini DEFAULT metadata_proxy_shared_secret 000000
[root@controller ~]# crudini --set /etc/neutron/metadata_agent.ini DEFAULT nova_metadata_port 8775
```

在 controller 节点中，通过对 Nova 主配置文件的修改来实现 Metadata 代理，配置命令如下：

```
[root@controller ~]# crudini --set /etc/nova/nova.conf neutron service_metadata_proxy true
[root@controller ~]# crudini --set /etc/nova/nova.conf neutron metadata_proxy_shared_secret 000000
```

注意，controller 节点和 compute 节点的预共享密钥需一致。

步骤 7：同步 Neutron 数据库。

```
[root@controller ~]# su -s /bin/sh -c "neutron-db-manage --config-file /etc/neutron/
neutron.conf --config-file /etc/neutron/plugins/ml2/ml2_conf.ini upgrade head" neutron
```

执行完成后，验证 neutron 数据库中是否有表，命令如下，结果如图 10.3.1 所示。

```
[root@controller ~]# mysql -uroot -p000000 -e "use neutron;show tables;"
```

```
+------------------------------------+
| Tables_in_neutron                  |
+------------------------------------+
| address_scopes                     |
| agents                             |
| alembic_version                    |
| allowedaddresspairs                |
| arista_provisioned_nets            |
| arista_provisioned_tenants         |
| arista_provisioned_vms             |
| auto_allocated_topologies          |
| bgp_peers                          |
| bgp_speaker_dragent_bindings       |
| bgp_speaker_network_bindings       |
| bgp_speaker_peer_bindings          |
| bgp_speakers                       |
| brocadenetworks                    |
+------------------------------------+
```

图 10.3.1　查询 neutron 数据表

步骤 8：启动 Neutron 相关服务。

配置完步骤 7，Neutron 的全部配置结束。接下来就可以启动 Neutron 相关服务了，本书中实验涉及的 Neutron 服务包括 neutron-server、neutron-openvswitch-agent、neutron-dhcp-agent、neutron-metadata-agent。

需要注意的是，在启动 compute 节点的 Neutron 服务前，需要重启 openstack-nova-compute 服务。

controller 节点命令如下：

```
[root@controller ~]# systemctl restart openstack-nova-api.service
[root@controller ~]# systemctl start neutron-server.service neutron-openvswitch-agent.service neutron-dhcp-agent.service neutron-metadata-agent.service neutron-l3-agent.service
```

compute 节点命令如下：

```
[root@compute ~]# systemctl restart openstack-nova-compute.service
[root@compute ~]# systemctl restart neutron-metadata-agent.service
[root@compute ~]# systemctl restart neutron-openvswitch-agent.service
```

启动完成后，将 Neutron 相关服务添加到开机启动项中。

controller 节点命令如下：

```
[root@controller ~]# systemctl enable neutron-server.service neutron-openvswitch-agent.service neutron-dhcp-agent.service neutron-metadata-agent.service neutron-l3-agent.service
```

compute 节点命令如下：

```
[root@compute ~]# systemctl enable neutron-metadata-agent.service
[root@compute ~]# systemctl enable neutron-openvswitch-agent.service
```

启动完成后，在 controller 节点可以查看 Neutron 相关服务的运行状态和位置信息，命令如下，结果如图 10.3.2 和图 10.3.3 所示。

```
[root@controller ~]# openstack network agent list
```

```
| ID                                   | Agent Type       |
| 83d871f3-bbaf-48e1-a3c6-bf24cabc2945 | DHCP agent       |
| 9a8a90e4-dc93-4db0-86cd-3190c6f242f9 | Metadata agent   |
| c236bafe-12c8-4952-b42b-1192bd158520 | Metadata agent   |
| db18a547-3ef0-4c51-b60d-146af092e22b | Open vSwitch agent |
| fab6188f-6c9a-4e05-9768-531799dc173f | Open vSwitch agent |
```

图 10.3.2　查看 Neutron 相关服务运行状态的反馈结果（1）

```
| Host       | Availability Zone | Alive | State | Binary                    |
| controller | nova              | :-)   | UP    | neutron-dhcp-agent        |
| controller | None              | :-)   | UP    | neutron-metadata-agent    |
| compute    | None              | :-)   | UP    | neutron-metadata-agent    |
| compute    | None              | :-)   | UP    | neutron-openvswitch-agent |
| controller | None              | :-)   | UP    | neutron-openvswitch-agent |
```

图 10.3.3　查看 Neutron 相关服务运行状态的反馈结果（2）

其中，Agent__Type 字段表示 Neutron 相关组件名；Host 表示服务运行的位置；Alive 表示运行状态信息，笑脸表示运行正常，xxx 表示运行不正常，可能相关服务未启动或配置文件存在问题，需要重新检查；State 表示该组件是否启用；UP 表示已启用。

至此，OpenStack 的 Neutron 组件配置已经完成。

步骤 9：创建 GRE 网络

通过 Neutron 创建一个小型 GRE 网络。

首先，将 service 项目的项目 ID 赋值给"projectID"变量，命令如下：

```
[root@controller ~]# projectID=$(openstack project list | grep service | awk '{print $2}')
```

接着，执行 Neutron 的 net-create 命令来创建虚拟网络，命令如下：

```
[root@controller ~]# openstack network create --project $projectID --share --provider-network-type gre --provider-segment 101 int-net1
```

创建完成后，可以看到创建成功的提示，如图 10.3.4 所示。

最后，创建子网，虚拟机在 192.168.200.0/24 网段，DHCP 地址为 192.168.200.100~192.168.200.200，命令如下：

```
[root@controller ~]# openstack subnet create --network int-net1 --subnet-range 10.0.0.0/24 --gateway 10.0.0.1 --allocation-pool start=10.0.0.100,end=10.0.0.200 int-subnet1
```

创建完成后，可以看到创建成功的提示，如图 10.3.5 所示。

同样，可以在 controller 节点执行"openstack network list"和"openstack subnet list"来查看网络和子网信息，如图 10.3.6 所示。

```
[root@controller ~]# openstack network create --project $projectID \
> --share --provider-network-type gre --provider-segment 101 int-net1
+---------------------------+--------------------------------------+
| Field                     | Value                                |
+---------------------------+--------------------------------------+
| admin_state_up            | UP                                   |
| availability_zone_hints   |                                      |
| availability_zones        |                                      |
| created_at                | 2019-11-16T09:06:53Z                 |
| description               |                                      |
| dns_domain                | None                                 |
| id                        | 516baea8-be9b-409f-a45a-50556b6cc5ad |
| ipv4_address_scope        | None                                 |
| ipv6_address_scope        | None                                 |
| is_default                | False                                |
| is_vlan_transparent       | None                                 |
| location                  | Munch({'project': Munch({'domain_name': None, |
| mtu                       | 1458                                 |
| name                      | int-net1                             |
| port_security_enabled     | True                                 |
| project_id                | b4224500ca3a464b8ac5adbad9233d13     |
| provider:network_type     | gre                                  |
| provider:physical_network | None                                 |
| provider:segmentation_id  | 101                                  |
| qos_policy_id             | None                                 |
| revision_number           | 1                                    |
| router:external           | Internal                             |
| segments                  | None                                 |
| shared                    | True                                 |
| status                    | ACTIVE                               |
| subnets                   |                                      |
| tags                      |                                      |
| updated_at                | 2019-11-16T09:06:54Z                 |
+---------------------------+--------------------------------------+
```

图 10.3.4　创建 GRE 网络的反馈结果

```
[root@controller ~]# openstack subnet create --network int-net1 --subnet-range 10.0.0.0/24 \
> --gateway 10.0.0.1 --allocation-pool start=10.0.0.100,end=10.0.0.200 int-subnet1
+-------------------+--------------------------------------+
| Field             | Value                                |
+-------------------+--------------------------------------+
| allocation_pools  | 10.0.0.100-10.0.0.200                |
| cidr              | 10.0.0.0/24                          |
| created_at        | 2019-11-16T09:11:36Z                 |
| description       |                                      |
| dns_nameservers   |                                      |
| enable_dhcp       | True                                 |
| gateway_ip        | 10.0.0.1                             |
| host_routes       |                                      |
| id                | 67a81386-9252-40c4-a4c0-0f5a043e484c |
| ip_version        | 4                                    |
| ipv6_address_mode | None                                 |
| ipv6_ra_mode      | None                                 |
| location          | Munch({'project': Munch({'domain_name': 'Default', 'domain_id': None, |
| name              | int-subnet1                          |
| network_id        | 516baea8-be9b-409f-a45a-50556b6cc5ad |
| prefix_length     | None                                 |
| project_id        | 23fb3921f1df4c1da0e4c1a7d59459c3     |
| revision_number   | 0                                    |
| segment_id        | None                                 |
| service_types     |                                      |
| subnetpool_id     | None                                 |
| tags              |                                      |
| updated_at        | 2019-11-16T09:11:36Z                 |
+-------------------+--------------------------------------+
```

图 10.3.5　创建子网成功的反馈结果

```
[root@controller ~]# openstack network list
+--------------------------------------+----------+--------------------------------------+
| ID                                   | Name     | Subnets                              |
+--------------------------------------+----------+--------------------------------------+
| 516baea8-be9b-409f-a45a-50556b6cc5ad | int-net1 | 67a81386-9252-40c4-a4c0-0f5a043e484c |
+--------------------------------------+----------+--------------------------------------+
[root@controller ~]# openstack subnet list
+--------------------------------------+-------------+--------------------------------------+-------------+
| ID                                   | Name        | Network                              | Subnet      |
+--------------------------------------+-------------+--------------------------------------+-------------+
| 67a81386-9252-40c4-a4c0-0f5a043e484c | int-subnet1 | 516baea8-be9b-409f-a45a-50556b6cc5ad | 10.0.0.0/24 |
+--------------------------------------+-------------+--------------------------------------+-------------+
```

图 10.3.6　查看网络和子网的反馈结果

10.4 Neutron 安装脚本及其解读

在介绍了手动安装 Neutron 服务后，下面介绍另一种安装 Neutron 服务的方法，即利用已经写好的脚本自动安装 Neutron 服务。

1. 需要安装的脚本内容

手动安装 Neutron 服务时，需要分别在 controller 和 compute 节点配置，所以脚本也需要分别在两个节点完成各自的配置内容。

controller 节点的脚本内容如下：

```bash
#!/bin/bash
source /etc/keystone/admin-openrc.sh
# 添加计算节点到单元数据库中
su -s /bin/sh -c "nova-manage cell_v2 discover_hosts --verbose" nova > /dev/null 2>&1
# 数据库配置
mysql -uroot -p000000 -e "create database IF NOT EXISTS neutron ;"
mysql -uroot -p000000 -e "GRANT ALL PRIVILEGES ON neutron.* TO 'neutron'@'localhost' IDENTIFIED BY '000000';"
mysql -uroot -p000000 -e "GRANT ALL PRIVILEGES ON neutron.* TO 'neutron'@'%' IDENTIFIED BY '000000';"
# Keystone 认证
openstack user create --domain default --password 000000 neutron
openstack role add --project service --user neutron admin
openstack service create --name neutron --description "OpenStack Networking" network
openstack endpoint create --region RegionOne network public http://controller:9696
openstack endpoint create --region RegionOne network internal http://controller:9696
openstack endpoint create --region RegionOne network admin http://controller:9696
# 安装 Neutron 需要的软件包
yum -y install openstack-neutron openstack-neutron-ml2 ebtables openstack-neutron-openvswitch haproxy
# 主配置文件设置
crudini --set /etc/neutron/neutron.conf DEFAULT core_plugin ml2
crudini --set /etc/neutron/neutron.conf DEFAULT service_plugins router
crudini --set /etc/neutron/neutron.conf DEFAULT allow_overlapping_ips true
crudini --set /etc/neutron/neutron.conf DEFAULT transport_url rabbit://openstack:000000@controller
crudini --set /etc/neutron/neutron.conf DEFAULT notify_nova_on_port_status_changes true
crudini --set /etc/neutron/neutron.conf DEFAULT notify_nova_on_port_data_changes true
# 配置数据库连接
crudini --set /etc/neutron/neutron.conf database connection mysql+pymysql://neutron:000000@controller/neutron
```

```
# 连接 Keystone
crudini --set /etc/neutron/neutron.conf DEFAULT auth_strategy keystone
crudini --set /etc/neutron/neutron.conf keystone_authtoken www_authenticate_uri http://controller:5000
crudini --set /etc/neutron/neutron.conf keystone_authtoken auth_url http://controller:5000
crudini --set /etc/neutron/neutron.conf keystone_authtoken memcached_servers controller:11211
crudini --set /etc/neutron/neutron.conf keystone_authtoken auth_type password
crudini --set /etc/neutron/neutron.conf keystone_authtoken project_domain_name default
crudini --set /etc/neutron/neutron.conf keystone_authtoken user_domain_name default
crudini --set /etc/neutron/neutron.conf keystone_authtoken project_name service
crudini --set /etc/neutron/neutron.conf keystone_authtoken username neutron
crudini --set /etc/neutron/neutron.conf keystone_authtoken password 000000
# 配置锁路径
crudini --set /etc/neutron/neutron.conf oslo_concurrency lock_path /var/lib/neutron/tmp
# 在 Neutron 中连接 Nova 到 Keystone
crudini --set /etc/neutron/neutron.conf nova auth_url http://controller:5000
crudini --set /etc/neutron/neutron.conf nova auth_type password
crudini --set /etc/neutron/neutron.conf nova project_domain_name default
crudini --set /etc/neutron/neutron.conf nova user_domain_name default
crudini --set /etc/neutron/neutron.conf nova region_name RegionOne
crudini --set /etc/neutron/neutron.conf nova project_name service
crudini --set /etc/neutron/neutron.conf nova username nova
crudini --set /etc/neutron/neutron.conf nova password 000000
# ML2 插件配置
crudini --set /etc/neutron/plugins/ml2/ml2_conf.ini ml2_type_gre tunnel_id_ranges 1:1000
crudini --set /etc/neutron/plugins/ml2/ml2_conf.ini ml2_type_flat flat_networks external
crudini --set /etc/neutron/plugins/ml2/ml2_conf.ini ml2 type_drivers flat,vlan,gre,vxlan,local
crudini --set /etc/neutron/plugins/ml2/ml2_conf.ini ml2 tenant_network_types gre
crudini --set /etc/neutron/plugins/ml2/ml2_conf.ini ml2 mechanism_drivers openvswitch,l2population
crudini --set /etc/neutron/plugins/ml2/ml2_conf.ini ml2 extension_drivers port_security
crudini --set /etc/neutron/plugins/ml2/ml2_conf.ini securitygroup enable_ipset true
crudini --set /etc/neutron/plugins/ml2/ml2_conf.ini securitygroup enable_security_group true
crudini --set /etc/neutron/plugins/ml2/ml2_conf.ini securitygroup firewall_driver iptables_hybrid
# DHCP 设置
```

```
crudini --set /etc/neutron/dhcp_agent.ini DEFAULT interface_driver neutron.agent.linux.interface.OVSInterfaceDriver
crudini --set /etc/neutron/dhcp_agent.ini DEFAULT dhcp_driver neutron.agent.linux.dhcp.Dnsmasq
crudini --set /etc/neutron/dhcp_agent.ini DEFAULT enable_isolated_metadata true
# 三层路由设置
crudini --set /etc/neutron/l3_agent.ini DEFAULT interface_driver neutron.agent.linux.interface.OVSInterfaceDriver
crudini --set /etc/neutron/l3_agent.ini DEFAULT ovs_integration_bridge br-int
crudini --set /etc/neutron/l3_agent.ini DEFAULT external_network_bridge br-ex
# Openvswitch 网桥设置
crudini --set /etc/neutron/plugins/ml2/openvswitch_agent.ini agent l2_population true
crudini --set /etc/neutron/plugins/ml2/openvswitch_agent.ini agent prevent_arp_spoofing true
crudini --set /etc/neutron/plugins/ml2/openvswitch_agent.ini ovs integration_bridge br-int
crudini --set /etc/neutron/plugins/ml2/openvswitch_agent.ini securitygroup firewall_driver iptables_hybrid
crudini --set /etc/neutron/plugins/ml2/openvswitch_agent.ini ovs bridge_mappings external:br-ex
crudini --set /etc/neutron/plugins/ml2/openvswitch_agent.ini agent tunnel_types gre
crudini --set /etc/neutron/plugins/ml2/openvswitch_agent.ini ovs local_ip 192.168.100.10
crudini --set /etc/neutron/plugins/ml2/openvswitch_agent.ini ovs enable_tunneling true
# 配置 Metadata 服务
crudini --set /etc/neutron/metadata_agent.ini DEFAULT nova_metadata_host controller
crudini --set /etc/neutron/metadata_agent.ini DEFAULT metadata_proxy_shared_secret 000000
crudini --set /etc/neutron/metadata_agent.ini DEFAULT nova_metadata_port 8775
# 在 Nova 中连接 Neutron 到 Keystone
crudini --set /etc/nova/nova.conf DEFAULT auto_assign_floating_ip true
crudini --set /etc/nova/nova.conf DEFAULT metadata_listen 0.0.0.0
crudini --set /etc/nova/nova.conf DEFAULT metadata_listen_port 8775
crudini --set /etc/nova/nova.conf neutron url http://controller:9696
crudini --set /etc/nova/nova.conf neutron auth_url http://controller:5000
crudini --set /etc/nova/nova.conf neutron auth_type password
crudini --set /etc/nova/nova.conf neutron project_domain_name default
crudini --set /etc/nova/nova.conf neutron user_domain_name default
crudini --set /etc/nova/nova.conf neutron region_name RegionOne
crudini --set /etc/nova/nova.conf neutron project_name service
crudini --set /etc/nova/nova.conf neutron username neutron
crudini --set /etc/nova/nova.conf neutron password 000000
crudini --set /etc/nova/nova.conf neutron service_metadata_proxy true
crudini --set /etc/nova/nova.conf neutron metadata_proxy_shared_secret 000000
```

```
# 配置内核转发
modprobe br_netfilter
echo "net.bridge.bridge-nf-call-iptables=1" >> /etc/sysctl.conf
echo "net.bridge.bridge-nf-call-ip6tables=1" >> /etc/sysctl.conf
echo "net.ipv4.ip_forward=1" >> /etc/sysctl.conf
echo "net.ipv4.conf.default.rp_filter=0" >> /etc/sysctl.conf
echo "net.ipv4.conf.all.rp_filter=0" >> /etc/sysctl.conf
sysctl -p

ln -s /etc/neutron/plugins/ml2/ml2_conf.ini /etc/neutron/plugin.ini
# 同步数据库
su -s /bin/sh -c "neutron-db-manage --config-file /etc/neutron/neutron.conf \
    --config-file /etc/neutron/plugins/ml2/ml2_conf.ini upgrade head" neutron
# 创建网桥
systemctl restart openvswitch
systemctl enable openvswitch
ovs-vsctl add-br br-int > /dev/null 2>&1
ovs-vsctl add-br br-ex > /dev/null 2>&1
ovs-vsctl add-port br-ex ens34 > /dev/null 2>&1
# 配置外网卡
cat > /etc/sysconfig/network-scripts/ifcfg-ens34 << EOF
DEVICE=ens34
TYPE=Ethernet
BOOTPROTO=none
ONBOOT=yes
EOF
ethtool -K ens34 gro off
systemctl restart network
# 落定服务
systemctl restart openstack-nova-api.service
systemctl enable neutron-server.service neutron-openvswitch-agent.service neutron-dhcp-agent.service neutron-metadata-agent.service neutron-l3-agent.service
systemctl start neutron-server.service neutron-openvswitch-agent.service neutron-dhcp-agent.service neutron-metadata-agent.service neutron-l3-agent.service
```

compute 节点的脚本内容如下：

```
#!/bin/bash
# 安装软件包
yum -y install openstack-neutron openstack-neutron-ml2 openstack-neutron-openvswitch ebtables ipset
# 连接消息队列
crudini --set /etc/neutron/neutron.conf DEFAULT transport_url rabbit://openstack:000000@controller
crudini --set /etc/neutron/neutron.conf DEFAULT core_plugin ml2
crudini --set /etc/neutron/neutron.conf DEFAULT service_plugins router
```

```
    crudini --set /etc/neutron/neutron.conf DEFAULT allow_overlapping_ips true
    # Keystone 认证
    crudini --set /etc/neutron/neutron.conf DEFAULT auth_strategy keystone
    crudini --set /etc/neutron/neutron.conf keystone_authtoken www_authenticate_uri
http://controller:5000
    crudini --set /etc/neutron/neutron.conf keystone_authtoken auth_url http://
controller:5000
    crudini --set /etc/neutron/neutron.conf keystone_authtoken memcached_servers
controller:11211
    crudini --set /etc/neutron/neutron.conf keystone_authtoken auth_type password
    crudini --set /etc/neutron/neutron.conf keystone_authtoken project_domain_name
default
    crudini --set /etc/neutron/neutron.conf keystone_authtoken user_domain_name default
    crudini --set /etc/neutron/neutron.conf keystone_authtoken project_name service
    crudini --set /etc/neutron/neutron.conf keystone_authtoken username neutron
    crudini --set /etc/neutron/neutron.conf keystone_authtoken password 000000

    crudini --set /etc/neutron/neutron.conf oslo_concurrency lock_path /var/lib/neutron/tmp
    # ML2 插件配置
    crudini --set /etc/neutron/plugins/ml2/ml2_conf.ini ml2 tenant_network_types gre
    crudini --set /etc/neutron/plugins/ml2/ml2_conf.ini ml2_type_gre tunnel_id_ranges
1:1000
    crudini --set /etc/neutron/plugins/ml2/ml2_conf.ini ml2 type_drivers
flat,vlan,gre,vxlan,local
    crudini --set /etc/neutron/plugins/ml2/ml2_conf.ini ml2 mechanism_drivers
openvswitch,l2population
    crudini --set /etc/neutron/plugins/ml2/ml2_conf.ini ml2 extension_drivers port_
security
    crudini --set /etc/neutron/plugins/ml2/ml2_conf.ini securitygroup enable_ipset true
    crudini --set /etc/neutron/plugins/ml2/ml2_conf.ini securitygroup enable_security_
group true
    crudini --set /etc/neutron/plugins/ml2/ml2_conf.ini securitygroup firewall_driver
iptables_hybrid
    # Openvswitch 网桥配置
    crudini --set /etc/neutron/plugins/ml2/openvswitch_agent.ini agent l2_population
true
    crudini --set /etc/neutron/plugins/ml2/openvswitch_agent.ini agent prevent_arp_
spoofing true
    crudini --set /etc/neutron/plugins/ml2/openvswitch_agent.ini ovs integration_bridge
br-int
    crudini --set /etc/neutron/plugins/ml2/openvswitch_agent.ini securitygroup firewall_
driver iptables_hybrid
    crudini --set /etc/neutron/plugins/ml2/openvswitch_agent.ini ovs bridge_mappings
external:br-ex
    crudini --set /etc/neutron/plugins/ml2/openvswitch_agent.ini agent tunnel_types gre
    crudini --set /etc/neutron/plugins/ml2/openvswitch_agent.ini ovs local_ip 192.168.100.20
```

```
crudini --set /etc/neutron/plugins/ml2/openvswitch_agent.ini ovs enable_tunneling true
# 在 Nova 连接 Neutron 到 Keystone
crudini --set /etc/nova/nova.conf neutron url http://controller:9696
crudini --set /etc/nova/nova.conf neutron auth_url http://controller:5000
crudini --set /etc/nova/nova.conf neutron auth_type password
crudini --set /etc/nova/nova.conf neutron project_domain_name default
crudini --set /etc/nova/nova.conf neutron user_domain_name default
crudini --set /etc/nova/nova.conf neutron region_name RegionOne
crudini --set /etc/nova/nova.conf neutron project_name service
crudini --set /etc/nova/nova.conf neutron username neutron
crudini --set /etc/nova/nova.conf neutron password 000000
crudini --set /etc/nova/nova.conf DEFAULT use_neutron true
crudini --set /etc/nova/nova.conf DEFAULT linuxnet_interface_driver nova.network.linux_net.LinuxOVSInterfaceDriver
crudini --set /etc/nova/nova.conf DEFAULT firewall_driver nova.virt.firewall.NoopFirewallDriver
crudini --set /etc/nova/nova.conf DEFAULT vif_plugging_is_fatal true
crudini --set /etc/nova/nova.conf DEFAULT vif_plugging_timeout 300
# 配置内核转发
modprobe br_netfilter
echo "net.bridge.bridge-nf-call-iptables=1" >> /etc/sysctl.conf
echo "net.bridge.bridge-nf-call-ip6tables=1" >> /etc/sysctl.conf
echo 'net.ipv4.ip_forward=1' >> /etc/sysctl.conf
echo "net.ipv4.conf.default.rp_filter=0" >> /etc/sysctl.conf
echo 'net.ipv4.conf.all.rp_filter=0' >> /etc/sysctl.conf
sysctl -p

ln -s /etc/neutron/plugins/ml2/ml2_conf.ini /etc/neutron/plugin.ini

# 创建网桥
systemctl restart openvswitch
systemctl enable openvswitch
ovs-vsctl add-br br-int > /dev/null 2>&1
ovs-vsctl add-br br-ex > /dev/null 2>&1
ovs-vsctl add-port br-ex ens34 > /dev/null 2>&1
# 配置外网网卡
cat > /etc/sysconfig/network-scripts/ifcfg-ens34 << EOF
DEVICE=ens34
TYPE=Ethernet
BOOTPROTO=none
ONBOOT=yes
EOF
ethtool -K ens34 gro off
systemctl restart network
```

```
# 落定服务
systemctl restart openstack-nova-compute.service
systemctl restart neutron-metadata-agent.service
systemctl enable neutron-metadata-agent.service
systemctl restart neutron-openvswitch-agent.service
systemctl enable neutron-openvswitch-agent.service
```

2. 脚本执行

刚写好的脚本文件，在核实过内容后，是不能直接使用的。要想使用脚本，首先需要赋予脚本执行权限，这里以 controller 节点的脚本为例进行讲解，命令如下：

```
[root@controller ~]# chmod a+x iaas-install-neutron-controller.sh
```

或者执行

```
[root@controller ~]# chmod 777 iaas-install-neutron-controller.sh
```

上述两种命令都可以实现赋予脚本执行权限的效果，这里的"iaas-install-neutron-controller.sh"是示范用的脚本名称，读者在实际使用时可自行命名脚本名称（说明：本次实验所使用的脚本均放在 root 用户的家目录里）。

脚本有了执行权限后就可以执行了。这里介绍两种执行脚本的方法。

方法 1：在脚本前加上脚本所在的路径，可以使用相对路径，也可以使用绝对路径，这取决于读者个人习惯，命令如下：

```
[root@controller ~]# ./iaas-install-neutron-controller.sh
```

"."表示当前目录，用户的脚本放在家目录中，而用户此时也在家目录中，所以为方便，这里使用的是当前目录。

方法 2：现将脚本所在路径加入环境变量 PATH 中，然后就可以像 ls 命令一样直接在命令中输入使用了，命令如下：

```
[root@controller ~]# echo $PATH
/usr/local/sbin:/usr/local/bin:/sbin:/bin:/usr/sbin:/usr/bin:/root/bin
[root@controller ~]# PATH=${PATH}:/root/
[root@controller ~]# echo $PATH
/usr/local/sbin:/usr/local/bin:/sbin:/bin:/usr/sbin:/usr/bin:/root/bin:/root/
[root@controller ~]# iaas-install-neutron-controller.sh
```

上述命令对比了将脚本所在目录加入环境变量 PATH 前后，环境变量 PATH 中的内容。最后可以直接在命令中像其他命令一样直接输入脚本名称。

说明：直接在命令中改变的环境变量只对当前 shell 和其子 shell 有效，当重启或另外打开新的 shell 时，当前修改的环境变量无效。但可以通过修改 /root/.bash_profile 配置文件中 PATH 的内容，达到环境变量 PATH 永久有效的效果，注意，修改 /root/.bash_profile 后需要通知系统内容变化，最直接的方法是重新打开新的 shell。

习题　Neutron 的安装及其配置

第 11 章　Cinder 的安装及其配置

本章导读：
　　本章首先介绍 Cinder 功能实现的技术基础、Linux 中的 LVM 卷，其后简单介绍 Cinder 服务的基本功能及其架构。作为云平台的块存储服务提供模块，Cinder 组件用于实现基本存储功能以及实例动态云硬盘的扩展功能。本章的实训项目 9 介绍如何通过命令行以及服务组件的参数配置安装 Cinder 服务，并解读 Cinder 在安装和使用过程中可能出现的问题。在以上基础上，本章最后还将介绍 Cinder 服务安装脚本的制作和使用。

电子资源：
　　电子教案　Cinder 的安装及其配置
　　PPT　Cinder 的安装及其配置
　　习题　Cinder 的安装及其配置

11.1 Cinder 功能简介

1. 逻辑卷管理

LVM（Logical Volume Manager，逻辑卷管理）是 Linux 环境下对磁盘分区进行管理的一种机制，它由 Heinz Mauelshagen 在 Linux 2.4 内核上实现，目前最新版本为稳定版 1.0.5、开发版 1.1.0–rc2，以及 LVM2 开发版。Linux 用户安装 Linux 操作系统时遇到的一个常见的难以决定的问题就是如何正确地评估各分区大小，以分配合适的硬盘空间。普通的磁盘分区管理方式在逻辑分区划分好之后就无法改变其大小，当一个逻辑分区存放不下某个文件时，该文件因受上层文件系统的限制，也不能跨越多个分区来存放，所以也不能同时放到别的磁盘上。而遇到出现某个分区空间耗尽时，解决的方法通常是使用符号链接，或者使用调整分区大小的工具，但这只是暂时的解决办法，没有从根本上解决问题。随着 Linux 的逻辑卷管理功能的出现，这些问题都迎刃而解了，用户在无须停机的情况下可以方便地调整各个分区大小。

LVM 是传统商业 UNIX 就带有的一项高级磁盘管理工具，异常强大。LVM 移植到了 Linux 操作系统上之后，可以在生产运行系统上面直接在线扩展硬盘分区，可以把分区 umount 以后收缩分区大小，还可以在系统运行过程中把一个分区从一块硬盘搬到另一块硬盘上面去等，可以在一个繁忙运行的系统上面直接操作，不会对系统运行产生任何影响，很安全。

与传统的磁盘与分区相比，LVM 为计算机提供了更高层次的磁盘存储。它使系统管理员可以更方便地为应用与用户分配存储空间。在 LVM 管理下的存储卷可以按需要随时改变大小与移除（可能需对文件系统工具进行升级）。LVM 也允许按用户组对存储卷进行管理，允许管理员用更直观的名称（如 sales、development）代替物理磁盘名（如 sda、sdb）来标识存储卷。

LVM 的一些术语是学习 LVM 之前必知的知识，内容如下。

① 物理存储介质（the physical media），这里指系统的存储设备——硬盘，如 /dev/hda1、/dev/sda 等，是存储系统最底层的存储单元。

② 物理卷（physical volume），是指硬盘分区或从逻辑上与磁盘分区具有同样功能的设备（如 RAID），是 LVM 的基本存储逻辑块，但和基本的物理存储介质（如分区、磁盘等）比较，却包含有与 LVM 相关的管理参数。

③ 卷组（volume group），LVM 卷组类似于非 LVM 系统中的物理硬盘，其由物理卷组成。可以在卷组上创建一个或多个"LVM 分区"（逻辑卷），LVM 卷组由一个或多个物理卷组成。

④ 逻辑卷（logical volume），LVM 的逻辑卷类似于非 LVM 系统中的硬盘分区，在逻辑卷之上可以建立文件系统（比如 /home 或者 /usr 等）。

⑤ PE（physical extent），每一个物理卷被划分为称为 PE(Physical Extents) 的基本单元（可理解为物理块，下面的 LE 则是对应逻辑块），具有唯一编号的 PE 是可以被 LVM 寻址的最小单元。PE 的大小是可配置的，默认为 4MB。

⑥ LE（logical extent），逻辑卷也被划分为被称为 LE(Logical Extents) 的可被寻址的基本单位。在同一个卷组中，LE 的大小和 PE 是相同的，并且一一对应。

PV、VG、LV 三者的对应关系如图 11.1.1 所示。

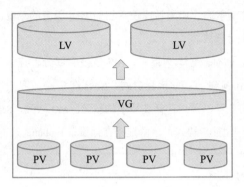

图 11.1.1　PV、VG、LV 三者的对应关系

2. Cinder 的基本架构

Cinder 是 OpenStack Block Storage 的项目名称；它为来宾虚拟机 (VM) 提供了持久块存储。对于可扩展的文件系统、最大性能、与企业存储服务的集成以及需要访问原生块级存储的应用程序而言，块存储通常是必需的。

系统可以暴露并连接设备，随后管理服务器的创建、附加到服务器和从服务器分离。应用程序编程接口 (API) 也有助于加强快照管理，这种管理可以备份大量块存储。

Cinder 从 OpenStack 的 Folsom 版本开始出现，用以替代 Nova-Volume 服务，Cinder 为 OpenStack 提供了管理卷（volunme）的基础设施。

按 OpenStack 官方文档的表述，Cinder 是受请求得到、自助化访问的块储存服务，即 Cinder 有两个显著的特点，第一，必须由用户提出请求，才能得到该服务；第二，用户可以自定义的半自动化服务。Cinder 实现 LVM(逻辑卷管理)，用以呈现存储资源给能够被 Nova 调用的端用户。简而言之，Cinder 虚拟化块存储设备池，提供端用户自助服务的 API 用以请求和使用这些块资源，并且不用了解存储的位置或设备信息。

LDAP（Lightweight Directory Access Protocol，轻量目录访问协议）是基于 X.500 标准的，但简单多了，并且可以根据需要定制。与 X.500 不同，LDAP 支持 TCP/IP，这对访问 Internet 是必需的。LDAP 的核心规范在 RFC 中都有定义。

Auth Manager：授权管理者。

AMQP：高级消息队列协议，应用于 MQ 中。

SCSI：小型计算机系统接口（Small computer System Interface，SCSI），是一种用于计算机和智能设备之间（硬盘、软驱、光驱、打印机、扫描仪等）系统级接口的独立处理器标准。SCSI 是一种智能的通用接口标准，也是各种计算机与外部设备之间的接口标准。

iSCSI：Internet 小型计算机系统接口（iSCSI）是一种基于 TCP/IP 的协议，用来建立和管理 IP 存储设备、主机和客户机等之间的相互连接，并创建存储区域网络（SAN）。SAN 使得 SCSI 协议应用于高速数据传输网络成为可能，这种传输以数据块级别（block-level）在多个数据存储网络间进行。

REST：表征状态转移（Representational State Transfer），定义了一种软件架构原则，它是一种针对网络应用的设计和开发方式，可以降低开发的复杂性，提高系统的可伸缩性。

根据这些原则设计以系统资源为中心的 Web 服务，包括使用不同语言编写的客户端如何通过 HTTP 处理和传输资源状态。如果考虑使用它的 Web 服务的数量，REST 近年来已经成为最主要的 Web 服务设计模式。

如图 11.1.2 所示，Cinder 整体构架中，API 是核心部分，连接了内部的授权管理，REST 请求，MQ 队列和前端的 Cinder 客户端，Nova 客户端。客户端通过 Web 界面显示 Cinder 信息，Nova 端用户可以调用存储在块中的资源。MQ 队列，Scheduler 调度 volume 上的存储信息。iSCSI 是一个接口协议，一端连接存储设备 (volume)，一端连接其他主机。

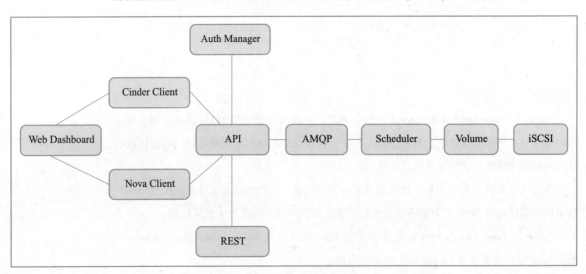

图 11.1.2　Cinder 架构

（1）Cinder 服务

API service：Cinder 构架图的核心部分，负责接收和处理 Rest 请求，并将请求放入 RabbitMQ 队列。

Scheduler service：处理任务队列的任务，并根据预定策略选择合适的 Volume Service 节点来执行任务。所以构架图中，Scheduler 一端连接 MQ 队列，处理 MQ 队列中的任务；一端连接 Volume 设备，任务在 Volume 节点上执行。

Volume service：服务运行在存储节点上，管理存储空间。每个存储节点都有一个 Volume Service，若干这样的存储节点联合起来可以构成一个存储资源池。

（2）Cinder 如何支持典型存储

从目前的实现来看，Cinder 对本地存储和 NAS 的支持比较好，可以提供完整的 Cinder API V2 支持，而对于其他类型的存储设备，Cinder 的支持会或多或少地受到限制。

① 对于本地存储，cinder-volume 可以使用 lvm 驱动，该驱动当前的实现需要在主机上事先用 lvm 命令创建一个 cinder-volumes 的 vg，当该主机接收到创建卷请求的时候，cinder-volume 在该 vg 上创建一个 LV，并且用 openiscsi 将这个卷当作一个 iscsi，target 给 export。当然还可以将若干主机的本地存储用 sheepdog 虚拟成一个共享存储，然后使用 sheepdog 驱动。

② EMC。图 11.1.3 是 EMC 的典型存储图。

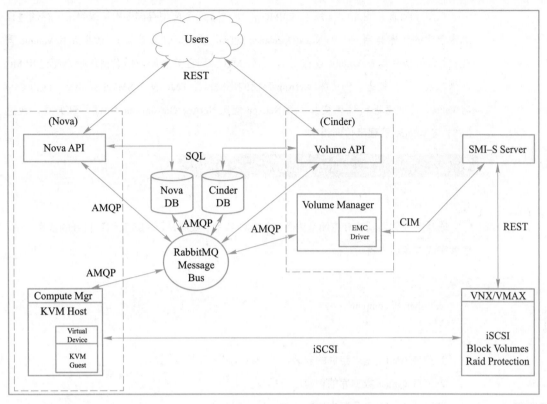

图 11.1.3　EMC 存储图

③ NetApp。图 11.1.4 是 NetApp 的典型存储图。

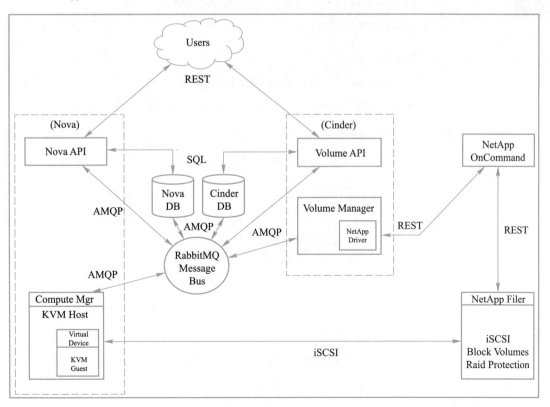

图 11.1.4　NetApp 存储图

在图 11.1.3 和图 11.1.4 所示的 EMC 和 NetApp 两个典型存储图中可以看出，存储过程必定包含的元素有用户，如 Nova(Openstack 的核心项目，计算功能)、数据库和 Volume 管理(资源就是存储在 Volume 设备上)，另外 MQ 队列至关重要，所有请求消息都汇集 MQ 队列，依次进行处理。EMC 和 NetApp 存储的不同点是 EMC 使用 SMI–S 服务器，通过 CIM 与 Volume 管理者进行信息交互，而 NetApp 使用 NetApp OnCommand 存储管理方式，通过 REST 与 Volume 管理者传递请求。

11.2 实训项目 9　Cinder 的手动安装与配置

电子教案　Cinder 的手动安装与配置

PPT　Cinder 的手动安装与配置

微课 12 Cinder 的手动安装与配置

1. 实训前提环境

成功完成实训项目 8 中所有内容后开始本实验，或者从已完成实训项目 8 的镜像开始，继续完成本实训内容。

2. 实训涉及节点

controller 和 compute。

3. 实训目标

① 完成 Cinder 基本组件的安装。
② 完成 Cinder 数据库的创建以及授权。
③ 完成 Cinder 主配置文件的修改。
④ 完成 Cinder 安全与认证配置。
⑤ 完成 Cinder 用户、项目、角色以及服务端点的创建。
⑥ 完成 Cinder 逻辑卷创建。

4. 实训步骤及其详解

步骤 1：在 controller 节点完成 Cinder 基本组件的安装。

通过 yum 命令安装 controller 节点 Cinder 服务所需要的依赖包，命令如下。

```
[root@controller~]#  yum -y install openstack-cinder
```

执行上述安装命令成功后，可以看到成功标志，如图 11.2.1 所示。

```
Installed:
  openstack-cinder.noarch 1:14.0.2-2.el7
Dependency Installed:
  glusterfs.x86_64 0:3.12.2-47.2.el7
  gperftools-libs.x86_64 0:2.6.1-1.el7
  librados2.x86_64 2:14.2.1-0.el7
  python-kmod.x86_64 0:0.9-4.el7
  python2-google-api-client.noarch 0:1.4.2-4.el7
  trousers.x86_64 0:0.3.14-2.el7

Complete!
[root@controller ~]#
```

图 11.2.1　Cinder 软件包安装完成的反馈结果

步骤 2：创建 Cinder 数据库并授权。

首先，用 root 用户登录 MySQL 数据库，命令如下。

```
[root@controller ~]# mysql -uroot -p000000
```

登录后，首先创建 Cinder 数据库，命令如下。

```
mysql>CREATE DATABASE cinder;
```

看到提示 Query OK, 1 row affected (0.00 sec)，表明数据库创建成功。

接着创建 MySQL 的 Cinder 用户，并赋予其 Cinder 数据库的操作权限，命令如下。

```
mysql> GRANT ALL PRIVILEGES ON cinder.* TO 'cinder' @ 'localhost'  IDENTIFIED BY  '000000';
mysql> GRANT ALL PRIVILEGES ON cinder.* TO 'cinder'@'%' IDENTIFIED BY  '000000';
mysql> exit
```

上述 SQL 语句与创建 Keystone 数据库的语句一致，请参考实训项目 3 中步骤 2 "创建 Keystone 数据库并授权"部分对创建数据库时 SQL 语句的说明，在此不做赘述。

然后，需要为 Cinder 服务创建数据库表。与其他 OpenStack 组件一样，在创建数据库表之前，需要先修改 Cinder 配置文件中用于数据库连接的内容，命令如下。

```
[root@controller ~]# crudini --set /etc/cinder/cinder.conf database connection mysql+pymysql://cinder:000000@controller/cinder
```

同样，也可以通过 vi/vim 编辑器，编辑 /etc/cinder/cinder.conf 文件中 [database] 字段与数据库连接的内容，以达到相同效果，编辑时请注意语法格式。

需要同步数据库，为 Cinder 服务创建数据库表，命令如下。

```
[root@controller ~]# su -s /bin/sh -c "cinder-manage db sync" cinder
```

与前面几个组件一样，也可以通过一条 SQL 命令来查看同步后创建的表，命令如下，结果如图 11.2.2 所示。

```
[root@controller ~]# mysql -u root -p000000 -e "use cinder;show tables;"
```

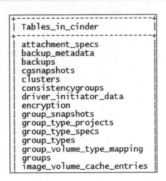

图 11.2.2　查看数据库同步的反馈结果

步骤 3：注册 Cinder 服务至 Keystone 服务器。

在 OpenStack 中，几乎所有的服务（包括 Keystone 服务）要想正常运行，都必须首先向 Keystone 服务器注册用户信息。

首先，注册 Cinder 用户信息，注册用户信息的目的是为了认证用户身份的。添加 Cinder 用户，并为用户设置密码，在本书实训过程中所有的密码统一使用 000000，命令如下，结果如图 11.2.3 所示。

```
[root@controller ~]# openstack user create --domain default --password 000000 cinder
```

```
+--------------------+----------------------------------+
| Field              | Value                            |
+--------------------+----------------------------------+
| domain_id          | default                          |
| enabled            | True                             |
| id                 | b56735032c174c8b9034fb6e66c5af25 |
| name               | cinder                           |
| options            | {}                               |
| password_expires_at| None                             |
+--------------------+----------------------------------+
```

图 11.2.3　创建 Cinder 用户的反馈结果

本书所有的服务都创建在 service 项目下。因此，Cinder 用户必须在 service 项目下具有 admin 权限。下面再为 Cinder 用户在 service 项目下分配 admin 权限，命令如下。

```
[root@controller ~]# openstack role add --project service --user cinder admin
```

上述命令用法在实训项目 3 步骤 7 中已经提到，在此不再赘述。

接下来，注册 Cinder 服务 (service) 和端点 (endpoint) 信息，注册服务和端点的信息目的是为了确保用户能够顺利访问 Cinder 服务，并执行一系列与 Cinder 服务有关的操作。需要注意，Cinder 与其他服务不同，Cinder 的服务和端点是两个版本并存的，所以，在进行 Cinder 服务和端点信息的注册操作时需要将两个版本的相关信息分别注册，命令如下，版本 2 即 v2 的结果如图 11.2.4 和图 11.2.5 所示，版本 3 即 v3 的结果如图 11.2.6 和图 11.2.7 所示。

```
[root@controller ~]# openstack service create --name cinderv2 --description "OpenStack Block Storage" volumev2
```

```
+-------------+----------------------------------+
| Field       | Value                            |
+-------------+----------------------------------+
| description | OpenStack Block Storage          |
| enabled     | True                             |
| id          | f6c8efe280c848d396bdb53c42eb4fac |
| name        | cinderv2                         |
| type        | volumev2                         |
+-------------+----------------------------------+
```

图 11.2.4　创建 Cinder 版本 2 服务的反馈结果

```
[root@controller ~]# openstack endpoint create --region RegionOne volumev2 public http://controller:8776/v2/%\(project_id\)s
    [root@controller ~]# openstack endpoint create --region RegionOne volumev2 internal http://controller:8776/v2/%\(project_id\)s
    [root@controller ~]# openstack endpoint create --region RegionOne volumev2 admin http://controller:8776/v2/%\(project_id\)s
```

```
[root@controller ~]# openstack endpoint list --service cinderv2
+----------------------------------+-----------+--------------+
| ID                               | Region    | Service Name |
+----------------------------------+-----------+--------------+
| 00fa6d1a08eb4e17ad25b5e4cfb416ec | RegionOne | cinderv2     |
| a1372e7ae3b84faaa28aae4ecf4a2540 | RegionOne | cinderv2     |
| f333961017324c49acce3e0e2ea9f90b | RegionOne | cinderv2     |
+----------------------------------+-----------+--------------+
```

图 11.2.5　创建 Cinder 版本 2 的 endpoint 的反馈结果

```
[root@controller ~]# openstack service create --name cinderv3 --description "OpenStack Block Storage" volumev3
```

```
| Field       | Value                            |
|-------------|----------------------------------|
| description | OpenStack Block Storage          |
| enabled     | True                             |
| id          | eb404769bb074277badae30b978d8849 |
| name        | cinderv3                         |
| type        | volumev3                         |
```

图 11.2.6　创建 Cinder 版本 3 服务的反馈结果

```
[root@controller ~]# openstack endpoint create --region RegionOne volumev3 public http://controller:8776/v3/%\(project_id\)s
[root@controller ~]# openstack endpoint create --region RegionOne volumev3 internal http://controller:8776/v3/%\(project_id\)s
[root@controller ~]# openstack endpoint create --region RegionOne volumev3 admin http://controller:8776/v3/%\(project_id\)s
```

```
[root@controller ~]# openstack endpoint list --service cinderv3
| ID                               | Region    | Service Name |
| 0bc606520e1641e6a0825edc2ed55bed | RegionOne | cinderv3     |
| 1d458dfb089b40d88c761873dab915e6 | RegionOne | cinderv3     |
| 9320ee2cbd824b8197587f235581693f | RegionOne | cinderv3     |
```

图 11.2.7　创建 Cinder 版本 3 的 endpoint 的反馈结果

步骤 4：配置 Cinder 使用消息队列服务。

消息队列仍然使用 RabbitMQ 队列，可以通过如下命令配置 controller 节点及 compute 节点的 Cinder 的消息队列。

```
[root@controller ~]# crudini --set /etc/cinder/cinder.conf DEFAULT transport_url rabbit://openstack:000000@controller
```

步骤 5：修改 Cinder 及 Nova 的配置文件。

① 修改 Cinder 的配置文件。

在完成 Cinder 用户信息、服务和端点信息的注册后，为了能让 Cinder 服务顺利通过 Keystone 的认证，需要在配置文件 /etc/cinder/cinder.conf 中记录 Keystone 的认证信息，即在配置文件的 [keystone_authtoken] 字段中修改 Keystone 的相关身份认证信息，同时在 [DEFAULT] 字段中添加 Keystone 支持。命令如下。

```
[root@controller ~]# crudini --set /etc/cinder/cinder.conf DEFAULT auth_strategy keystone
[root@controller ~]# crudini --set /etc/cinder/cinder.conf keystone_authtoken www_authenticate_uri http://controller:5000
[root@controller ~]# crudini --set /etc/cinder/cinder.conf keystone_authtoken auth_url http://controller:5000
[root@controller ~]# crudini --set /etc/cinder/cinder.conf keystone_authtoken memcached_servers controller:11211
[root@controller ~]# crudini --set /etc/cinder/cinder.conf keystone_authtoken auth_type password
[root@controller ~]# crudini --set /etc/cinder/cinder.conf keystone_authtoken project_domain_name default
[root@controller ~]# crudini --set /etc/cinder/cinder.conf keystone_authtoken user_
```

```
domain_name default
    [root@controller ~]# crudini --set /etc/cinder/cinder.conf keystone_authtoken
project_name service
    [root@controller ~]# crudini --set /etc/cinder/cinder.conf keystone_authtoken
username cinder
    [root@controller ~]# crudini --set /etc/cinder/cinder.conf keystone_authtoken
password 000000
    [root@controller ~]# crudini --set /etc/cinder/cinder.conf DEFAULT my_ip
192.168.100.10
    [root@controller ~]# crudini --set /etc/cinder/cinder.conf oslo_concurrency lock_
path /var/lib/cinder/tmp
```

对于上述命令，也可以通过 vi/vim 直接编辑配置文件 /etc/cinder/cinder.conf，在相应的地方做上述内容的修改，以达到相同的效果。修改完成后使用 grep 命令过滤 /etc/cinder/cinder.conf 文件中的内容来检查已完成的配置。命令和结果如下所示。

```
[root@controller ~]# grep -Ev "^#|^$" /etc/cinder/cinder.conf
[DEFAULT]
transport_url = rabbit://openstack:000000@controller
auth_strategy = keystone
my_ip = 192.168.100.10
[backend]
[backend_defaults]
[barbican]
[brcd_fabric_example]
[cisco_fabric_example]
[coordination]
[cors]
[database]
connection = mysql+pymysql://cinder:000000@controller/cinder
[fc-zone-manager]
[healthcheck]
[key_manager]
[keystone_authtoken]
www_authenticate_uri = http://controller:5000
auth_url = http://controller:5000
memcached_servers = controller:11211
auth_type = password
project_domain_name = default
user_domain_name = default
project_name = service
username = cinder
password = 000000
[nova]
[oslo_concurrency]
lock_path = /var/lib/cinder/tmp
```

② 修改 Nova 的配置文件。

```
[root@controller ~]# crudini --set /etc/nova/nova.conf cinder os_region_name RegionOne
```

步骤 6：启动服务并将其加入开机自启。

在 controller 节点启动 openstack-cinder-api 和 openstack-cinder-scheduler 服务并设置为开机自动启动，命令如下。

```
[root@controller ~]# systemctl restart openstack-nova-api.service
[root@controller ~]# systemctl start openstack-cinder-api.service openstack-cinder-scheduler.service
[root@controller ~]# systemctl enable openstack-cinder-api.service openstack-cinder-scheduler.service
```

步骤 7：在 compute 节点安装 Cinder 软件包。

通过 yum 命令安装 compute 节点 Cinder 服务所需要的依赖包，命令如下。

```
[root@compute ~]# yum -y install lvm2 device-mapper-persistent-data openstack-cinder targetcli python-keystone
```

步骤 8：创建 LVM 物理卷和 cinder-volumes 卷组。

将事先创建好的分区初始化为 LVM 物理卷，本书采用的是 sdb 硬盘，命令如下。

```
[root@compute ~]# pvcreate /dev/sdb
```

使用 vgcreate 命令创建 cinder-volumes 卷组，命令如下。

```
[root@compute ~]# vgcreate cinder-volumes /dev/sdb
```

步骤 9：在 compute 节点修改 Cinder 配置文件。

可以通过 openstack-config 命令配置 Cinder 与数据库的连接，命令如下。

```
[root@compute ~]# crudini --set /etc/cinder/cinder.conf database connection mysql+pymysql://cinder:000000@controller/cinder
```

修改 [DEFAULT] 段落和 [keystone_authtoken] 段落中 Keystone 相关认证信息命令如下。

```
[root@compute ~]# crudini --set /etc/cinder/cinder.conf DEFAULT auth_strategy keystone
[root@compute ~]# crudini --set /etc/cinder/cinder.conf keystone_authtoken www_authenticate_uri http://controller:5000
[root@compute ~]# crudini --set /etc/cinder/cinder.conf keystone_authtoken auth_url http://controller:5000
[root@compute ~]# crudini --set /etc/cinder/cinder.conf keystone_authtoken memcached_servers controller:11211
[root@compute ~]# crudini --set /etc/cinder/cinder.conf keystone_authtoken auth_type password
[root@compute ~]# crudini --set /etc/cinder/cinder.conf keystone_authtoken project_domain_name default
[root@compute ~]# crudini --set /etc/cinder/cinder.conf keystone_authtoken user_domain_name default
```

```
[root@compute ~]# crudini --set /etc/cinder/cinder.conf keystone_authtoken project_name service
[root@compute ~]# crudini --set /etc/cinder/cinder.conf keystone_authtoken username cinder
[root@compute ~]# crudini --set /etc/cinder/cinder.conf keystone_authtoken password 000000
```

在 [DEFAULT] 段落中修改消息队列服务相关信息,命令如下。

```
[root@ compute ~]# crudini --set /etc/cinder/cinder.conf DEFAULT transport_url rabbit://openstack:000000@controller
```

在 [DEFAULT] 段落中修改 glance 服务的主机名为 controller 节点,命令如下。

```
[root@compute ~]# crudini --set /etc/cinder/cinder.conf DEFAULT glance_api_servers http://controller:9292
```

在 [oslo_concurrency] 段落中配置锁定路径,命令如下。

```
[root@compute ~]# crudini --set /etc/cinder/cinder.conf oslo_concurrency lock_path /var/lib/cinder/tmp
```

配置计算节点及存储节点的管理接口 IP 地址,命令如下。

```
[root@compute ~]# crudini --set /etc/cinder/cinder.conf DEFAULT my_ip 192.168.100.20
```

步骤 10:配置 cinder 开启支持 lvm,命令如下。

```
[root@compute ~]# crudini --set /etc/cinder/cinder.conf lvm volume_driver cinder.volume.drivers.lvm.LVMVolumeDriver
[root@compute ~]# crudini --set /etc/cinder/cinder.conf lvm volume_group cinder-volumes
[root@compute ~]# crudini --set /etc/cinder/cinder.conf lvm target_protocol iscsi
[root@compute ~]# crudini --set /etc/cinder/cinder.conf lvm target_helper lioadm
[root@compute ~]# crudini --set /etc/cinder/cinder.conf DEFAULT enabled_backends lvm
```

步骤 11:在 compute 节点启动 openstack-cinder-volume 服务和 target 服务并设置为开机自动启动,命令如下。

```
[root@compute ~]# systemctl start openstack-cinder-volume.service target.service
[root@compute ~]# systemctl enable openstack-cinder-volume.service target.service
```

5. 常见错误及调试排错

使用 cinder create 命令创建云硬盘时,出现以下报错,如图 11.2.8 所示。

图 11.2.8 创建卷失败

解决方法:

通过 vi 命令进入 Cinder 配置文件,命令如下。

```
[root@controller ~]# vi /etc/cinder/cinder.conf
```

查看 [database] 段落下数据库的连接配置是否正确，应改为以下内容，如图 11.2.9 所示。

```
[database]
connection = mysql+pymysql://cinder:000000@controller/cinder
```

图 11.2.9　数据库配置选项

修改完成后在 controller 节点重启相关服务，命令如下。

```
[root@controller ~]# systemctl restart openstack-cinder-api
[root@controller ~]# systemctl restart openstack-cinder-scheduler
```

重启 compute 节点相关服务，命令如下。

```
[root@compute ~]# systemctl restart openstack-cinder-volume
[root@compute ~]# systemctl restart target
```

完成后，重新使用 cinder create 命令创建云硬盘，命令如下，结果如图 11.2.10 所示。

```
[root@controller ~]# cinder create --display-name myvolume 1
+---------------------------------+--------------------------------------+
| Property                        | Value                                |
+---------------------------------+--------------------------------------+
| attachments                     | []                                   |
| availability_zone               | nova                                 |
| bootable                        | false                                |
| consistencygroup_id             | None                                 |
| created_at                      | 2019-11-15T12:11:58.000000           |
| description                     | None                                 |
| encrypted                       | False                                |
| id                              | 04819fe2-4ee2-4d72-b0d0-f9f145a81780 |
| metadata                        | {}                                   |
| migration_status                | None                                 |
| multiattach                     | False                                |
| name                            | myvolume                             |
| os-vol-host-attr:host           | None                                 |
| os-vol-mig-status-attr:migstat  | None                                 |
| os-vol-mig-status-attr:name_id  | None                                 |
| os-vol-tenant-attr:tenant_id    | 23fb3921f1df4c1da0e4c1a7d59459c3     |
| replication_status              | None                                 |
| size                            | 1                                    |
| snapshot_id                     | None                                 |
| source_volid                    | None                                 |
| status                          | creating                             |
| updated_at                      | None                                 |
| user_id                         | 9932b64573f0422993f572f020785e5c     |
| volume_type                     | None                                 |
+---------------------------------+--------------------------------------+
```

图 11.2.10　创建卷反馈

11.3　Cinder 安装脚本及其解读

在介绍了手动安装 Cinder 服务后，接下来将介绍另一种安装 Cinder 服务的方法，即利用已经写好的脚本自动安装 Cinder 服务。

电子教案　Cinder 安装脚本及其解读

PPT　Cinder 安装脚本及其解读

1. 需要安装的脚本内容

Cinder 服务手工安装时，需要分别在 controller 和 compute 节点配置，所以脚本也需要分别在两个节点写各自的配置内容。

controller 节点的脚本内容如下。

```
#!/bin/bash
source /etc/keystone/admin-openrc.sh
# 安装软件包
yum -y install openstack-cinder
# 配置数据库
mysql -uroot -p000000 -e "create database IF NOT EXISTS cinder ;"
mysql -uroot -p000000 -e "GRANT ALL PRIVILEGES ON cinder.* TO 'cinder'@'localhost'
```

```
IDENTIFIED BY '000000' ;"
    mysql -uroot -p000000 -e "GRANT ALL PRIVILEGES ON cinder.* TO 'cinder'@'%' IDENTIFIED BY '000000' ;"
    # Keystone 认证
    openstack user create --domain default --password 000000 cinder
    openstack role add --project service --user cinder admin
    openstack service create --name cinderv2 --description "OpenStack Block Storage" volumev2
    openstack service create --name cinderv3 --description "OpenStack Block Storage" volumev3
    openstack endpoint create --region RegionOne volumev2 public http://controller:8776/v2/%\(project_id\)s
    openstack endpoint create --region RegionOne volumev2 internal http://controller:8776/v2/%\(project_id\)s
    openstack endpoint create --region RegionOne volumev2 admin http://controller:8776/v2/%\(project_id\)s
    openstack endpoint create --region RegionOne volumev3 public http://controller:8776/v3/%\(project_id\)s
    openstack endpoint create --region RegionOne volumev3 internal http://controller:8776/v3/%\(project_id\)s
    openstack endpoint create --region RegionOne volumev3 admin http://controller:8776/v3/%\(project_id\)s
    # 主配置文件设置
    crudini --set /etc/cinder/cinder.conf database connection mysql+pymysql://cinder:000000@controller/cinder
    crudini --set /etc/cinder/cinder.conf DEFAULT transport_url rabbit://openstack:000000@controller
    crudini --set /etc/cinder/cinder.conf DEFAULT auth_strategy keystone
    crudini --set /etc/cinder/cinder.conf keystone_authtoken www_authenticate_uri http://controller:5000
    crudini --set /etc/cinder/cinder.conf keystone_authtoken auth_url http://controller:5000
    crudini --set /etc/cinder/cinder.conf keystone_authtoken memcached_servers controller:11211
    crudini --set /etc/cinder/cinder.conf keystone_authtoken auth_type password
    crudini --set /etc/cinder/cinder.conf keystone_authtoken project_domain_name default
    crudini --set /etc/cinder/cinder.conf keystone_authtoken user_domain_name default
    crudini --set /etc/cinder/cinder.conf keystone_authtoken project_name service
    crudini --set /etc/cinder/cinder.conf keystone_authtoken username cinder
    crudini --set /etc/cinder/cinder.conf keystone_authtoken password 000000
    crudini --set /etc/cinder/cinder.conf DEFAULT my_ip 192.168.100.10
    crudini --set /etc/cinder/cinder.conf oslo_concurrency lock_path /var/lib/cinder/tmp
    # 同步数据库
    su -s /bin/sh -c "cinder-manage db sync" cinder
    crudini --set /etc/nova/nova.conf cinder os_region_name RegionOne
```

```bash
# 落定服务
systemctl restart openstack-nova-api.service
systemctl enable openstack-cinder-api.service openstack-cinder-scheduler.service
systemctl start openstack-cinder-api.service openstack-cinder-scheduler.service
```

compute 节点的脚本内容如下。

```bash
#!/bin/bash
# 加载环境变量
source openrc.sh
# 安装需要的软件包
yum -y install lvm2 device-mapper-persistent-data
systemctl enable lvm2-lvmetad.service
systemctl start lvm2-lvmetad.service
# 创建卷组
pvcreate -f /dev/sdb
vgcreate cinder-volumes /dev/sdb
# 安装需要的软件包
yum -y install openstack-cinder targetcli python-keystone
# 配置数据库连接
crudini --set /etc/cinder/cinder.conf database connection mysql+pymysql://cinder:000000@controller/cinder
# 连接消息队列
crudini --set /etc/cinder/cinder.conf DEFAULT transport_url rabbit://openstack:000000@controller
# Keystone 认证
crudini --set /etc/cinder/cinder.conf DEFAULT auth_strategy keystone
crudini --set /etc/cinder/cinder.conf keystone_authtoken www_authenticate_uri http://controller:5000
crudini --set /etc/cinder/cinder.conf keystone_authtoken auth_url http://controller:5000
crudini --set /etc/cinder/cinder.conf keystone_authtoken memcached_servers controller:11211
crudini --set /etc/cinder/cinder.conf keystone_authtoken auth_type password
crudini --set /etc/cinder/cinder.conf keystone_authtoken project_domain_name default
crudini --set /etc/cinder/cinder.conf keystone_authtoken user_domain_name default
crudini --set /etc/cinder/cinder.conf keystone_authtoken project_name service
crudini --set /etc/cinder/cinder.conf keystone_authtoken username cinder
crudini --set /etc/cinder/cinder.conf keystone_authtoken password 000000
crudini --set /etc/cinder/cinder.conf DEFAULT my_ip 192.168.100.20
# 卷存储配置
crudini --set /etc/cinder/cinder.conf lvm volume_driver cinder.volume.drivers.lvm.LVMVolumeDriver
crudini --set /etc/cinder/cinder.conf lvm volume_group cinder-volumes
crudini --set /etc/cinder/cinder.conf lvm target_protocol iscsi
crudini --set /etc/cinder/cinder.conf lvm target_helper lioadm
```

```
crudini --set /etc/cinder/cinder.conf DEFAULT enabled_backends lvm
crudini --set /etc/cinder/cinder.conf DEFAULT glance_api_servers http://controller:9292
crudini --set /etc/cinder/cinder.conf oslo_concurrency lock_path /var/lib/cinder/tmp
# 落定服务
systemctl enable openstack-cinder-volume.service target.service
systemctl start openstack-cinder-volume.service target.service
```

2. 脚本的执行

刚写好的脚本文件，在核实过内容后，是不能直接使用的。要想使用脚本，首先需要赋予脚本执行权限，这里以 controller 节点的脚本为例进行讲解，命令如下。

```
[root@controller ~]# chmod a+x iaas-install-cinder-controller.sh
```

或者执行

```
[root@controller ~]# chmod 777 iaas-install-cinder-controller.sh
```

上述两种命令都可以实现赋予脚本执行权限的效果，这里的 "iaas-install-cinder-controller.sh" 是示范用的脚本名称，读者在实际使用时可自行命名脚本名称（说明：本次实验所使用的脚本均放在 root 用户的家目录里）。

脚本有了执行权限后就可以执行了。这里介绍两种执行脚本的方法。

方法1：在脚本前加上脚本所在的路径，可以使相对路径，也可以是绝对路径，这取决于读者个人习惯，命令如下。

```
[root@controller ~]# ./iaas-install-cinder-controller.sh
```

"."表示当前目录，我们的脚本放在家目录中，而用户此时也在家目录中，所以为方便，这里使用的是当前目录。

方法2：现将脚本所在路径加入环境变量 PATH 中，然后就可以像 ls 命令一样直接在命令中输入使用了，命令如下。

```
[root@controller ~]# echo $PATH
/usr/local/sbin:/usr/local/bin:/sbin:/bin:/usr/sbin:/usr/bin:/root/bin
[root@controller ~]# PATH=${PATH}:/root/
[root@controller ~]# echo $PATH
/usr/local/sbin:/usr/local/bin:/sbin:/bin:/usr/sbin:/usr/bin:/root/bin:/root/
[root@controller ~]# iaas-install-cinder-controller.sh
```

上述命令中对比了将脚本所在目录加入环境变量 PATH 前后，环境变量 PATH 中的内容，最后可以直接在命令中像其他命令一样直接输入脚本名称了。

说明：直接在命令中改变的环境变量只对当前 shell 和其子 shell 有效，当重启或另外打开新的 shell 时，当前修改的环境变量无效。然而可以通过修改 /root/.bash_profile 配置文件中 PATH 的内容，达到环境变量 PATH 永久有效的效果，注意修改 /root/.bash_profile 后需要通知系统内容变化，最直接的方法是重新打开新的 shell。

习题 Cinder 的安装及其配置

第 12 章　Dashboard 的安装及其配置

 本章导读：

　　本章首先介绍 Dashboard 在云平台中的作用。安装并实现该模块基本功能后可以通过 Web 页面的形式对 OpenStack 进行 GUI 管理。在本章的实训项目 10 中介绍如何通过命令行以及服务组件的参数配置安装 Dashboard，本章还将解读启动虚拟机实例过程中可能会出现的问题以及原因分析。

电子资源：
　　　电子教案　Dashboard 的安装及其配置
　　　PPT　Dashboard 的安装及其配置
　　　习题　Dashboard 的安装及其配置

第 12 章　Dashboard 的安装及其配置

12.1　Dashboard 功能简介

OpenStack 在 2012 年的 Essex 版本中引入了 Dashboard 组件，Dashboard 组件的项目命名为 Horizon，旨在通过 Web 页面的形式对 OpenStack 进行 GUI 管理。

Dashboard 采用 Python 作为后台语言进行开发，使用 Python 的 Django 架构，运行相当流畅稳定，各模块布局清晰。

Dashborad 几乎可以完成 Openstack 的所有管理操作，包括创建用户、启动、终止虚拟机实例，管理域等操作，Dashboard 将原本复杂的 CLI 操作转化为图形化页面，使 OpenStack 逐渐走进大众的眼睛中。

电子教案　Dashboard 功能简介及安装与配置

PPT　Dashboard 功能简介及安装与配置

12.2　实训项目 10　Dashboard 的安装与配置

1. 实训前提环境

完成实训项目 9 所有内容，实训项目 9 成功，或者从已完成实训项目 9 镜像开始，继续完成本实训内容。

2. 实训涉及节点

controller

微课 13 Dashboard 的安装与配置

3. 实训目标

① 完成 Dashboard 基本组件的安装。
② 完成 Dashboard 本地配置的修改。
③ 成功登录 Dashboard 页面创建虚拟机实例。

4. 实训步骤及其详解

步骤 1：在 controller 节点完成 Dashboard 基本组件的安装。

通过 yum 命令在 controller 节点安装 Dashboard 服务所需要的依赖包，命令如下。

```
[root@controller ~]# yum -y install openstack-dashboard
```

执行上述安装命令成功后，可以看到成功标志，所有 controller 节点 Dashboard 依赖包都安装完成，如图 12.2.1 所示。

```
Installed:
  openstack-dashboard.noarch 1:15.1.1-1.el7
Dependency Installed:
  XStatic-Angular-common.noarch 1:1.5.8.0-1.el7
  fontawesome-fonts.noarch 0:4.4.0-1.el7
  mdi-fonts.noarch 0:1.4.57.0-4.el7
  python-XStatic-Bootstrap-Datepicker.noarch 0:1.3.1.0-1.el7
```

图 12.2.1　Dashboard 依赖包安装完成反馈结果

步骤 2：配置 Dashboard 本地策略。

在 Dashboard 中，需要指定能够通过访问 Dashboard 页面的主机名或 IP 地址，和 OpenStack 自己的主机名，可以通过修改 controller 节点下 /etc/openstack-dashboard/local_settings 文件来实现，命令如下。

```
[root@controller ~]# vi /etc/openstack-dashboard/local_settings
```

修改内容如下。

```
ALLOWED_HOSTS = ['*','localhost']
OPENSTACK_HOST = "controller"
SESSION_ENGINE = 'django.contrib.sessions.backends.cache'
CACHES = {
    'default': {
        'BACKEND': 'django.core.cache.backends.memcached.MemcachedCache',
        'LOCATION': 'controller:11211',
    },
}
OPENSTACK_KEYSTONE_MULTIDOMAIN_SUPPORT = True
OPENSTACK_API_VERSIONS = {
    "identity": 3,
    "image": 2,
    "volume": 3,
}
OPENSTACK_KEYSTONE_DEFAULT_DOMAIN = 'Default'
OPENSTACK_KEYSTONE_DEFAULT_ROLE = "user"
[root@controller ~]# vi /etc/httpd/conf.d/openstack-dashboard.conf
```

在该配置文件中添加如下一行内容。

```
WSGIApplicationGroup %{GLOBAL}
```

需要注意，在生产环境中，可以根据实际情况修改允许访问的主机和 OpenStack 主机名。

步骤 3：启动相关服务。

Dashboard 组件的运行依赖 Apache 和 Memcached 服务，故 Dashboard 配置完成后需要重新启动相关服务。Apache 服务对应的守护进程为 httpd，Memcache 对应的守护进程为 memcached。

重启服务，命令如下。

```
[root@controller ~]# systemctl restart httpd
[root@controller ~]# systemctl restart memcached
```

至此，Dashboard 安装及配置已经完成。

步骤 4：登录 Dashboard 页面创建虚拟机实例。

首先登录 Dashboard，在浏览器地址栏输入 http://192.168.100.10/dashboard，按 Enter 键，打开 Dashboard 的登录界面如图 12.2.2 所示。

在用户名和密码框中输入用户名和密码信息，这里的用户名和密码在 Keystone 的安装配置时已经建立，使用管理员的账号登录，用户名为 admin，密码为 000000，之后单击"登录"按钮即可进入 Dashboard 管理页面，结果如图 12.2.3 所示。

之后，在左侧选择"项目"→"compute"→"实例"标签栏，如图 12.2.4 所示。

在界面右侧上方单击"创建实例"按钮，弹出"创建实例"子窗口，选择云主机名称为 demo，云主机类型为 m1.small，镜像名称选择 cirros，其他标签选项使用默认参数即可，如图 12.2.5 所示。

图 12.2.2　Dashboard 的登录界面

图 12.2.3　Dashboard 管理页面

图 12.2.4　Dashboard 中实例标签栏

图 12.2.5　创建实例

单击"运行"按钮，等待虚拟机实例启动，第一次启动实例一般时间较长，电源将显示"无状态"，耐心等待 5 分钟左右，如图 12.2.6 所示。

图 12.2.6　创建实例过程

如果其他组件的配置都没有问题，将成功看到实例启动，"电源状态"变为"运行中"，如图 12.2.7 所示。

单击实例 demo 后的"更多"按钮▼，在弹出的下拉列表中选择"控制台"选项，如图 12.2.8 所示。

此时便可以看到云主机的 VNC 控制台，可以通过该控制台对云主机进行操作，如图 12.2.9 所示。

12.2 实训项目 10 Dashboard 的安装与配置

至此，Dashboard 的云主机实例创建已经全部完成。

为了使得外部终端软件可以成功接入所创建的"实例"，必须进行安全规则的有关配置，首先在左侧"项目"选择"网络"→"安全组"选项进入如图 12.2.10 所示界面。

单击 default 安全组后的"管理规则"按钮，在管理安全组规则 default 页面中添加规则，如图 12.2.11 所示。

单击"添加规则"按钮，弹出"添加规则"子页面，如图 12.2.12 所示。

通过上述界面，依次添加 ALL TCP、ALL UDP、ALL ICMP，在入口和出口两个方向上共 6 条规则，添加完成后如图 12.2.13 所示。

图 12.2.7　实例成功创建

图 12.2.8　实例更多操作列表

图 12.2.9　实例控制台

图 12.2.10　安全组界面

图 12.2.11　default 管理安全组

图 12.2.12　"添加规则"子页面

219

☐	出口	IPv4	ICMP	任何	0.0.0.0/0	-	-	删除规则
☐	入口	IPv4	ICMP	任何	0.0.0.0/0	-	-	删除规则
☐	出口	IPv4	TCP	1 - 65535	0.0.0.0/0	-	-	删除规则
☐	入口	IPv4	TCP	1 - 65535	0.0.0.0/0	-	-	删除规则
☐	出口	IPv4	UDP	1 - 65535	0.0.0.0/0	-	-	删除规则
☐	入口	IPv4	UDP	1 - 65535	0.0.0.0/0	-	-	删除规则

图 12.2.13　添加完成所需规则反馈结果

12.3　启动虚拟机实例及其排错案例

1. Dashboard 创建虚拟网络报错

问题描述：

用户在通过 Dashboard 页面创建虚拟网络时，会报提示信息"危险：在提交表单的时候出现错误，请再次尝试"，如图 12.3.1 所示。

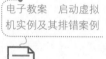
电子教案　启动虚拟机实例及其排错案例

PPT　启动虚拟机实例及其排错案例

图 12.3.1　创年网络失败错误反馈

问题原因：

数据库连接出现故障。

解决方法：

重新启动数据库配置文件，并且检查 neutron 数据库是否同步成功。

2. Dashboard 可用主机报错

问题描述：

用户在通过 Dashboard 页面创建虚拟机时，系统提示错误，如图 12.3.2 所示。

问题原因：

"No valid host was found"报错是云平台的常见报错，可以查看云平台日志来定位具体错误，查看云平台 nova 主机调度相关日志的命令如下。

```
[root@controller ~]# tail -f /var/log/nova/scheduler.log
```

12.3 启动虚拟机实例及其排错案例

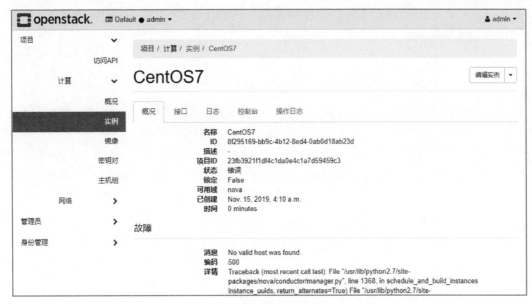

图 12.3.2 错误反馈结果

（1）出错的可能有很多，一般可以归结为以下几种

① 云主机实例资源分配不足。

如果在日志文件中出现：raise exception.FlavorDiskTooSmall()\n', u"FlavorDiskTooSmall: Flavor's disk is too small for requested image.\n" 则表示云主机实例资源分配不足。

② 云平台资源不足。

云平台硬盘资源不足在 Nova 日志中不会出现明显特征：

WARNING nova.scheduler.driver [req-60dd3dd1-51a7-4b42-9b52-bf77fb50ea74 admin admin] [instance: fbd9763d-ec5a-4aba-8e73-f832c8c29951] Setting instance to ERROR state.

可以通过查看云平台中的主机来定位错误，如云平台中其他主机实例可以正常运行，如图 12.3.3 所示。

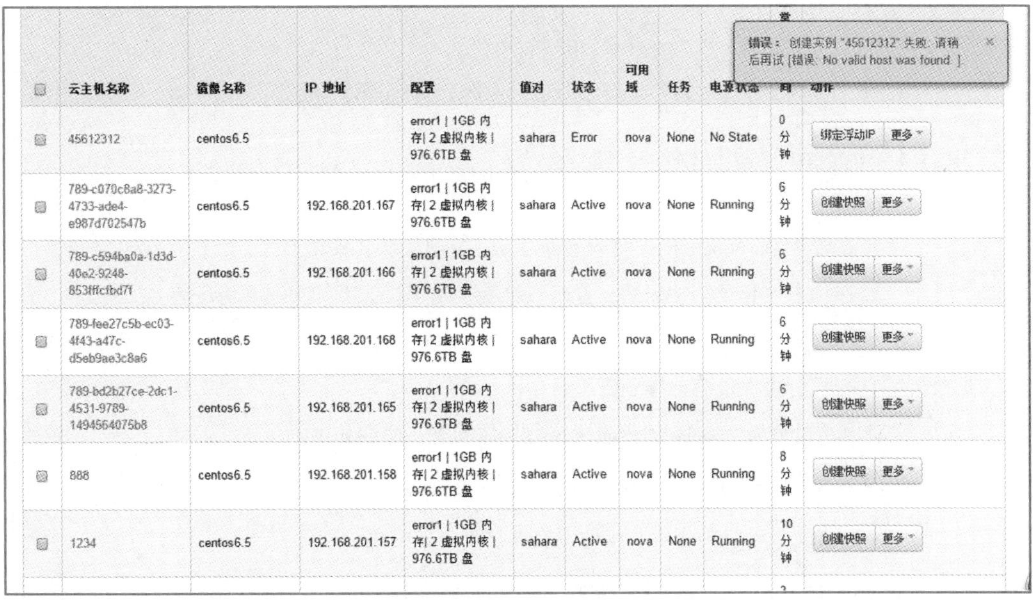

图 12.3.3 比较实例间运行状态

而刚创建的主机实例在很短的时间内（一般少于 2 秒）报错，基本可以确定是云平台资源不足问题。

③ 配置文件错误。

查看 nova 的主机调度文件，发现如下报错：

```
    File "/usr/lib/python2.6/site-packages/nova/virt/libvirt/vif.py", line 384, in get_
config\n        _("Unexpected vif_type=%s") % vif_type)\n', u'NovaException: Unexpected vif_
type=binding_failed\n']
```

关键字：vif_type=binding_failed，一般可以定位为配置文件加载错误。

④ 其他原因。

如果在排查以上原因后仍然不能定位云平台问题，可以先通过以下命令：

```
[root@controller ~]# openstack-status
```

该命令用于查看云平台各个组件的运行状况，通过这个命令可以定位问题组件，再进一步排查错误。

（2）解决方法

① 云主机实例资源分配不足，可以通过修改云主机类型（flavor）的方式来解决，一般可以将云主机类型值改大。在启动 centos 系统时，可以使用系统默认的 small 类型来创建。

② 云平台资源不足，如果定位问题是云平台资源问题，可以将部分云主机类型值改小（在保证启动的情况下），或者增加云平台计算资源。

③ 配置文件错误，针对配置文件问题，只能通过 grep 命令来对配置文件进行核对操作，一般来说，配置文件的问题集中在 Neutron 配置中，注意大小写。

④ 其他原因，由于云平台的自由性较高，在运行时难免会发生很多日志文件没有记录的错误，通过查看云平台整体状态可以对某一组件进行针对排错，即通过可以在云平台命令后跟上 --debug 参数，来进行 Debug 跟踪，例如查看 Neutron 组件运行状态时进行 Debug 跟踪的命令如下。

```
[root@controller1 ~]# neutron --debug agent-list
```

通过该参数可以对组件运行细节进行跟踪，排查具体问题。

3. Dashboard 登录报错

（1）问题描述

① 在登录 Dashboard 时，提示登录失败，返回服务器报错页面，如图 12.3.4 所示。

② 在单击 "登录" 按钮后，提示认证发生错误，如图 12.3.5 所示。

（2）问题原因

① 云平台组件没有连接到云平台中。

可以通过查看 Apache 的日志来定位具体问题，命令如下。

```
[root@controller ~]# tail -f /var/log/httpd/error_log
```

通过该命令可以判断哪些组件没有连接到云平台中，例如：

```
[Sun Oct 11 08:33:20 2015] [error]         raise exceptions.ConnectionFailed(reason=_
```

```
("Maximum attempts reached"))
    [Sun Oct 11 08:33:20 2015] [error] ConnectionFailed: Connection to neutron failed:
Maximum attempts reached
```

发现关键字：Connection to neutron failed，确定是 Neutron 组件问题，有时可能不会明显提示具体组件名，需要通过 openstack-status 命令来查看云平台组件的具体状态来进行进一步的问题定位。

图 12.3.4 登录失败（1）   图 12.3.5 登录失败（2）

② 认证服务出现问题。

如果报错页面也没有提示，一般是 Keystone 配置问题，可以通过 openstack user list 等命令来查看 Keystone 的状态，同样，也可以通过查看 Apache 的报错日志：

```
[root@controller ~]# tail -f /var/log/httpd/error_log
```

查看报错：

```
[Sun Oct 11 08:33:55 2015] [error] Logging out user "admin".
[Sun Oct 11 08:33:55 2015] [error] Could not delete token
[Sun Oct 11 08:34:00 2015] [error] DeprecationWarning: BaseException.message has
been deprecated as of Python 2.6Login failed for user "admin".
[Sun Oct 11 08:42:13 2015] [error] Login failed for user "admin".
```

判断用户认证方面出现错误。

（3）解决方法

① 组件连接问题。通过 openstack-status 命令查看各组件状态后，启动相应组件，连接到平台，如果启动失败，则可以对该组件进行进一步排查，需要注意，在 Keystone 中注册的组件都会影响 Dashboard 的启动，需要对每个注册过的组件都进行排查工作。

② 认证服务配置问题。认证服务 Keystone 出现问题，可以通过查看 Keystone 的日志问题进行排查，具体命令如下。

```
[root@controller ~]# tail -f /var/log/keystone/keystone.log
```

另外，在排查前，需要确保 Keystone 服务已经正常运行，如果没有正常运行，需要首先启动 Keystone 服务。

4. Dashboard 中看不到用户标签

问题描述：

Dashboard 通过 admin 用户登录但看不到用户标签，不能通过管理员对用户进行操作，如图 12.3.6 所示。

图 12.3.6　看不到用户标签

问题原因：

Keystone 的 Endpoint 创建错误，可以通过 openstack endpoint list --service keystone 命令来查看端点信息，结果如图 12.3.7 所示。

可以发现 Keystone 的 adminurl 与 internalurl、publicurl 不同，配置成了 http://controller:35357/v3，导致连接到 Keystone adminurl 时出现问题，返回报错信息。

解决方法：

可以重新创建 Keystone 的 Endpoint，注意，需要先将错误的 Endpoint 删除，创建时通过令牌创建（详见 Keystone 章节）。另外，创建完成后，需要重启浏览器，以保证浏览器读取缓存继续报错。

图 12.3.7　查看 Endpoint 反馈结果

第三部分 原生 OpenStack 云平台运维详解

第 13 章 原生 OpenStack 云平台各组件运维

本章导读：
　　本章所有内容均为实训项目，通过不同的实训项目介绍云平台不同模块的基本运维操作，包括通过命令行进行运维以及通过 Dashboard 界面进行运维。

电子资源：
　　电子教案原生　OpenStack 云平台各组件运维
　　PPT 原生　OpenStack 云平台各组件运维
　　习题原生　OpenStack 云平台各组件运维

第 13 章 原生 OpenStack 云平台各组件运维

电子教案 Keystone 基本运维命令及其应用

PPT Keystone 基本运维命令及其应用

微课 14 Keystone 基本运维命令及其应用

13.1 实训项目 11 Keystone 基本运维命令及其应用

1. 实训前提环境

本地主机安装终端软件，通过终端软件连接到安装成功的 OpenStack 云平台中的控制节点，通过命令进行运维操作。本地主机需安装浏览器，推荐使用谷歌浏览器或者火狐浏览器，通过浏览器登录云平台 Dashboard 界面进行运维的操作。

2. 实训涉及节点

controller

3. 实训目标

① 熟悉云平台用户、用户组、角色、项目和域查看的命令。

② 熟悉云平台用户、用户组、角色、项目和域创建与删除的命令。

③ 能够熟练使用命令以及云平台 Dashboard 界面进行用户、用户组、角色、项目和域的管理。

④ 能够使用命令查看以及修改云平台的服务端点。

4. 实训内容

步骤 1：查看用户、用户组、角色、项目和域查看的命令。

在使用 openstack user 命令前，首先要执行环境变量的文件，该文件在实训项目 3 中已经创建过，找到其路径并通过 source 命令执行。之后通过 openstack user list 查看当前云平台上有哪些用户，命令如下，结果如图 13.1.1 所示。

```
[root@controller ~]# openstack user list
```

ID	Name
0165a8c49c4647118053a06ce2fbce68	admin
132a3f8ef8934480887436a802e3f808	glance
39ef92e66e584bb8b8bdf509910bb5f9	placement
577cfebf19cf4ba18ba3c6602d0db8b3	neutron
8a5897353b7043e1a801d6a2397baf4d	demo
cc6f950c440f46a5b215d67acf3580f9	nova

图 13.1.1 查看用户的反馈结果

从结果中可以看到目前平台上所有的用户，表中第 1 列的 id 字段表明每个用户都有一个唯一的与之相对应的 id 值，第 2 列 name 字段显示了当前已存在用户的用户名。

可以进一步用 keystone user-get 命令查看某一个特定用户的信息，例如，想查看 admin 用户的信息，可以在上述命令之后直接跟 admin 用户名，或直接跟 admin 用户所对应的 id，命令如下，结果如图 13.1.2 所示。domain_id 表示用户属于哪个域。enabled 字段的值表明所对应的用户是否被激活，显然，"True"为激活状态，非激活状态这里的值应该是"False"只有激活的用户才能登录并使用云平台。

```
[root@controller ~]# openstack user show admin
```

或者

```
[root@controller ~]# openstack user show bc23d4f89c314521a030f620bf548a7e
```

```
+---------------------+----------------------------------+
| Field               | Value                            |
+---------------------+----------------------------------+
| domain_id           | default                          |
| enabled             | True                             |
| id                  | 0165a8c49c4647118053a06ce2fbce68 |
| name                | admin                            |
| options             | {}                               |
| password_expires_at | None                             |
+---------------------+----------------------------------+
```

图 13.1.2　查看特定用户的反馈结果

查看云平台域信息的命令与查看用户信息的命令类似为 openstack domain list，命令如下，结果如图 13.1.3 所示。

```
[root@controller ~]# openstack domain list
```

```
+----------------------------------+---------+---------+--------------------+
| ID                               | Name    | Enabled | Description        |
+----------------------------------+---------+---------+--------------------+
| dd50fd9b36384c9ab429a00353ec615b | demo    | True    | My Domain          |
| default                          | Default | True    | The default domain |
+----------------------------------+---------+---------+--------------------+
```

图 13.1.3　查看域的反馈结果

从反馈的结果中可以看到目前平台上所有的域，和查看用户的结果类似，第 1 列的 id 字段表明每个域也有一个唯一的与之对应的 id 值，第 2 列 name 字段显示了当前已存在的域的名称，第 3 列 enabled 字段的值表明所对应的域是否被激活，"True"表明所对应的域已激活，相反的状态为"False"。

和查看特定用户的操作类似，如果想查看某一个特定域的信息，可以使用 openstack domain show 命令，在命令后可以直接跟域的名称或 id，例如要单独查看 demo 域的信息，命令如下，结果如图 13.1.4 所示。

```
[root@controller ~]#openstack domain show demo
```

或

```
[root@controller ~]#openstack domain show dd50fd9b36384c9ab429a00353ec615b
```

```
+-------------+----------------------------------+
| Field       | Value                            |
+-------------+----------------------------------+
| description | My Domain                        |
| enabled     | True                             |
| id          | dd50fd9b36384c9ab429a00353ec615b |
| name        | demo                             |
| tags        | []                               |
+-------------+----------------------------------+
```

图 13.1.4　查看特定域的反馈结果

查看用户组的操作与查看用户类似，可以通过 openstack group list 查看所有用户组，也可以通过 openstack group show 查看单个用户组的信息，命令如下，结果如图 13.1.5 和图 13.1.6 所示。

```
[root@controller ~]# openstack group list
```

```
+----------------------------------+------+
| ID                               | Name |
+----------------------------------+------+
| b8f97f51b1bb47a885e0f2466e52dcfc | test |
+----------------------------------+------+
```

图 13.1.5　查看用户组的反馈结果

```
[root@controller ~]# openstack group show test
```

```
+-------------+----------------------------------+
| Field       | Value                            |
+-------------+----------------------------------+
| description |                                  |
| domain_id   | default                          |
| id          | b8f97f51b1bb47a885e0f2466e52dcfc |
| name        | test                             |
+-------------+----------------------------------+
```

图 13.1.6　查看特定用户组的反馈结果

查看用户是否属于某个组，可以通过命令 openstack group contains user 来查看。如果 demo 属于 test 组，则会出现 "demo in group test"。

```
[root@controller ~]# openstack group contains user test demo
demo in group test
```

查看角色的操作和查看用户和域类似，可以通过 openstack role list 查看所有角色，也可以通过 openstack role show 查看单个角色的信息，命令如下，结果如图 13.1.7 和图 13.1.8 所示。

```
[root@controller ~]# openstack role list
```

```
+----------------------------------+--------+
| ID                               | Name   |
+----------------------------------+--------+
| 940922970e384ed5930b5937482f69a7 | member |
| 9c34246705c445a3856bb53cd8cdfa00 | user   |
| d4aa683be0b24aa5aa23f2a53f923a27 | reader |
| e44a6ba4bff74cbe924be9daaeba94f6 | admin  |
+----------------------------------+--------+
```

图 13.1.7　查看角色的反馈结果

```
[root@controller ~]# openstack role show admin
[root@controller ~]# keystone role show e44a6ba4bff74cbe924be9daaeba94f6)
```

```
+-------------+----------------------------------+
| Field       | Value                            |
+-------------+----------------------------------+
| description | None                             |
| domain_id   | None                             |
| id          | e44a6ba4bff74cbe924be9daaeba94f6 |
| name        | admin                            |
+-------------+----------------------------------+
```

图 13.1.8　查看特定角色的反馈结果

查看角色的分配情况，可以通过 openstack role assignment list 查看角色授予的情况，包括角色、用户、组、项目以及域和系统。结果如图 13.1.9 所示。

```
[root@controller ~]# openstack role assignment list
```

```
+--------------------------------+--------------------------------+-------+--------------------------------+--------+--------+-----------+
| Role                           | User                           | Group | Project                        | Domain | System | Inherited |
+--------------------------------+--------------------------------+-------+--------------------------------+--------+--------+-----------+
| e44a6ba4bff74cbe924be9daaeba94f6 | 0165a8c49c4647118053a06ce2fbce68 |       | dab3559d05534ff9815746bb3bf20864 |        |        | False     |
| e44a6ba4bff74cbe924be9daaeba94f6 | 132a3f8ef8934480887436a802e3f808 |       | a823ad0ccc144975b5300e7c20ba3d7b |        |        | False     |
| e44a6ba4bff74cbe924be9daaeba94f6 | 39ef92e66e584bb8b8bdf509910bb5f9 |       | a823ad0ccc144975b5300e7c20ba3d7b |        |        | False     |
| e44a6ba4bff74cbe924be9daaeba94f6 | 577cfebf19cf4ba18ba3c6602d0db8b3 |       | a823ad0ccc144975b5300e7c20ba3d7b |        |        | False     |
| 9c34246705c445a3856bb53cd8cdfa00 | 8a5897353b7043e1a801d6a2397baf4d |       | 5804d6b60d854cf282a0d35ecf8b598c |        |        | False     |
| e44a6ba4bff74cbe924be9daaeba94f6 | cc6f950c440f46a5b215d67acf3580f9 |       | a823ad0ccc144975b5300e7c20ba3d7b |        |        | False     |
| e44a6ba4bff74cbe924be9daaeba94f6 | 0165a8c49c4647118053a06ce2fbce68 |       |                                |        | all    | False     |
+--------------------------------+--------------------------------+-------+--------------------------------+--------+--------+-----------+
```

图 13.1.9 查看角色分配情况的反馈结果

查看项目的操作和查看用户和域类似，可以通过 openstack project list 查看所有项目，也可以通过 openstack project show 查看单个项目的信息，命令如下，结果如图 13.1.10 和图 13.1.11 所示。

```
[root@controller ~]# openstack project list
+----------------------------------+---------+
| ID                               | Name    |
+----------------------------------+---------+
| 5804d6b60d854cf282a0d35ecf8b598c | demo    |
| a823ad0ccc144975b5300e7c20ba3d7b | service |
| dab3559d05534ff9815746bb3bf20864 | admin   |
+----------------------------------+---------+
```

图 13.1.10 查看项目的反馈结果

```
[root@controller ~]# openstack project show admin
```

或

```
[root@controller ~]# keystone project show dab3559d05534ff9815746bb3bf20864
+-------------+----------------------------------------------+
| Field       | Value                                        |
+-------------+----------------------------------------------+
| description | Bootstrap project for initializing the cloud.|
| domain_id   | default                                      |
| enabled     | True                                         |
| id          | dab3559d05534ff9815746bb3bf20864             |
| is_domain   | False                                        |
| name        | admin                                        |
| parent_id   | default                                      |
| tags        | []                                           |
+-------------+----------------------------------------------+
```

图 13.1.11 查看特定项目的反馈结果

步骤 2：熟悉云平台用户、用户组、角色、项目和域创建与删除的命令。

可以使用 openstack 加上一些参数来实现云平台用户、角色、项目和域的管理，为 OpenStack 服务提供认证服务。

系统登录时，除了要输入用户名和密码之外，首先需要输入的是域。下面使用 openstack domain create 创建域，如图 13.1.12 所示，由于系统里面已经有了 demo 和 default，所以创建一个不同名的，如 example，具体命令如下所示：

```
[root@controller ~]# openstack domain create example
+-------------+----------------------------------+
| Field       | Value                            |
+-------------+----------------------------------+
| description |                                  |
| enabled     | True                             |
| id          | 9acd032a835e4fb59ebe75be0a6e8d2e |
| name        | example                          |
| tags        | []                               |
+-------------+----------------------------------+
```

图 13.1.12 创建 example 域

然后在这个域下面创建一个 project 名为 myproject,如图 13.1.13 所示,具体命令如下:

```
[root@controller ~]# openstack project create --domain example myproject
```

```
+-------------+----------------------------------+
| Field       | Value                            |
+-------------+----------------------------------+
| description |                                  |
| domain_id   | 9acd032a835e4fb59ebe75be0a6e8d2e |
| enabled     | True                             |
| id          | 3bc320cc8bba4ee7b7782b56e4933046 |
| is_domain   | False                            |
| name        | myproject                        |
| parent_id   | 9acd032a835e4fb59ebe75be0a6e8d2e |
| tags        | []                               |
+-------------+----------------------------------+
```

图 13.1.13　创建 myproject 项目

首先,使用 keystone user-create 创建使用云平台的用户 user1、user2,两个用户创建时所跟的参数略有不同,首先 user1 用户创建时设定密码但不指定所属的项目,user2 用户创建时既设定密码也指定所属的项目 myproject,密码统一设置为"000000",如图 13.1.14 和图 13.1.15,两条命令如下所示。

```
[root@controller ~]# openstack user create user1 --password 000000
```

```
+---------------------+----------------------------------+
| Field               | Value                            |
+---------------------+----------------------------------+
| domain_id           | default                          |
| enabled             | True                             |
| id                  | e4d0692dcf1e4a74b53a8d55421d12ad |
| name                | user1                            |
| options             | {}                               |
| password_expires_at | None                             |
+---------------------+----------------------------------+
```

图 13.1.14　创建 user1 用户

```
[root@controller ~]# openstack user create user2 -password 000000 -project myproject -domain example
```

```
+---------------------+----------------------------------+
| Field               | Value                            |
+---------------------+----------------------------------+
| default_project_id  | 3bc320cc8bba4ee7b7782b56e4933046 |
| domain_id           | 9acd032a835e4fb59ebe75be0a6e8d2e |
| enabled             | True                             |
| id                  | b50a83755bbb4c85bf5a808c816926ca |
| name                | user2                            |
| options             | {}                               |
| password_expires_at | None                             |
+---------------------+----------------------------------+
```

图 13.1.15　创建 user2 用户

创建成功之后,由于没有给 user1 指定域,所以系统设置了一个默认的域 default,打开浏览器,在地址栏上输入云平台 Dashboard 登录界面的 url: http://192.168.100.10/Dashboard,使用用户 user1 登录,会发现,user1 因为没有分配角色,不能登录云平台,如图 13.1.16 所示。

把用户加入组,具体命令如下:

```
[root@controller ~]# openstack group add user test user1
[root@controller ~]# openstack group add user test user2
```

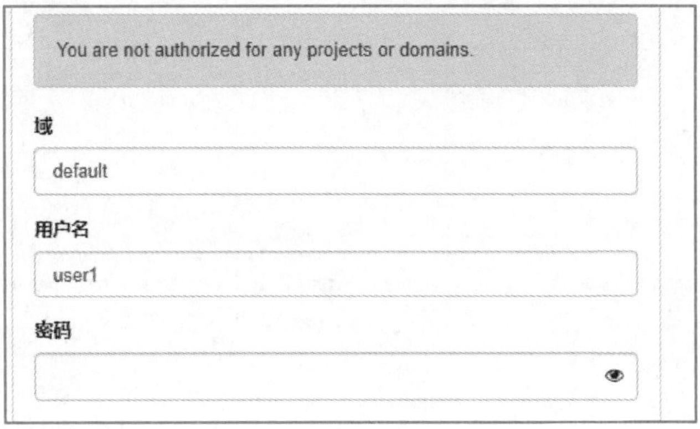

图 13.1.16　user2 用户登录失败提示

这样在 test 组中又多了两个用户 user1、user2。

给用户 user1、user2 授予 user 角色，具体命令如下：

```
[root@controller ~]# openstack role add user --user user1 --project demo
[root@controller ~]# openstack role add user --user user2 --project myproject
```

很多时候可以为用户组授予角色，其优点是设置了用户组的权限，所有用户也同时具备了相应的权限。

```
[root@controller ~]# openstack role add user --group test -project demo
```

接下来，使用 user1 用户、域 example 及其所对应的密码尝试登录云平台，会发现，由于 user1 默认属于域 default，系统也是提示"Invalid credentials"，如图 13.1.17 所示。

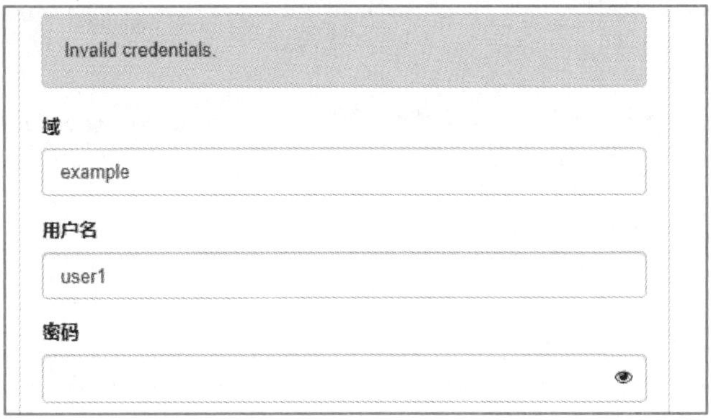

图 13.1.17　user2 用户登录失败提示

最后，使用 user1 用户、域 default、密码 000000 登录云平台，成功打开并使用平台中的资源，如图 13.1.18 所示。

从图 13.1.8 中可以发现，user1 登录后并没有管理的功能，因为 user1 用户并不具备管理的权限，如果想让 user1 成为和 admin 用户一样的管理员，可以进一步使用 keystone role add 命令，赋予 user1 用户 admin 角色的权限，具体命令如下。

```
[root@controller ~]# openstack role add admin -user user1 -project admin
```

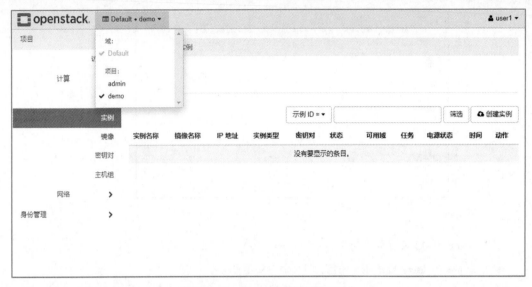

图 13.1.18　user1 用户成功登录云平台

执行完上述命令后，退出当前的登录，然后再次用 user1 用户登录云平台就会发现，左侧的菜单中多了"管理员"的项目，如图 13.1.19 所示，所以如果想让其他普通用户具备云平台的管理权限，只需要通过上述命令将用户加入 admin 组中并赋予 admin 角色的权限。

图 13.1.19　赋予 user1 用户 admin 角色权限后的登录界面

如果仅想让 user1 用户能使用云平台，则要使用 keystone user-role-remove 命令删除刚刚赋予 user1 用户 admin 角色的权限，其命令如下。

```
[root@controller ~]# keystone role remove --user user1 --project myproject
```

删除权限后使用之前的 role add 重新加上管理权限。然后通过 user1 用户登录后，选择 admin 项目，发现之前已经创建过了一个实例 cirros。而如果选择原来的 demo 项目，则看不到该实例。通过上述的操作，由于 user1 是属于 admin 项目和 demo 项目，可以看出，一个用户可以属于多个不同的项目，并赋予多个不同的角色权限。可以进一步推断，属于项目 1 的用户创建的实例，项目 2 当中的用户是无法访问并操作的，即使是用的同一个用户。

下面创建一个 user3，加入到这个 admin 项目，然后创建 demo 角色，并赋予 demo 角色权限，其命令如下。

```
[root@controller ~]# openstack user create user3 -password 000000 -project admin
[root@controller ~]# openstack role create demo
[root@controller ~]# openstack role add user --user demo --project admin
```

完成后，通过 user3 登录后，选择实例，会看到有个 cirros 实例，并没有管理员权限，如图 13.1.20 所示。

图 13.1.20　实例界面

过上述的操作，由于 user3 属于 admin 项目，可以看出，不同用户不同角色，只要属于同一个域，就能管理相同实例。

下面，可以尝试启动一个虚拟机实例，名称为 cirros，内存为 1GB，硬盘大小为 20GB，具体的启动过程在实训项 6 中已经详细介绍过，启动结果如图 13.1.21 所示。

图 13.1.21　启动名为 cirros 虚拟机实例

接下来，登录 user1，在云平台 Dashboard 界面上创建一个项目 project2，具体操作步骤如下。

登录成功后，选择 admin 项目，然后在左侧选择"身份管理"→"项目"选项，在右侧则可以看到之前创建的所有的项目，如图 13.1.22 所示。

图 13.1.22　从 Dashboard 界面查看项目信息

单击第 1 个"创建项目"按钮，出现创建项目的子页面，在"名称"栏中输入 project2，因为现在只创建一个 project2 项目，项目中暂时没有任何成员，所以不需要在"项目成员"和"项目组"两个子页中进行设置，直接单击"创建项目"按钮即可，如图 13.1.23 所示。

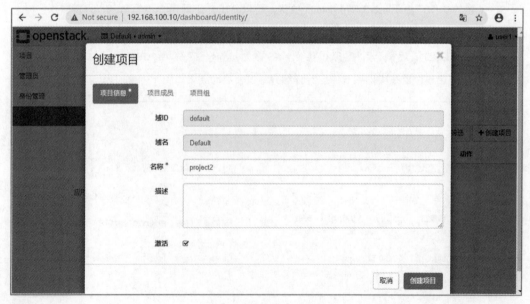

图 13.1.23　创建 project2 项目

此时,在"项目"页面中,可以看到 project2 的信息,如图 13.1.24 所示。

图 13.1.24　创建好的 project2 项目的信息

其后,创建一个用户 user4,属于项目 project2,并具有 admin 角色的权限,具体操作步骤如下:

在"身份管理"栏的"用户"选项中,可以看到之前创建的所有用户信息,如图 13.1.25 所示。

图 13.1.25　从 Dashboard 界面查看用户信息

单击"创建用户"按钮,在弹出的子页面中,将带"*"的栏目填写完整,用户名 user4,密码 000000,在"主项目"的下拉菜单中,选择 project2 选项,角色选择 admin,然后单击"创建用户"按钮即可,如图 13.1.26 所示。

此时,可以在如图 13.1.26 所示的用户界面看到刚刚创建的 user4 用户,在以上的操作过程中一共创建了 4 个用户,这 4 个用户可以列表稍做比较,今后在云平台的使用过程中,可以参考表 13.1.1 创建具有不同权限的用户。

图 13.1.26 从 Dashboard 界面查看用户信息

表 13.1.1 不同类的用户项目权限比较

用户名	所属域	所属项目	所属用户组	绑定角色	云平台登录使用权限	云平台管理权限
user1	default	demo,admin	test	user,admin	有	有
user2	example	myproject	test	user	有	无
user3	default	admin		demo	有	无
user4	default	project2		admin	有	有

与用户的创建类似，可以通过命令删除某个用户、项目或是角色。使用 user delete、project delete 和 role delete 三个 OpenStack 子命令，在后面直接跟上要删除的名称或 id 号。下面分别删除用户 user4、项目 project2 和角色 demo，具体操作命令如下。

```
[root@controller ~]# openstack user delete user4
```

或

```
[root@controller ~]# openstack user delete b3a3f99b728e490a9e1beeb66982da05
[root@controller ~]# openstack project delete project2
```

或

```
[root@controller ~]# openstack project delete 81d4a4c829c64553abefb6376123b002
[root@controller ~]# openstack role delete demo
```

或

```
[root@controller ~]# openstack role delete 8820becacc964196978a42fede49373f
```

上述操作没有任何返回值，可以分别通过 user list、project list 和 role list 查看操作后的结果。上述命令中，id 号可以通过前面提到的查看用户、项目，以及角色信息的命令进行查看。

为了不影响下面实验的使用，利用前面讲到的用户、项目和角色的创建命令将删除的用户 user4、项目 project2 和角色 demo 重新创建起来，具体操作参照步骤 2 中云平台用户、项目和角色的创建以及角色的分配。

删除用户和项目的操作还可以在云平台 Dashboard 界面上完成，同样需要管理员权限。

使用 admin 用户登录云平台，在左侧选择"身份管理"→"项目"选项，可以看到已经创建的所有项目，选中希望删除的项目名称前的复选框，单击界面右上角的"删除项目"按钮即可删除，同时也可以在"更多"按钮▼的下拉菜单中选择删除项目，如图 13.1.27 所示。

图 13.1.27　通过 Dashboard 界面删除项目

删除用户与删除项目方法一样，在"身份管理"→"用户"选项中进行操作即可，如图 13.1.28 所示。

步骤 3：使用命令以及云平台 Dashboard 界面进行用户、角色、项目和域的管理。

在云平台的使用过程中，新建的用户、角色、项目和域都处于默认的激活状态，如果想禁用某个用户，可以通过命令将其状态修改为非激活状态。例如，通过命令将 user4 的状态改为非激活状态，命令如下。

图 13.1.28　通过 Dashboard 界面删除用户

```
[root@controller ~]# openstack user set user4  --disable
```

　　需要注意，如果将一个用户的状态改为非激活状态，那么这个用户将不能登录使用云平台，此时使用 user4 登录云平台，将出现如图 13.1.29 所示的提示。

图 13.1.29　用户处于非激活状态无法登录 Dashboard 界面使用云平台

　　上述命令执行后 user4 将暂时不能登录使用或管理云平台，如果想恢复 user4 的激活状态，只需要执行上述命令将 disable 的参数改为 enable，命令如下。

```
[root@controller ~]# openstack user set user4 --enable
```

以上的操作同样可以在云平台 Dashboard 界面上完成，但前提条件仍然是需要具有管理员权限的用户登录云平台，具体操作步骤如下：

登录管理账号后，在左侧选择"身份管理"→"用户"选项，其右侧列出了之前创建的所有用户，如图 13.1.30 所示，默认情况下，所有用户的激活状态都为"Yes"，在用户信息右侧"更多"▼按钮下拉菜单中，有"禁用用户"选项，如图 13.1.30 所示。

图 13.1.30　Dashboard 界面中的"禁用用户"选项

在需要禁用的用户右侧单击▼按钮，在弹出的下拉列表中选择"禁用用户"选项即可，将 user4 用户禁用，禁用后的用户在以上界面上的激活状态变为了"No"，如图 13.1.31 所示。

图 13.1.31　Dashboard 界面中 user4 用户的状态改变

想要恢复被禁用的用户，只需使用具有管理权限的用户登录云平台，单击用户右侧的▼按钮，在弹出的下拉列表中选择"激活用户"选项即可。

同样，可以用命令修改项目的状态，例如，将名为 project2 的项目禁用，只需要将 project2 的状态改为非激活，命令如下。

```
[root@controller ~]# openstack project set project2 -disable
```

需要注意，如果将一个项目的状态改为非激活状态，那么这个项目下所有的用户将不能登录使用云平台。这里尝试使用属于 project2 项目的 user4 用户登录云平台，出现如图 13.1.32 所示的提示。

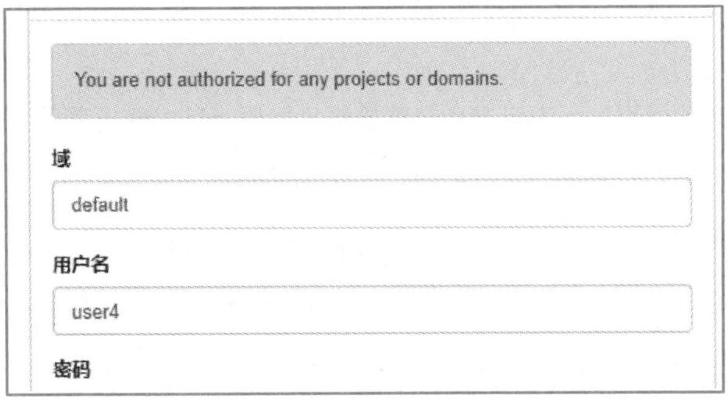

图 13.1.32　用户所属项目被禁用后该用户无法登录

如果一个用户同时属于两个或两个以上项目，其中一个项目的状态被改为非激活并不会影响这个用户登录并使用云平台，因为这个用户所属的其他项目仍然处于激活状态，除非同时将这个用户所属的其他所有项目禁用。禁用项目的操作同样可以在云平台 Dashboard 界面上完成，这里通过界面将项目 demo 禁用，具体如下。

使用具有管理员的用户登录云平台，在左侧选择"身份管理"→"项目"选项，在列表中选择想要禁用的项目，在其右侧单击▼按钮，在弹出的下拉列表中选择"编辑项目"选项，如图 13.1.33 所示。

图 13.1.33　通过 Dashboard 界面打开编辑项目的子页面

弹出一个编辑项目的子页面，默认情况下"激活"复选框是选中的，如图 13.1.34 所示。只需要取消选中复选框，然后单击"保存"按钮即可。

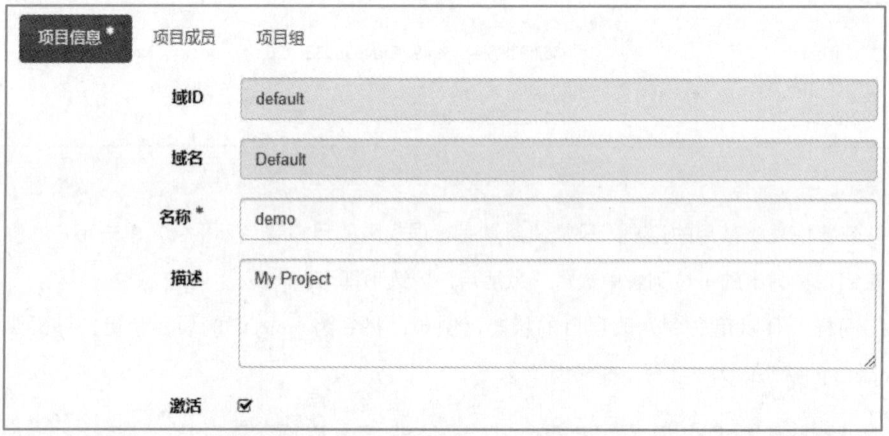

图 13.1.34　通过编辑项目的子页面禁用项目

禁用后，可以尝试使用属于 demo 的用户 user1 登录云平台，这时会发现登录操作依然成功，因为 user1 用户不仅仅属于项目 demo 还属于项目 admin，而项目 admin 并没有被禁用，所以得到如下结论，只有当一个用户所属的所有项目都被禁用时，该用户才不能登录云平台进行操作。

在使用云平台的过程中，不可避免地需要修改密码，可以使用命令 openstack user set 修改指定用户密码，例如将 user2 的密码修改为 123456，命令如下。

```
[root@controller ~]# openstack user set user2 --password 123456
```

修改后，成功使用新密码登录云平台，上述命令中，也可以使用"--password-prompt"参数，此时系统会出现"User Password"的提示，输入新密码，然后重复输入一次也可以完成密码修改。

同样，上述命令也可以在云平台中进行修改，使用 user2 登录云平台，单击界面右上角用户名右侧的▼下拉按钮，在弹出的下拉列表中选择"设置"选项，如图 13.1.35 所示，进入用户设置子页面。

图 13.1.35　进入用户设置子页面

在窗口左侧选择"修改密码"选项后，在右侧可以看到修改密码的内容，输入当前密码，新密码，然后确认新密码后，单击界面右下角的"修改"按钮，如图 13.1.36 所示。

图 13.1.36　通过用户设置子界面修改当前用户密码

例如，将 user2 用户的密码改回 000000，密码修改完成后，系统会自动跳到云平台登录界面，要求用户重新登录，如图 13.1.37 所示。

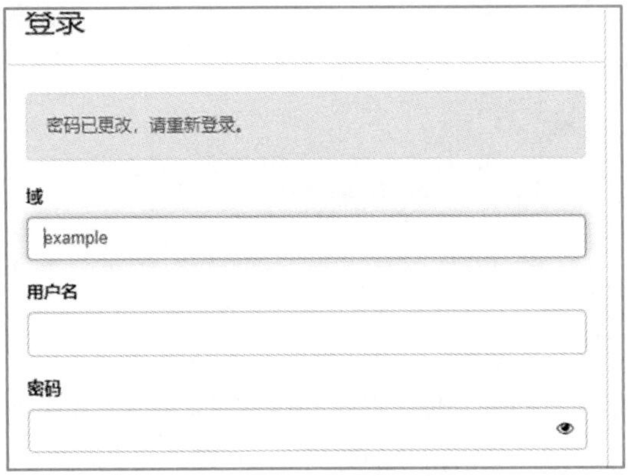

图 13.1.37　修改密码后提示重新登录

此时，输入刚刚修改过密码的 user2 和其新密码，可以成功登录云平台。

上述在云平台操作的内容为普通用户修改自己密码的过程，而管理员则拥有修改所有用户密码的权限。使用拥有管理员权限的 admin 用户登录云平台，在左侧选择"身份管理"→"用户"选项，在右侧列表中选择需要修改密码的用户，单击其右侧的"更多"▼按钮，在弹出的下拉列表中选择"修改密码"选项，如 user4 用户，如图 13.1.38 所示。

图 13.1.38　在界面中找到 user4 用户

在弹出的"修改密码"子页面中，输入新密码，然后确认密码，单击"保存"按钮即可修改密码，如图 13.1.39 所示。修改完成后，会出现修改成功的提示。

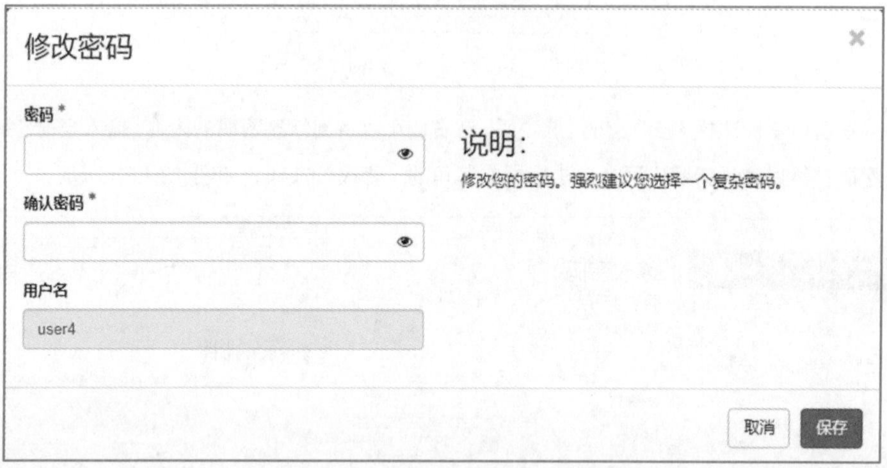

图 13.1.39　通过"更新用户"子界面修改用户密码

用户组是一个域中部分用户的集合。用户组不存在激活还是未激活，主要作用是可以批量对组中用户进行授权操作。使用管理员登录云平台，在左侧选择"身份管理"→"组"选项后，在其右侧不但可以创建、删除、编辑组，还可以管理组成员，如图 13.1.40 所示。

图 13.1.40　用户组管理

角色如果要生效，必须映射到每个 OpenStack 服务配置文件目录下的 policy.json 文件中。默认策略是给予 admin 角色大部分服务的管理访问权限。

步骤 4：云平台服务和端点的查看和修改。

服务和端点的创建在本书第二部分的实训项目中每个组件的安装都用到过，这里不再赘述。可以通过 openstack service list 命令查看 OpenStack 的服务和其对应的端点，命令如下，结果如图 13.1.41 所示。

```
[root@controller ~]# openstack service list
```

```
+----------------------------------+-----------+-----------+
| ID                               | Name      | Type      |
+----------------------------------+-----------+-----------+
| 14a1dd7eb7bc434eb706db5b639f15ed | glance    | image     |
| 8248dee59ccd4154b3e003468726d45d | keystone  | identity  |
| 9d228582eacc4b7b883c3ca08285607a | nova      | compute   |
| e1730e94a0d44b6898240a79474788c2 | neutron   | network   |
| e2776b886fac4eada8a76ddf0f5ce669 | placement | placement |
+----------------------------------+-----------+-----------+
```

图 13.1.41　查看服务命令的反馈结果

如果需要查看指定服务的具体信息，可以通过 openstack service show 命令查看。如果想查看 keystone 服务的具体信息，只要在上述命令之后直接跟 keystone 服务名或 ID，命令如下，其结果如图 13.1.42 所示。

```
[root@controller ~]# openstack service show keystone
（或 [root@controller ~]# openstack service show 8248dee59ccd4154b3e003468726d45d）
```

```
+---------+----------------------------------+
| Field   | Value                            |
+---------+----------------------------------+
| enabled | True                             |
| id      | 8248dee59ccd4154b3e003468726d45d |
| name    | keystone                         |
| type    | identity                         |
+---------+----------------------------------+
```

图 13.1.42　查看服务具体信息命令的反馈结果

如果要查看具体某个服务端点的信息，可以通过 openstack endpoint list --service 来查看，可以跟服务名称或者端点 ID。具体命令如下，其结果如图 13.1.43 所示。

```
[root@controller ~]# openstack endpoint list --service keystone
（或 [root@controller ~]# openstack endpoint list --service 8248dee59ccd4154b3e003468726d45d）
```

```
+----------------------------------+-----------+--------------+--------------+---------+-----------+---------------------------+
| ID                               | Region    | Service Name | Service Type | Enabled | Interface | URL                       |
+----------------------------------+-----------+--------------+--------------+---------+-----------+---------------------------+
| 28b19f4b006d4e7f8a18bd4298e034c4 | RegionOne | keystone     | identity     | True    | public    | http://controller:5000/v3/|
| 4993ef92e6454baebc611c1f77c9e833 | RegionOne | keystone     | identity     | True    | internal  | http://controller:5000/v3/|
| 961e21645360460697a94196c5ed8bc5 | RegionOne | keystone     | identity     | True    | admin     | http://controller:5000/v3/|
+----------------------------------+-----------+--------------+--------------+---------+-----------+---------------------------+
```

图 13.1.43　查看某一特定服务的端点信息反馈结果

如果要查看某个端点的具体信息，可以通过 openstack endpoint show 命令来查看，可以跟端点 ID，具体命令如下，其结果如图 13.1.44 所示。

同时，服务和端点的删除和用户、项目、角色的删除方式一样，需要注意，在删除服务时，参数可以是服务的名称或 id，命令如下。

```
[root@controller ~]# openstack service delete glance
（或 [root@controller ~]# openstack service delete glance 14a1dd7eb7bc434eb706db5b639f15ed）
```

```
| Field        | Value                            |
| enabled      | True                             |
| id           | 28b19f4b006d4e7f8a18bd4298e034c4 |
| interface    | public                           |
| region       | RegionOne                        |
| region_id    | RegionOne                        |
| service_id   | 8248dee59ccd4154b3e003468726d45d |
| service_name | keystone                         |
| service_type | identity                         |
| url          | http://controller:5000/v3/       |
```

图 13.1.44　查看某一特定端点的反馈结果

但删除端点时，参数只能是端点的 id，所以先要通过查询命令查询对应端点的 id 后才能将其删除，例如，删除 glance 服务下某个端点，其命令如下。

```
[root@controller ~]# openstack endpoint delete 2bb539a9f30d420d9076cf9cf6df071e
```

注意，为不影响后面实验的正常使用，删除后，再根据前面的方法分别创建 glance 服务和其端点。

13.2　实训项目 12　Glance 基本运维命令及其应用

1. 实训前提环境

电子教案　Glance 基本运维命令及其应用

本地主机安装终端软件，通过终端软件连接到安装成功的 OpenStack 云平台中的控制节点，通过命令进行运维操作。本地主机需安装浏览器，推荐使用谷歌浏览器或者火狐浏览器，通过浏览器登录云平台 Dashboard 界面进行运维的操作。本地主机准备最小化的 cirros-0.3.1-x86_64-disk.img 镜像文件用于本实训项目中镜像上传以及修改的实验。

PPT　Glance 基本运维命令及其应用

2. 实训涉及节点

controller 节点

3. 实训目标

① 熟悉镜像的查看命令。

② 能够通过命令上传镜像。

③ 能够通过命令以及 Dashboard 界面修改已上传镜像的有关参数。

微课 15　Glance 基本运维命令及其应用

④ 能够通过命令以及 Dashboard 界面下载云平台中的镜像。

⑤ 能够通过命令以及 Dashboard 界面删除云平台中的镜像。

4. 实训内容

步骤 1：查看镜像和其详细信息。

在使用 Glance 命令前首先要执行环境变量的文件，之后查看当前云平台上目前已有镜像的信息，命令如下，结果如图 13.2.1 所示。

```
[root@controller ~]# openstack image list
```

13.2 实训项目 12　Glance 基本运维命令及其应用

```
| ID                                   | Name   | Status |
| 2d698a77-ffb8-4a42-8890-1b839f67f55e | cirros | active |
```

图 13.2.1　查看云平台镜像的反馈结果

从反馈信息中可以看到目前平台上有一个镜像，这是之前上传用于检验 Glance 服务是否正常运行，同时也是为虚拟机实例创建提供的可用镜像。图 13.2.1 中第 1 列的 ID 字段表明每个用户都有一个唯一的与之相对应的 ID 值，第 2 列 Name 字段显示当前已存在镜像的镜像名，这是用户在上传时定义的名称。第 3 列 Status 字段的值表明所对应的镜像状态是否可用，显然，"active"为可用状态，当变为"inactive"时表示为不可用状态，此时是无法用状态为"inactive"的镜像创建实例的。

可以进一步用 openstack image show 命令查看某一特定镜像的详细信息，例如，想查看已经上传的这个 cirros 镜像的详细信息，可以在上述命令之后直接跟上 cirros 镜像名，或直接跟上 cirros 镜像所对应的 ID，具体命令如下，反馈结果如图 13.2.2 所示。

[root@controller ~]# openstack image show cirros
（或 [root@controller ~]# openstack image show 2d698a77-ffb8-4a42-8890-1b839f67f55e）

```
| Field            | Value                                                              |
| checksum         | 443b7623e27ecf03dc9e01ee93f67afe                                   |
| container_format | bare                                                               |
| created_at       | 2019-11-03T08:11:38Z                                               |
| disk_format      | qcow2                                                              |
| file             | /v2/images/2d698a77-ffb8-4a42-8890-1b839f67f55e/file               |
| id               | 2d698a77-ffb8-4a42-8890-1b839f67f55e                               |
| min_disk         | 0                                                                  |
| min_ram          | 0                                                                  |
| name             | cirros                                                             |
| owner            | dab3559d05534ff9815746bb3bf20864                                   |
| properties       | os_hash_algo='sha512', os_hash_value='6513f21e44aa3da349f248188a44bc304a3653a0... |
| protected        | False                                                              |
| schema           | /v2/schemas/image                                                  |
| size             | 12716032                                                           |
| status           | active                                                             |
| tags             |                                                                    |
| updated_at       | 2019-11-03T08:11:38Z                                               |
| virtual_size     | None                                                               |
| visibility       | public                                                             |
```

图 13.2.2　查看某一特定镜像的反馈结果

从反馈结果中可以看到一些更详细的信息。Container_format 字段是镜像的容器格式，这里定义的是 bare，OpenStack 也支持其他容器格式（如 docker、ova、ovf 等），具体可以通过命令 openstack image create --help 进行查看。尽管现在定义 Container Format 并没有任何的 OpenStack 服务使用，但该字段是必需的，同时定义为 bare 也是安全的。Created_at 字段表示上传时间。Disk_format 字段显示的是上传镜像的磁盘格式，OpenStack 推荐使用 qcow2，它也支持其他格式（如 ami、ari、vhd、vmdk、raw、vhdx、vdi、iso 等），具体可以通过命令 openstack image create --help 进行查看。File 字段显示文件路径，实际存储路径在 /var/lib/glance/images 下。Owner 字段表示所属项目的 id。Size 字段显示的是上传镜像大小。Updated_at 字段表示更改时间。

步骤 2：云平台镜像的上传。

在步骤 1 中，已经查看过前面配置 Glance 服务时上传过的镜像，下面通过使用一些参数来实现云平台镜像的上传。这里使用 cirros 镜像文件作为测试镜像来为例来进行详细的讲解。cirros 由于镜像文件很小，非常适合做实验，可以通过 wget 去 download.cirros-cloud.net 下载最新的 cirros 镜像。同时使用前面 keystone 服务部分创建的普通用户和 admin 用户进行对比。

首先，在 root 用户中下载 cirros 镜像，然后通过 openstack image create 命令上传一个除必要参数外其他都默认的镜像，并命名为 cirros-d190515，命令如下，其反馈结果如图 13.2.3 所示。

```
[root@controller ~]# wget https://download.cirros-cloud.net/contrib/nowster/20190515/cirros-d190515-x86_64-disk.img
[root@controller ~]# openstack image create cirros-d190515 --file cirros-d190515-x86_64-disk.img --disk-format qcow2 --container-format bare
```

```
+------------------+------------------------------------------------------+
| Field            | Value                                                |
+------------------+------------------------------------------------------+
| checksum         | 58ff5732c5ac290a70d7977d4b542fd2                     |
| container_format | bare                                                 |
| created_at       | 2019-12-02T02:58:23Z                                 |
| disk_format      | qcow2                                                |
| file             | /v2/images/da18f4c8-3131-4fb7-8b0a-015ec1605010/file |
| id               | da18f4c8-3131-4fb7-8b0a-015ec1605010                 |
| min_disk         | 0                                                    |
| min_ram          | 0                                                    |
| name             | cirros-d190515                                       |
| owner            | dab3559d05534ff9815746bb3bf20864                     |
| properties       | os_hash_algo='sha512', os_hash_value='78948fbe9a335fbcc4db752f2822b4d48e |
| protected        | False                                                |
| schema           | /v2/schemas/image                                    |
| size             | 14222336                                             |
| status           | active                                               |
| tags             |                                                      |
| updated_at       | 2019-12-02T02:58:24Z                                 |
| virtual_size     | None                                                 |
| visibility       | shared                                               |
+------------------+------------------------------------------------------+
```

图 13.2.3　上传镜像的进度指示和上传后的反馈结果

从反馈信息中可以看到默认情况下，上传的镜像还是属于 admin 项目（owner 值是 admin 项目的 ID，可以通过前面所学命令查询 admin 项目的 ID 进行验证），同时默认情况下上述命令不加任何参数上传后的镜像是共享的和不受保护的（图 13.2.3 中 visibility 的值为 shared，protected 的值为 False），可以使用 admin 用户（该用户属于 admin 项目并具有 admin 角色的权限）通过浏览器登录 Dashboard 界面更直观地查看云平台中已上传的镜像，如图 13.2.4 所示。

从图 13.2.4 中可以看到，在通过命令行新添加了 cirros-d190515 镜像后，admin 用户可以看到该镜像，因为 cirros-d190515 默认是属于 admin 项目的，admin 用户也是属于 admin 项目的，所以在"镜像"选项中有两个镜像。然而 cirros-d190515 的"可见性"参数是 shared 即是共享的，而之前实训项目中上传的 cirros 镜像是公共的。

下面再上传一个名为 ownerdemo 的镜像，其磁盘格式和容器格式都不变，visibility 设置为 public，protected 设置为 True，同时定义其属于 demo 项目，命令如下，其反馈结果如图 13.2.5 所示。

```
[root@controller ~]# openstack image create ownerdemo --file cirros-d190515-x86_64-disk.img --disk-format qcow2 --container-format bare --protected --public --project 5804d6b60d854cf282a0d35ecf8b598c
```

13.2 实训项目 12　Glance 基本运维命令及其应用

镜像

所有者	名称	类型	状态	可见性	受保护的	
admin	cirros	镜像	运行中	公有	否	启动
admin	cirros-d190515	镜像	运行中	共享的	否	启动

图 13.2.4　Dashboard 界面查看已上传镜像

以上命令中 --project 后跟 demo 项目的 id，可以通过 openstack project list 命令来查询。

```
| Field            | Value                                                          |
| checksum         | 58ff5732c5ac290a70d7977d4b542fd2                               |
| container_format | bare                                                           |
| created_at       | 2019-12-02T05:35:04Z                                           |
| disk_format      | qcow2                                                          |
| file             | /v2/images/3347a41a-85d9-4998-8633-1f79d9a545d9/file            |
| id               | 3347a41a-85d9-4998-8633-1f79d9a545d9                           |
| min_disk         | 0                                                              |
| min_ram          | 0                                                              |
| name             | ownerdemo                                                      |
| owner            | 5804d6b60d854cf282a0d35ecf8b598c                               |
| properties       | os_hash_algo='sha512', os_hash_value='78948fbe9a335fbcc4db752f2822 |
| protected        | True                                                           |
| schema           | /v2/schemas/image                                              |
| size             | 14222336                                                       |
| status           | active                                                         |
| tags             |                                                                |
| updated_at       | 2019-12-02T05:35:04Z                                           |
| virtual_size     | None                                                           |
| visibility       | public                                                         |
```

图 13.2.5　上传镜像 ownerdemo 的反馈结果

为了清晰地比较观察 visibility 和 protected 参数变化的区别，分别用属于 admin 项目的 admin 用户和属于 demo 项目的 demo 用户登录 Dashboard 进行查看。用 admin 用户登录后进入 admin 项目，看到的镜像如图 13.2.6 所示。

用 admin 用户登录后可以发现有 3 个镜像，在云平台的"项目"选项中可以看到属于当前登录用户所在项目的所属镜像和所有可见性为 public 的镜像。由于 ownerdemo 镜像 visibility 参数是 public，所以可以看到刚刚创建的 ownerdemo 镜像。ownerdemo 镜像的 protected 参数是 True 即为受保护的，在云平台中对于属性为公开但处于受保护状态的镜像，只有同属于该镜像所属项目的成员用户可以对其进行编辑，而非该镜像所属项目的用户只可以使用运行该镜像，但不可以编辑。

然后在"管理员"栏中的镜像选项中，可以看到上传的 ownerdemo 镜像，如图 13.2.7 所示，通过与"项目"栏中看到的镜像比较，可以得知处于受保护状态的镜像即使是拥有管理员权限的用户依然无法直接对镜像执行删除操作。若要执行删除操作，只要先去除保护即可。

图 13.2.6　admin 用户 admin 项目下镜像子页面中的镜像列表

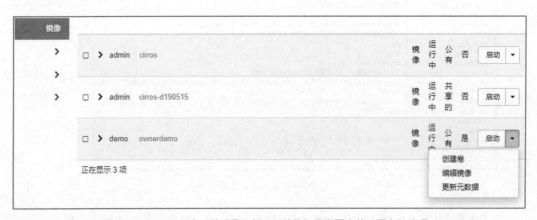

图 13.2.7　admin 用户"管理员"栏下"镜像"子页面中的"更多"选项

作为对比，再用 demo 用户重新登录云平台，如图 13.2.8 所示。可以发现 demo 用户的项目选项中出现了刚刚上传的属于 demo 项目的 ownerdemo 镜像，与上面 admin 用户登录的图 13.2.6 相比较会发现少了个 admin 项目"共享"的镜像，因为共享是指项目内部都可以看到的。另外在"更多"选项里面，依然不能删除镜像，只能通过"编辑镜像"修改 protected 属性为 false。

同时可以发现，在 demo 用户界面，无法看到默认创建属于 admin 项目的 def 镜像，在上述实验中，比较了 cirros、cirros-d190515 和 ownerdemo 这 3 个参数不同的镜像，同时通过分别使用 admin 和 demo 用户登录 Dashboard 的方式，直观对比了 owner、visibility 和 protected 三个参数对用户和镜像带来的影响，可以得到以下两条结论：

① 在 Dashboard 镜像栏中，如果选择"项目"选项，只能看到属于当前用户所属项目的镜像和所有公共镜像。

13.2 实训项目 12 Glance 基本运维命令及其应用

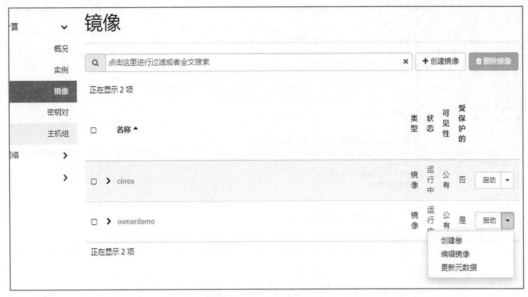

图 13.2.8 demo 用户"项目"栏下镜像子页面中的镜像列表

② 对于属性为公开但处于受保护状态的镜像,镜像所属项目的成员用户可以对其进行编辑,而非镜像所属项目的用户只可以使用运行,对于拥有管理员权限的用户来说,只有在执行管理员权限时才能对非本项目且处于受保护的镜像进行编辑。

步骤 3:云平台镜像的修改。

通过对步骤 2 中上传镜像的修改,讲解如何使用 openstack image set 这个修改已经上传镜像的命令。可以通过 openstack image set 命令来查看这个命令的具体使用方法。以下主要讲解 --name、--project、--disk-format、--container-format、--public、--protected 这 6 个参数的修改。

首先,通过 openstack image list 命令查看已经上传的镜像,这里为了可以通过 admin 用户查看所有已经上传的镜像。命令如下,反馈信息如图 13.2.9 所示。

```
[root@controller ~]# openstack image list
```

```
| ID                                   | Name          | Status |
| 2d698a77-ffb8-4a42-8890-1b839f67f55e | cirros        | active |
| da18f4c8-3131-4fb7-8b0a-015ec1605010 | cirros-d190515| active |
| 3347a41a-85d9-4998-8633-1f79d9a545d9 | ownerdemo     | active |
```

图 13.2.9 查询所有镜像的反馈信息

以修改 ownerdemo 镜像的参数为例,首先看一下 ownerdemo 镜像的详细信息,命令如下,反馈信息如图 13.2.10 所示。

```
[root@controller ~]# openstack image show ownerdemo
 (或 [root@controller ~]# glance image-show  3347a41a-85d9-4998-8633-1f79d9a545d9)
```

从图 13.2.10 中的反馈信息可以看出没做过任何修改的镜像 created_at 和 updated_at 的值是一样的。

openstack image set 命令的语法格式如下所示。

```
| Field            | Value                                                                        |
+------------------+------------------------------------------------------------------------------+
| checksum         | 58ff5732c5ac290a70d7977d4b542fd2                                             |
| container_format | bare                                                                         |
| created_at       | 2019-12-02T05:35:04Z                                                         |
| disk_format      | qcow2                                                                        |
| file             | /v2/images/3347a41a-85d9-4998-8633-1f79d9a545d9/file                         |
| id               | 3347a41a-85d9-4998-8633-1f79d9a545d9                                         |
| min_disk         | 0                                                                            |
| min_ram          | 0                                                                            |
| name             | ownerdemo                                                                    |
| owner            | 5804d6b60d854cf282a0d35ecf8b598c                                             |
| properties       | os_hash_algo='sha512', os_hash_value='78948fbe9a335fbcc4db752f2822b4d48e218  |
| protected        | True                                                                         |
| schema           | /v2/schemas/image                                                            |
| size             | 14222336                                                                     |
| status           | active                                                                       |
| tags             |                                                                              |
| updated_at       | 2019-12-02T05:35:04Z                                                         |
| virtual_size     | None                                                                         |
| visibility       | public                                                                       |
```

图 13.2.10 查询 ownerdemo 镜像的反馈信息

```
usage: openstack image set [optional arguments] <IMAGE>
```

其中 [optional arguments] 是它支持的一些可选参数，<IMAGE> 代指需要修改的镜像的名字和 ID。

将 ownerdemo 镜像的名字修改为 ownerdemo2，将 visibility 的值修改为 shared，将 protected 的值修改为 False。命令如下，反馈信息如图 13.2.11。

```
[root@controller ~]# openstack image set ownerdemo --name ownerdemo2 --shared --unprotected
[root@controller ~]# openstack image show ownerdemo2
```

```
| Field            | Value                                                                        |
+------------------+------------------------------------------------------------------------------+
| checksum         | 58ff5732c5ac290a70d7977d4b542fd2                                             |
| container_format | bare                                                                         |
| created_at       | 2019-12-02T05:35:04Z                                                         |
| disk_format      | qcow2                                                                        |
| file             | /v2/images/3347a41a-85d9-4998-8633-1f79d9a545d9/file                         |
| id               | 3347a41a-85d9-4998-8633-1f79d9a545d9                                         |
| min_disk         | 0                                                                            |
| min_ram          | 0                                                                            |
| name             | ownerdemo2                                                                   |
| owner            | 5804d6b60d854cf282a0d35ecf8b598c                                             |
| properties       | os_hash_algo='sha512', os_hash_value='78948fbe9a335fbcc4db752f2822b4d48e2187e|
| protected        | False                                                                        |
| schema           | /v2/schemas/image                                                            |
| size             | 14222336                                                                     |
| status           | active                                                                       |
| tags             |                                                                              |
| updated_at       | 2019-12-02T06:55:46Z                                                         |
| virtual_size     | None                                                                         |
| visibility       | shared                                                                       |
```

图 13.2.11 修改 ownerdemo 镜像参数后的反馈信息

从图 13.2.12 的反馈信息中可以看到已经被修改的内容，name 发生了更新，同时可以看到修改时间（updated_at）已经发生变化，创建时间（created_at）还是一样，同样，修改后的内容可以通过 demo 用户登录 Dashboard 界面，在 Dashboard 界面上看到修改后的镜像有何变化。

从图 13.2.12 可以看到，ownerdemo2 镜像由公共变为共享，同时 ownerdemo2 镜像不再为受保护的状态，随后的"更多"按钮中多了一个"删除镜像"的功能。

13.2 实训项目 12 Glance 基本运维命令及其应用

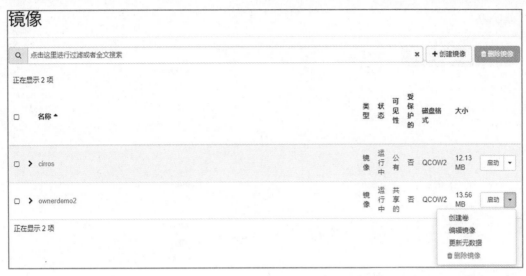

图 13.2.12　通过 Dashboard 界面查看修改后的镜像

在 Dashboard 上选择如图 13.2.13 所示的镜像名称 ownerdemo2，即可查看镜像的详细信息，如图 13.2.13 所示。

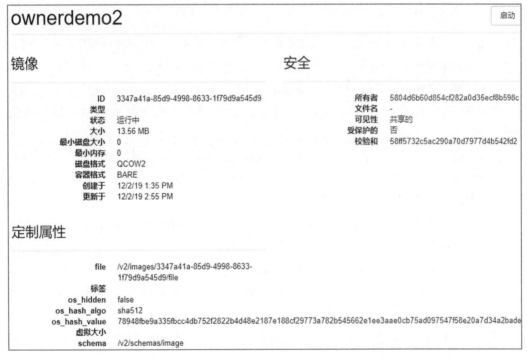

图 13.2.13　Dashboard 界面查看镜像的详细信息

当然上述修改镜像参数的命令在 Dashboard 界面也可以完成，在 Dashboard 上单击如图 13.2.12 中所示的"编辑"按钮，打开如图 13.2.14 所示的上传及修改镜像子页面。

目前在 Dashboard 上只能修改镜像的名称，以及修改镜像是否公有、是否受保护，修改后单击"更新镜像"按钮即可。

图 13.2.14 上传以及修改镜像子页面

目前版本中，镜像其他参数还需要依赖命令修改，如所在的项目和磁盘格式等。由于磁盘格式只有在没有上传镜像文件时修改，也就是处于 queued 状态。所以，新建一个不带镜像文件的镜像 ownerdemo，然后将 ownerdemo 镜像的所属项目修改为 user4 用户所在的 project2 项目，同时将磁盘格式（disk_format）修改为 raw 格式。命令如下，反馈信息如图 13.2.15 所示。

```
[root@controller ~]# openstack image create ownerdemo --disk-format qcow2 --container-format bare --protected --public --project 5804d6b60d854cf282a0d35ecf8b598c
[root@controller ~]# openstack image set ownerdemo --project 81d4a4c829c64553abefb6376123b002 --disk-format raw --private
```

以上命令中 --project 后的 ID 值 81d4a4c829c64553abefb6376123b002 为 project2 项目的 ID，可以通过之前学习到的命令查询得到该 ID。

需要注意，由于 ownerdemo 镜像可见性是私有，属于 project2 项目，并不属于 admin 项目，在 demo 用户登录后也不会看到 ownerdemo 镜像。使用 user4 用户登录 Dashboard，如图 13.2.16 和图 13.2.17 所示。

从两张图中可以清晰地看到，user4 用户的"项目"选项中出现了 ownerdemo 镜像，由于没有上传任何镜像文件，所以没有操作按钮。同时从详细信息可以看到，镜像的格式由 qcow2 变为了 RAW。

步骤 4：云平台镜像的下载和删除。

云平台中不仅能可以上传新镜像，同时还可以将已经上传到云平台上的镜像下载到本地，将前面步骤中上传的 ownerdemo2 镜像下载到本地并保存到 image-ownerdemo2 的文件中，命令如下。

```
+------------------+----------------------------------------------------------+
| Field            | Value                                                    |
+------------------+----------------------------------------------------------+
| checksum         | None                                                     |
| container_format | bare                                                     |
| created_at       | 2019-12-02T14:26:19Z                                     |
| disk_format      | raw                                                      |
| file             | /v2/images/f96663d3-d249-4888-85cb-f51b595c98f4/file     |
| id               | f96663d3-d249-4888-85cb-f51b595c98f4                     |
| min_disk         | 0                                                        |
| min_ram          | 0                                                        |
| name             | ownerdemo                                                |
| owner            | 81d4a4c829c64553abefb6376123b002                         |
| properties       | os_hash_algo='None', os_hash_value='None', os_hidden='False' |
| protected        | True                                                     |
| schema           | /v2/schemas/image                                        |
| size             | None                                                     |
| status           | queued                                                   |
| tags             |                                                          |
| updated_at       | 2019-12-02T14:37:26Z                                     |
| virtual_size     | None                                                     |
| visibility       | private                                                  |
+------------------+----------------------------------------------------------+
```

图 13.2.15 在此修改镜像 ownerdemo 后的反馈信息

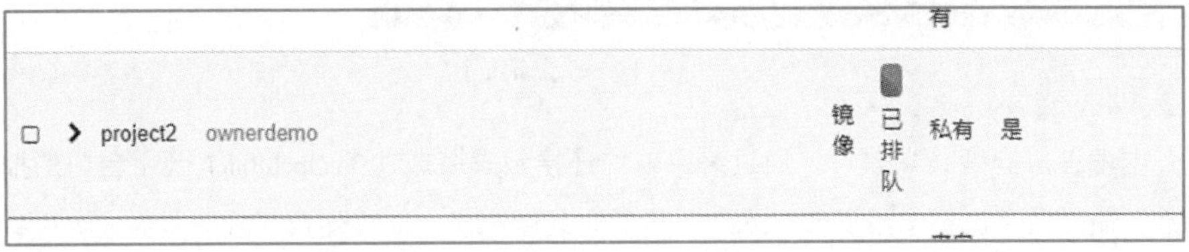

图 13.2.16 user4 用户登录后项目下的镜像列表

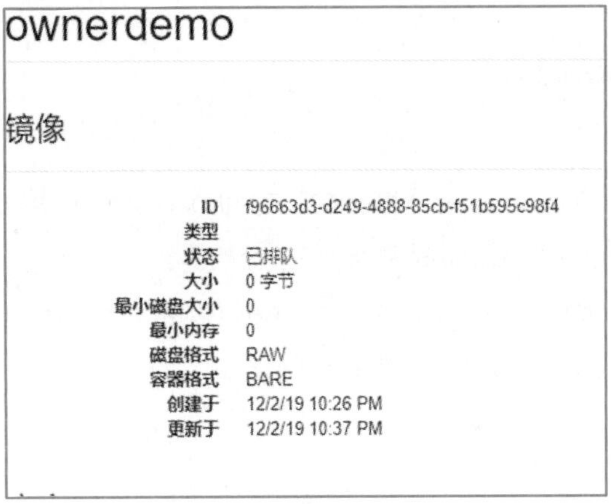

图 13.2.17 user4 用户登录后 ownerdemo 的详细信息

```
[root@controller ~]# openstack image save ownerdemo2 --file image-ownerdemo2
```

在上述 --file 参数后跟的是下载完成后本地存储镜像的镜像名，该镜像名之前还可以加上完整的存储路径，若不加，该镜像默认保存在当前路径下。下载完成后没有返回结果，可以到下载文件存放的目录下查看。

在前面几个步骤中，在 Dashboard 中可以看到处于不受保护状态的镜像可以直接被镜像所属组成员用户和拥有管理员权限的用户在管理员镜像界面删除。对于处于保护状态的镜像，则需要先通过"编辑"按钮取消受保护的状态，然后才可以删除指定镜像。

如果通过命令删除镜像，对于处于受保护状态的镜像，需要先修改其 protected 参数值为 False，然后才可以执行删除命令，下面在命令中删除 ownerdemo 镜像，命令如下。

```
[root@controller ~]# openstack image set ownerdemo -unprotected
[root@controller ~]# openstack image delete ownerdemo
```

执行命令后没有任何返回结果,可以通过 openstack image list 命令确认该镜像是否被删除,命令如下,确认结果如图 13.2.18 所示。

```
[root@controller ~]# openstack image list
```

ID	Name	Status
2d698a77-ffb8-4a42-8890-1b839f67f55e	cirros	active
da18f4c8-3131-4fb7-8b0a-015ec1605010	cirros-d190515	active
3347a41a-85d9-4998-8633-1f79d9a545d9	ownerdemo2	active

图 13.2.18 确认镜像 demo2pfalse 是否被删除

从反馈信息中已经看不到 ownerdemo 镜像了,删除成功。

13.3 实训项目 13 Nova 基本运维命令及其应用

1. 实训前提环境

本地主机安装终端软件,通过终端软件连接到安装成功的 OpenStack 云平台中的控制节点,通过命令进行运维操作。本地主机需安装浏览器,推荐使用谷歌浏览器或者火狐浏览器,通过浏览器登录云平台 Dashboard 界面进行运维的操作。

2. 实训涉及节点

controller 节点和 compute 节点

3. 实训目标

① 掌握通过命令以及 Dashboard 界面管理云主机(实例)类型模板。
② 掌握通过命令以及 Dashboard 界面管理安全组。
③ 掌握通过命令以及 Dashboard 界面对云主机(实例)创建和管理。
④ 掌握通过命令以及 Dashboard 界面管理配额。

4. 实训内容

步骤 1:云主机(实例)类型模板(flavor)的管理。

在实训项目 10 中,已经成功通过 admin 用户登录 Dashboard 界面并创建实例。在创建实例的界面可以直观地了解到创建实例的要求,如图 13.3.1 所示。

从图 13.3.1 中可以知道启动一个实例需要指明"可用域""实例名称""镜像源""实例类型"等参数,同时还需要定义"网络""安全组"两部分参数。启动云主机还可以用命令来完成,下面讲解如何通过相关命令启动实例。在命令中启动实例,命令的基本格式如下。

```
# openstack server start [-h] <server> [<server> ...]
```

注意,用命令启动实例有许多可选参数可以被指定,但上述命令是一个基础命令。server 参数后需要指明使用镜像的名称或者是 ID。

13.3 实训项目 13 Nova 基本运维命令及其应用

图 13.3.1 创建实例界面

此外，可以使用 openstack flavor 命令进行查看主机类型模板，命令如下，其反馈信息如图 13.3.2 所示。

```
[root@controller ~]# openstack flavor list
```

```
+--------------------------------------+-------+------+------+-----------+-------+-----------+
| ID                                   | Name  | RAM  | Disk | Ephemeral | VCPUs | Is Public |
+--------------------------------------+-------+------+------+-----------+-------+-----------+
| 98ce29c5-8963-4db3-a19e-6a8aa9ece574 | small | 1024 |   20 |         0 |     1 | True      |
+--------------------------------------+-------+------+------+-----------+-------+-----------+
```

图 13.3.2 查看主机类型模板的反馈信息

从反馈的信息可知，系统中有 1 个实例类型模板可供选择。第 3 列"RAM"字段从名称上可以知道表示的是实例内存大小，注意它的单位是 MB。第 4 列"Disk"字段表示的是实例硬盘大小，它的单位是 GB，这是个装载启动虚拟机的临时的硬盘。当从一个永久使用的硬盘启动时就不需要了。当它的大小为 0 时，是一个特殊的大小，表示采用和启动实例镜像相同的大小。第 5 列字段"Ephemeral"表示指定第 2 个临时硬盘的大小。0 表示这块硬盘不存在。第 6 列"VCPUs"字段表示实例中虚拟 CPU 的核数。第 7 列"Is_Public"字段表示是否公开，即某一指定类型模板是否可以给其他项目中的用户使用，默认为 True，即公开。

下面定义自己需要的类型，创建自定义的类型模板之前，首先通过 openstack flavor create -h 命令查看一下创建类型模板的具体命令要求，反馈信息如图 13.3.3 所示。

其中包括 --ephemeral <size-gb>、--swap <size-mb>、--rxtx-factor <factor>、--public|--private 这 4 个表中出现的字段。从帮助中可以清晰地知道创建类型模板需要 1 个必选参数"<flavor-name>"，表示创建的类型模板的名称。像 id 之类的信息，系统会自行分配。

```
usage: openstack flavor create [-h] [-f {json,shell,table,value,yaml}]
                                [-c COLUMN] [--max-width <integer>]
                                [--fit-width] [--print-empty] [--noindent]
                                [--prefix PREFIX] [--id <id>] [--ram <size-mb>]
                                [--disk <size-gb>] [--ephemeral <size-gb>]
                                [--swap <size-mb>] [--vcpus <vcpus>]
                                [--rxtx-factor <factor>] [--public | --private]
                                [--property <key=value>] [--project <project>]
                                [--description <description>]
                                [--project-domain <project-domain>]
                                <flavor-name>
```

图 13.3.3　创建 flavor 帮助信息

现在定义一个名为 myflavor 的类型模板，其内存（Memory_MB）为 1024MB，硬盘（Disk）为 15GB，1 个虚拟 CPU，命令如下，其结果如图 13.3.4 所示。

```
[root@controller ~]# openstack flavor create myflavor --ram 1024 --disk 15 --vcpus 1
```

```
+----------------------------+--------------------------------------+
| Field                      | Value                                |
+----------------------------+--------------------------------------+
| OS-FLV-DISABLED:disabled   | False                                |
| OS-FLV-EXT-DATA:ephemeral  | 0                                    |
| disk                       | 15                                   |
| id                         | 0086f4af-ea70-4a57-8920-eb90f2829dcc |
| name                       | myflavor                             |
| os-flavor-access:is_public | True                                 |
| properties                 |                                      |
| ram                        | 1024                                 |
| rxtx_factor                | 1.0                                  |
| swap                       |                                      |
| vcpus                      | 1                                    |
+----------------------------+--------------------------------------+
```

图 13.3.4　创建新类型模板的反馈结果

从反馈信息中可以看到，"ID"字段如果没有给出，结果会是很长的随机值，"Ephemeral"字段默认是 0，"Swap"字段默认没有，"RXTX_Factor"字段默认为 1.0，"Is_Public"字段默认为 True 即是公开的。

如果在创建的过程中不小心将一些参数定义错误或者不喜欢随机给出的 ID 值，无法通过命令修改，命令能修改的是——project、——description、——project-domain。然而在 Dashboard 界面中即使是拥有管理员权限的用户无法做出一些修改，只能删除重建。如图 13.3.5 所示通过 admin 用户登录云平台 Dashboard 界面并打开管理员下的"实例类型"栏，在右侧选择需要编辑的实例类型。

实例类型名称	VCPU数量	内存	根磁盘	临时磁盘	Swap磁盘	RX/TX因子	ID	公有	元数据	动作
myflavor	1	1GB	15 GB	0 GB	0 MB	1.0	0086f4af-ea70-4a57-8920-eb90f2829dcc	Yes	No	更新元数据
small	1	1GB	20 GB	0 GB	0 MB	1.0	98ce29c5-8963-4db3-a19e-6a8aa9ece574	Yes	No	更新元数据

图 13.3.5　Dashboard 界面中的实例类型

选择后即可看到如图 13.3.5 所示的实例类型管理界面，从图中可以看到，只能重新创建"实例类型"，原因是这些"实例类型"是 public 的，所有用户都可以使用。接下来通过 openstack flavor set 来设置 myflavor 的所属项目，命令用法如下：

```
[root@controller ~]# openstack flavor set myprivateflavor --project project2 -project demo
```

执行上述命令后，会发现如下的反馈信息。

```
Failed to set flavor access to project: Cannot set access for a public flavor
```

从反馈的信息可以看到系统报错，中文翻译为"不能设置公共实例类型的可访问性"。

如果想控制"实例类型"的使用权，可以重新创建一个 private 属性的"实例类型"，然后分配给指定的项目使用。为什么需要对"实例类型"进行访问权限的控制，其原因是可能会遇到这样的情况，某个用户需要一个特定的类型模板来协调他正在进行的工程，例如，用户可能需要 128GB 的内存，如果按照之前的方式创建一个新的 public "实例类型"，该用户当然可以访问这个特定的"实例类型"，但在云中，其他组的用户也可以访问使用。允许所有用户可以访问使用这个 128GB 内存的"实例类型"，可能会使云容量迅速达到饱和状态甚至可能会出现更危险的情况。作为管理员，为了防止这种情况的发生，不能共享所有的"实例类型"，可以在创建"实例类型"时使用"--private"参数来限制对特定"实例类型"的访问使用，具体使用命令如下，结果如图 13.3.6 所示。

```
[root@controller ~]# openstack flavor create myprivateflavor --private
```

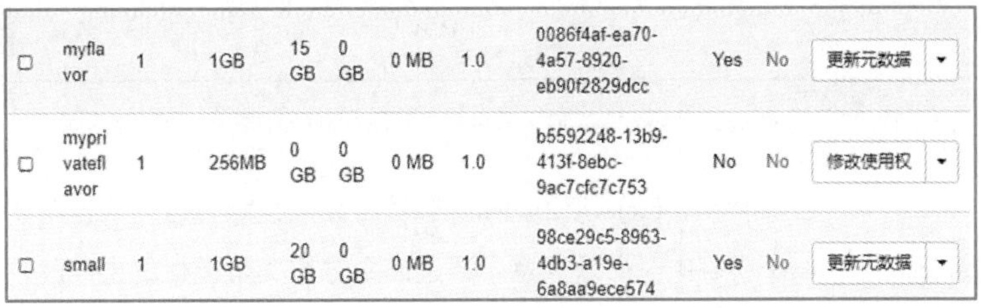

图 13.3.6　私有"实例类型"的区别

在图 13.3.6 中，可以发现两个 public 的"实例类型"都是只能更新元数据，而 myprivate flavor 可以修改使用权。

将刚刚创建的 myprivateflavor "实例类型"加入 user4 用户所在的 project2 项目和 demo 所在的 demo 项目，使用 openstack flavor set 命令如下。

```
[root@controller ~]# openstack flavor set myprivateflavor --project project2 -project demo
```

上述命令中 myprivateflavor 后面的字符串是 user4 用户所在的 project2 项目的名称和 demo 项目的名称。以此类推，可以把"实例类型"加到很多项目中去。可以通过执行 openstack flavor show myprivateflavor，得到反馈出结果，如图 13.3.7 所示。Access_project_ids 中有两个 project 的 id。

```
| Field                      | Value                                                      |
|----------------------------|------------------------------------------------------------|
| OS-FLV-DISABLED:disabled   | False                                                      |
| OS-FLV-EXT-DATA:ephemeral  | 0                                                          |
| access_project_ids         | 5804d6b60d854cf282a0d35ecf8b598c, 81d4a4c829c64553abefb6376123b002 |
| disk                       | 0                                                          |
| id                         | b5592248-13b9-413f-8ebc-9ac7cfc7c753                       |
| name                       | myprivateflavor                                            |
| os-flavor-access:is_public | False                                                      |
| properties                 |                                                            |
| ram                        | 256                                                        |
| rxtx_factor                | 1.0                                                        |
| swap                       |                                                            |
| vcpus                      | 1                                                          |
```

图 13.3.7　将"实例类型"加入某项目的反馈信息

接下来用 admin 用户登录 Dashboard 查看 myprivateflavor 的状态，打开如图 13.3.6 所示界面，选中 myprivateflavor 实例类型进行编辑，切换到编辑实例类型的界面中，如图 13.3.8 所示。

图 13.3.8　myflavor 实例类型中显示 project2 项目和 demo 项目为选中的项目

从图 13.3.8 中也可以看到，project2 和 demo 项目已经被移到右边的选框中，表明已经成功加入 project2 和 demo 项目中。必须注意，通过命令进行类型模板的操作时，只有私有类型的"实例类型"可以成功加入指定项目中。可以随时单击 + 或 – 按钮来添加和删除与项目的关联。

分别用 project2 项目下的 user4 用户和 admin 项目的 admin 用户，登录 Dashboard 创建实例，看一看对实例类型修改后的效果，如图 13.3.9 和图 13.3.10 所示。

图 13.3.9 是 user4 用户在创建"实例类型"的可选项，图 13.3.10 是 admin 用户启动"实例类型"的可选项，图中清晰表明了两者之间的区别。所以，一旦一个"实例类型"使用权被限制，除了被授权项目内的用户可以使用，其他任何用户在创建实例时都无法使用，包括拥有 admin 项目（管理员组）中的用户。所以作为管理员，必须解决"实例类型"的使用权问题。

作为管理员有时还需要删除废弃的"实例类型"或者将某一"实例类型"从指定的项目中去除。两者方法都很简单，删除"实例类型"的命令用法如下。

13.3 实训项目 13 Nova 基本运维命令及其应用

图 13.3.9　user4 登录后可使用的实例类型

图 13.3.10　admin 登录后可使用的实例类型

```
[root@controller ~]# openstack help flavor delete
usage: openstack flavor delete [-h] <flavor> [<flavor> ...]
```

上述命令中，flavor 的值是要被删除的 flavor 的 ID 或名称。

将某一类型模板从指定的项目中去除的命令用法如下：

```
[root@controller ~]# openstack help flavor set
usage: openstack flavor set [--project <project>] <flavor>
```

其实就是设置新的 project 即可。

步骤 2：安全组的管理。

对于新用户来说，在 OpenStack 中经常会遇到这样的问题，在云平台中成功启动了一个虚拟机，但是却无法通过终端软件连接访问网络上的虚拟机。这是因为没有设置适当的安全组。本书在 10.3 节的步骤 2 中介绍过，在 OpenStack 中使用虚拟网络插件 Openvswitch 提供的防火墙和安全组功能来管理云平台的网络。安全组是一组应用于一个虚拟机实例的网络 IP 过滤规则，是基于具体项目的，项目成员可以编辑默认的规则，或者添加属于他们组的新规则。在 OpenStack 中，如果所有的项目都没有定义其他安全组，都会有一个"默认"安全组，并将应用于虚拟机。需要注意，在不做任何修改的情况下，默认的安全组会阻挡所有的进入流量。

图 13.3.11　查看安全组相关命令的反馈结果

在讲解安全组（secgroup）的用法之前，先了解关于安全组的命令选项，命令如下，其结果如图 13.3.11 所示。

```
[root@controller ~]# openstack security group | grep group
```

从反馈信息中可以看到 security group 安全组的所有子命令。

首先可以通过 openstack security group list 命令查看所有项目中的所有安全组，命令如下，其反馈信息如图 13.3.12 所示。

```
[root@controller ~]# openstack security group list
```

从反馈信息中可以看到每个项目里面只有一个默认的 default 安全组。

```
+--------------------------------------+---------+------------------------+--------------------------------------+------+
| ID                                   | Name    | Description            | Project                              | Tags |
+--------------------------------------+---------+------------------------+--------------------------------------+------+
| 2d448383-eaa2-47ae-86b5-4226a4036138 | default | Default security group | dab3559d05534ff9815746bb3bf20864     | []   |
| 3196537b-6b53-4210-9881-2ed9c036d38a | default | Default security group | 5804d6b60d854cf282a0d35ecf8b598c     | []   |
| 6294692a-cc5e-4b19-89a7-67db109263d7 | default | Default security group | 3bc320cc8bba4ee7b7782b56e4933046     | []   |
| 766ff948-b2fe-4a13-a892-cebe5bc42449 | default | Default security group | a823ad0ccc144975b5300e7c20ba3d7b     | []   |
| a8938b77-5e85-4af7-8290-43daa9da71d8 | default | Default security group | 81d4a4c829c64553abefb6376123b002     | []   |
| bf47bec2-0995-490a-8db0-2f627790d13b | default | Default security group |                                      | []   |
+--------------------------------------+---------+------------------------+--------------------------------------+------+
```

图 13.3.12 查看已有安全组命令的反馈信息

通过 admin 用户登录后，在 Dashboard 界面中，可以"网络"选项下"安全"页面查看安全组的条目，如图 13.3.13 所示。

当用户想知道指定安全组的具体内容时，可以通过 security group list 命令查看所有安全组内的规则，通过以下命令查看 default 安全组的具体规则，其结果如图 13.3.14 所示。

```
[root@controller ~]# openstack security group rule list 2d448383-eaa2-47ae-86b5-4226a4036138
```

图 13.3.13 Dashboard 界面中"安全组"页面

```
+--------------------------------------+-------------+----------+------------+--------------------------------------+
| ID                                   | IP Protocol | IP Range | Port Range | Remote Security Group                |
+--------------------------------------+-------------+----------+------------+--------------------------------------+
| 402301cb-0c1a-4d6a-970e-382fbbcecdaa | None        | None     |            | None                                 |
| 87fffe99-e29c-4c23-81ab-ff541dd9a83f | None        | None     |            | None                                 |
| 8fe2bd63-b538-404b-8636-a4a8dd1a132c | None        | None     |            | 2d448383-eaa2-47ae-86b5              |
| ef08af9a-5352-4c27-8d74-c96a48793762 | None        | None     |            | 2d448383-eaa2-47ae-86b5              |
+--------------------------------------+-------------+----------+------------+--------------------------------------+
```

图 13.3.14 查看 default 安全组具体规则的反馈结果

在 Dashboard 界面中，单击"管理规则"按钮，进入 default 子页面时便可以看到 default 安全组的规则，如图 13.3.15 所示。

方向	以太网类型 (EtherType)	IP协议	端口范围	远端IP前缀	远端安全组	Description	动作
出口	IPv4	任何	任何	0.0.0.0/0	-	-	删除规则
出口	IPv6	任何	任何	::/0	-	-	删除规则
入口	IPv4	任何	任何	-	default	-	删除规则
入口	IPv6	任何	任何	-	default	-	删除规则

图 13.3.15 default 安全组默认规则

在 Dashboard 界面可以获得更加详细的规则信息，当然在命令行也可以通过 openstack security group rule show 命令来获得详细规则信息。第 1 条是允许通过 IPv4 访问外部；第 2 条是允许通过 IPv6 访问外部；第 3 条是允许同一组内主机通过 IPv4 访问；第 4 条是允许同一组内主机通过 IPv6 访问。

在实训项目 10 中介绍过，为了可以正常访问云平台中创建的实例，需要额外添加一些访问规则。一般情况下，在对实例的操作中，会涉及 3 种协议，即 TCP 协议、ICMP 协议和 UDP 协议。添加 TCP 协议可以通过 SSH 登录到虚拟机，添加 ICMP 协议可以使用 ping 命令测试实例的连通性，添加 UDP 协议可以在实例中部署 DNS 服务时，确保 DNS 服务可以正常运行。本次实训中不涉及 UDP 协议，这里不做详细说明。

在使用默认安全组的情况下，无法 ping 通云平台中的实例，也无法使用 SSH 连接到实例，在 Dashboard 上创建一个 test-ext 实例进行验证，该实例使用默认安全组，创建成功后的实例如图 13.3.16 所示。

图 13.3.16 使用默认安全组实例创建 test-ext 虚拟机实例

接下来，通过本地的主机去 ping 该虚拟机实例的地址 192.168.200.83，发现该实例不能 ping 通，如图 13.3.17 所示。

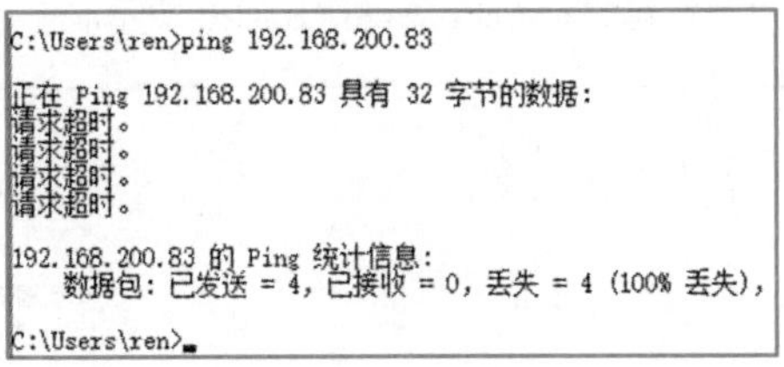

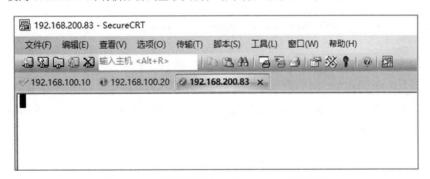

图 13.3.17 本地主机无法 ping 通 test-ext 虚拟机实例

使用 SecureCRT 终端软件尝试登录实例，结果仍然登录不上，如图 13.3.18 所示。

图 13.3.18 无法通过终端软件连接 test 虚拟机实例

为确保可以正常访问操作云平台的实例，需要在启动实例前将其添加到指定的安全组内。关于添加规则，有两种方式，第一，在原有的 default 安全组内直接添加新规则；第二，创建新的安全组然后添加新规则。第 2 种方式的操作内容包含第 1 种操作方式中的操作，以下重点介绍第 2 种方式。

首先，需要创建一个新的安全组，在创建之前，先了解创建安全组的语法，命令如下。

```
[root@controller ~]# openstack help security group create
usage: openstack security group create [--description <description>] <name>
Create a security group.
```

从反馈信息中可以看到，创建一个新安全组的命令非常简单，只需要定义名称和描述信息即可。接下来定义一个名为 testsec，描述信息为 test security group 的安全组，命令如下。

```
[root@controller ~]# openstack security group create testsec --description "test security group"
```

```
+----------------+---------------------------------------------------------------------------+
| Field          | Value                                                                     |
+----------------+---------------------------------------------------------------------------+
| created_at     | 2019-12-14T08:04:47Z                                                      |
| description    | test security group                                                       |
| id             | eb609f49-70d5-4637-8ac9-b538c2fa4995                                      |
| location       | Munch({'project': Munch({'domain_name': 'Default', 'domain_id': None, 'name': |
| name           | testsec                                                                   |
| project_id     | dab3559d05534ff9815746bb3bf20864                                          |
| revision_number| 1                                                                         |
| rules          | created_at='2019-12-14T08:04:47Z', direction='egress', ethertype='IPv4', id=' |
|                | created_at='2019-12-14T08:04:47Z', direction='egress', ethertype='IPv6', id=' |
| tags           | []                                                                        |
| updated_at     | 2019-12-14T08:04:47Z                                                      |
+----------------+---------------------------------------------------------------------------+
```

图 13.3.19 创建 testsec 安全组的反馈信息

从图 13.3.19 中可以看出默认有两条出接口的规则，分别是 IPv4 和 IPv6。Dashboard 界面上安全组中会多出刚刚新建的 testsec，如图 13.3.20 所示。

图 13.3.20 Dashboard 界面上显示新建安全组 testsec

可以使用命令 openstack security group rule list testsec 查看，看到新创建的安全组中是有两条规则的，命令如下，其反馈结果如图 13.3.21 所示。

```
[root@controller ~]# nova secgroup-list-rules testsec
```

```
+--------------------------------------+-------------+----------+------------+-----------------------+
| ID                                   | IP Protocol | IP Range | Port Range | Remote Security Group |
+--------------------------------------+-------------+----------+------------+-----------------------+
| 72c6f323-5331-446c-99dd-57065ec9cc0c | None        | None     | None       | None                  |
| cf092f6f-a565-41e3-82e6-6c7c678af38b | None        | None     | None       | None                  |
+--------------------------------------+-------------+----------+------------+-----------------------+
```

图 13.3.21 查看 testsec 安全组具体规则的反馈结果

具体的规则可以通过命令 openstack security group rule show 来查看规则的具体信息。

首先需要查看如何向指定安全组内添加规则，通过 help 命令查询下 security group rule add 命令具体的参数，命令如下。

```
[root@controller ~]# openstack help security group rule create
usage: openstack security group rule create
[--remote-ip <ip-address> | --remote-group <group>]
[--dst-port <port-range>]
[--protocol <protocol>]
[--ethertype <ethertype>]
[--ingress | --egress]……
<group>
Create a new security group rule.
```

从反馈信息中可以看出为新规则主要定义如下内容，即安全组 <group>、协议 <protocol>、远程地址 <ip-address>、远程端口 <port-range>。这里需要说明一下，对于需要添加的 3 个 IP 协议来说，作为管理员，可以按照需要添加指定内容，同时，也可以通过直接添加全部的方式一次将所有端口和 IP 地址加入规则。有两种方式提供参考。

下面分别添加协议并通过结果观察其带来的影响。先添加 ICMP 协议，可以 ping 通云平台中的实例，命令如下，反馈信息如图 13.3.22 示。

```
[root@controller ~]# openstack security group rule create testsec --protocol icmp
```

```
+-------------------+--------------------------------------------------------------+
| Field             | Value                                                        |
+-------------------+--------------------------------------------------------------+
| created_at        | 2019-12-15T01:21:55Z                                         |
| description       |                                                              |
| direction         | ingress                                                      |
| ether_type        | IPv4                                                         |
| id                | e5397543-4908-444d-9e91-2c38c350c7f0                         |
| location          | Munch({'project': Munch({'domain_name': 'Default', 'domain_  |
| name              | None                                                         |
| port_range_max    | None                                                         |
| port_range_min    | None                                                         |
| project_id        | dab3559d05534ff9815746bb3bf20864                             |
| protocol          | icmp                                                         |
| remote_group_id   | None                                                         |
| remote_ip_prefix  | 0.0.0.0/0                                                    |
| revision_number   | 0                                                            |
| security_group_id | eb609f49-70d5-4637-8ac9-b538c2fa4995                         |
| tags              | []                                                           |
| updated_at        | 2019-12-15T01:21:55Z                                         |
+-------------------+--------------------------------------------------------------+
```

图 13.3.22 添加所有 IP 地址 ICMP 协议允许访问规则的反馈结果

从反馈信息中可以看到成功向 testsec 安全组中添加 ICMP 规则，"IP Range"字段为"0.0.0.0/0"，这样就允许所有 IP 地址 ping 通云平台的实例，如果需要指定 IP 地址能 ping 通，则在 <ip-address> 选项处填写指定的 IP 地址即可，示例命令如下。

```
[root@controller ~]# openstack security group rule create testsec --protocol icmp
--remote-ip 192.168.200.0/24
```

这里指只允许 192.168.200.0 子网掩码 24 位的主机 ping 通 test 实例。

添加 ICMP 协议后，给 test-ext 主机添加安全组 testsec。然后通过在本地 ping 云平台中的 test-ext 主机查看影响，如图 13.3.23 示。

成功 ping 通 test-ext 实例。但依然无法通过终端仿真软件登录 test-ext，因为终端软件 SSH 是通过 TCP22 端口访问虚拟机实例，还需要添加 TCP 协议。

目前是需要通过终端仿真软件远程登录 test-ext，如果仅仅是实现这个目的，只需要放行 22 端口即可，命令如下。

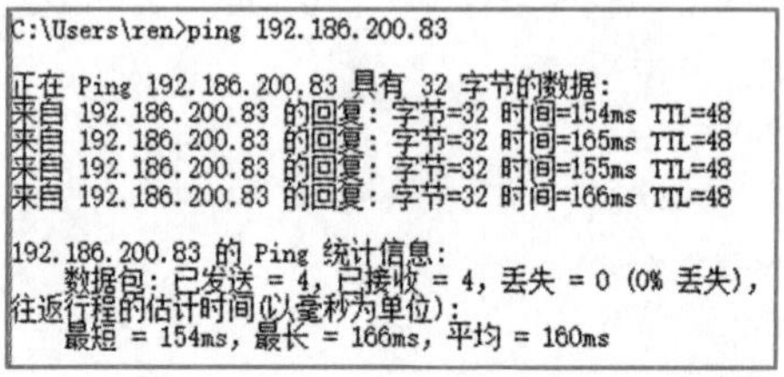

图 13.3.23　规则后本地主机 ping 通 test-ext 虚拟机实例

```
[root@controller ~]# openstack security group rule create testsec   --protocol tcp --dst-port 22
```

现在，端口 22 就为所有 IP 地址的请求开放了。验证结果如图 13.3.24 所示。

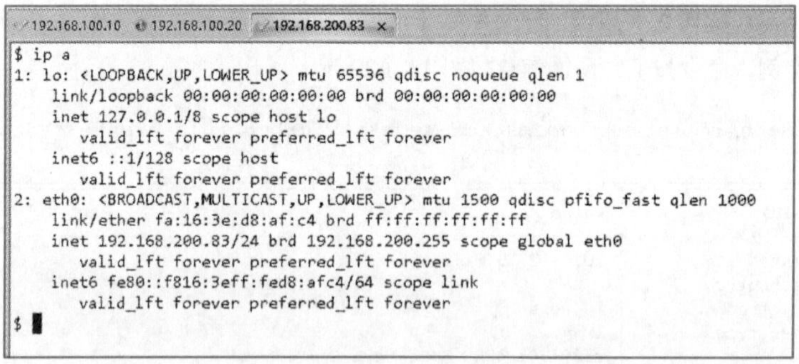

图 13.3.24　规则后本地主机可以通过终端连接 test-ext 虚拟机实例

同样，如果有其他需求，可以通过放行所有 TCP 端口的形式添加规则，命令如下。

```
[root@controller ~]# openstack security group rule create testsec   --protocol tcp --dst-port 1:65535
```

这样所有 TCP 协议端口就为所有 IP 地址的请求开放了。现在使用属于 project2 项目的 user4 用户登录云平台，查看"安全组"页面的内容，如图 13.3.25 所示。

图 13.3.25　user4 用户登录后安全组页面内容

从图 13.3.25 中可以看到，属于 project2 项目的 user4 用户的安全组并没有任何变化，之前建立的 testsec 安全组，在 user4 用户登录后并不能看到，这说明安全组是基于项目定义的，不同项目间不共享自定义的安全组。

在 OpenStack 云平台中，安全组的定义也可以在 Dashboard 界面中也可以实现，实训项目 10 步骤 4 中已经做过介绍。以下是简单回顾：用 admin 用户登录 Dashboard 界面，在"项目"→"网络"选项的"安全组"页中看到系统默认的安全组和已经创建的属于 admin 项目的安全组，同时单击如图 13.3.26 所示右上方的"创建安全组"按钮，按提示可以创建新的安全组。

图 13.3.26　Dashboard 界面创建安全组

单击指定安全组的"管理规则"按钮，即可进入子页面，查看已经添加的规则，如图 13.3.27 所示。

单击"删除规则"可以将已经存在的规则删除，单击如图 13.3.28 所示右上方的"添加规则"按钮，即可为指定安全组添加需要的规则。

在如图 13.3.28 所示的"规则"下拉列表中会显示所有支持的协议，"方向"选择"入口"或"出口"，再选择端口或是端口范围，然后，输入具体端口内容和 CIDR 的内容，单击"添加"按钮，即可添加成功。

在 Dashboard 中只需要单击"删除安全组"或"删除规则"按钮即可将相关内容删除，作为管理员还需要知道如何在命令中删除安全组和指定安全组中的相应规则。删除指定安全组，查看该命令参数如下。

```
[root@controller ~]# openstack help security group delete
usage: openstack security group delete [-h] <group> [<group> ...]
Delete security group(s)
positional arguments:
  <group>       Security group(s) to delete (name or ID)
```

图 13.3.27　通过 Dashboard 界面添加删除安全组规则

图 13.3.28　添加规则具体操作

从反馈信息中可以看到删除安全组的命令很简单，需要指定安全组组名或者 ID。

删除指定安全组中的相应规则，通过 help 命令查看 security group rule delete 所跟参数如下。

```
[root@controller ~]# openstack help security group rule delete
usage: openstack security group rule delete [-h] <rule> [<rule> ...]

Delete security group rule(s)

positional arguments:
  <rule>        Security group rule(s) to delete (ID only)
```

从反馈信息中发现删除一条规则仅仅是需要提供规则 ID。

安全组是确保用户可以正常访问操作云平台中虚拟机的重要保障，作为管理员需要定义合适的安全组。

步骤 3：云主机（实例）的创建和管理。

在本实训项目的步骤 1 中用命令创建实例，命令如下。

```
# openstack server create <server-name> --flavor <flavor> --image <image> <name>
```

用命令创建实例有许多可选参数可以被指定，但上述命令是一个基础命令。其他具体参数可以通过命令 openstack help server create 查看。以下使用最基本的命令参数来尝试启动一个实例，首先需要列出可用镜像和可用的实例类型模板，命令如下，其结果如图 13.3.29 和图 12.3.30 所示。

```
[root@controller ~]# openstack image list

+--------------------------------------+----------------+--------+
| ID                                   | Name           | Status |
+--------------------------------------+----------------+--------+
| 2d698a77-ffb8-4a42-8890-1b839f67f55e | cirros         | active |
| da18f4c8-3131-4fb7-8b0a-015ec1605010 | cirros-d190515 | active |
| 3347a41a-85d9-4998-8633-1f79d9a545d9 | ownerdemo2     | active |
+--------------------------------------+----------------+--------+
```

图 13.3.29　查看可用镜像的反馈结果

```
[root@controller ~]# openstack flavor list
```

13.3 实训项目 13 Nova 基本运维命令及其应用

```
+--------------------------------------+----------+------+------+-----------+-------+-----------+
| ID                                   | Name     | RAM  | Disk | Ephemeral | VCPUs | Is Public |
+--------------------------------------+----------+------+------+-----------+-------+-----------+
| 0086f4af-ea70-4a57-8920-eb90f2829dcc | myflavor | 1024 | 15   | 0         | 1     | True      |
| 98ce29c5-8963-4db3-a19e-6a8aa9ece574 | small    | 1024 | 20   | 0         | 1     | True      |
+--------------------------------------+----------+------+------+-----------+-------+-----------+
```

图 13.3.30　查看可用实例类型模板的反馈结果

以下是用命令创建的一个简单实例，定义其名称为 test，使用 myflavor 实例类型模板和 cirros-d190515 的镜像，命令如下，反馈信息如图 13.3.31 所示。

```
[root@controller ~]# openstack server create test --flavor myflavor --image cirros-d190515 --network ext-net
```

从反馈的信息中可以看到它的 ID、名称、镜像信息、所属用户 ID、所属项目 ID 等，同时也可以看到 status 的值是 BUILD，正在创建，此时可以用 admin 用户登录到 Dashboard 上查看一下实例页中的变化，如图 13.3.32 所示。

从图 13.3.32 中看到，当登录到 Dashboard 界面后，刚刚创建的实例电源状态已经为"运行中"，状态为"运行"，这表明它已经成功启动。也可以在命令中查看，命令如下，其结果如图 13.3.33 所示，其中第 1 行为 test 实例所对应的值。

```
+-------------------------------------+-------------------------------------------------------+
| Field                               | Value                                                 |
+-------------------------------------+-------------------------------------------------------+
| OS-DCF:diskConfig                   | MANUAL                                                |
| OS-EXT-AZ:availability_zone         |                                                       |
| OS-EXT-SRV-ATTR:host                | None                                                  |
| OS-EXT-SRV-ATTR:hypervisor_hostname | None                                                  |
| OS-EXT-SRV-ATTR:instance_name       |                                                       |
| OS-EXT-STS:power_state              | NOSTATE                                               |
| OS-EXT-STS:task_state               | scheduling                                            |
| OS-EXT-STS:vm_state                 | building                                              |
| OS-SRV-USG:launched_at              | None                                                  |
| OS-SRV-USG:terminated_at            | None                                                  |
| accessIPv4                          |                                                       |
| accessIPv6                          |                                                       |
| addresses                           |                                                       |
| adminPass                           | oMX3qH4KfLVF                                          |
| config_drive                        |                                                       |
| created                             | 2019-12-15T03:57:49Z                                  |
| flavor                              | myflavor (0086f4af-ea70-4a57-8920-eb90f2829dcc)       |
| hostId                              |                                                       |
| id                                  | 3585fd55-8803-4dbf-b8f3-d3b7409631cc                  |
| image                               | cirros-d190515 (da18f4c8-3131-4fb7-8b0a-015ec1605010) |
| key_name                            | None                                                  |
| name                                | test                                                  |
| progress                            | 0                                                     |
| project_id                          | dab3559d05534ff9815746bb3bf20864                      |
| properties                          |                                                       |
| security_groups                     | name='default'                                        |
| status                              | BUILD                                                 |
| updated                             | 2019-12-15T03:57:49Z                                  |
| user_id                             | 0165a8c49c4647118053a06ce2fbce68                      |
| volumes_attached                    |                                                       |
+-------------------------------------+-------------------------------------------------------+
```

图 13.3.31　用命令启动 test 虚拟机实例的反馈信息

实例名称	镜像名称	IP 地址	实例类型	密钥对	状态	可用域	任务	电源状态	时间	动作
test	cirros-d190515	192.168.200.78	myflavor	-	运行	nova	无	运行中	15 minutes	创建快照

正在显示 4 项

图 13.3.32　从 Dashboard 界面看到刚启动的 test 虚拟机实例

```
[root@controller ~]# openstack server list
```

```
+--------------------------------------+----------+---------+--------------------------+---------------+----------+
| ID                                   | Name     | Status  | Networks                 | Image         | Flavor   |
+--------------------------------------+----------+---------+--------------------------+---------------+----------+
| 3585fd55-8803-4dbf-b8f3-d3b7409631cc | test     | ACTIVE  | ext-net=192.168.200.78   | cirros-d190515| myflavor |
| 9a8887c0-1a63-48a8-99df-764a122dc0ee | cirros2  | ACTIVE  | int-net=10.0.0.187       | cirros-d190515| myflavor |
| 0a65606c-89c1-43d9-b875-436a03a11d83 | test-ext | ACTIVE  | ext-net=192.168.200.83   | cirros-d190515| myflavor |
| 01ebb576-88ec-46e8-b28e-314d873a573c | cirros   | SHUTOFF | int-net=10.0.0.71        | cirros        | small    |
+--------------------------------------+----------+---------+--------------------------+---------------+----------+
```

图 13.3.33　查看当前虚拟机实例的反馈结果

看到用命令创建的实例成功启动之后，还可以通过查看其详细信息，来观察哪些参数系统会自动添加分配，命令如下，结果如图 13.3.34 所示。

```
[root@controller ~]# openstack server show test
```

上述命令中使用的是 test 实例的名称，也可以直接跟实例 ID。从反馈信息中可以了解，除 IP 外，系统还会给实例分配安全组，这里是 default 安全组。Dashboard 界面中同样可以查看实例的详细信息，如图 13.3.35 所示。

原生 OpenStack 云平台支持多种方式创建和启动实例，可以在 Dashboard 界面上直观地看到，如图 13.3.36 所示。

```
+-------------------------------------+----------------------------------------------------------+
| Field                               | Value                                                    |
+-------------------------------------+----------------------------------------------------------+
| OS-DCF:diskConfig                   | MANUAL                                                   |
| OS-EXT-AZ:availability_zone         | nova                                                     |
| OS-EXT-SRV-ATTR:host                | compute                                                  |
| OS-EXT-SRV-ATTR:hypervisor_hostname | compute                                                  |
| OS-EXT-SRV-ATTR:instance_name       | instance-00000018                                        |
| OS-EXT-STS:power_state              | Running                                                  |
| OS-EXT-STS:task_state               | None                                                     |
| OS-EXT-STS:vm_state                 | active                                                   |
| OS-SRV-USG:launched_at              | 2019-12-15T03:57:57.000000                               |
| OS-SRV-USG:terminated_at            | None                                                     |
| accessIPv4                          |                                                          |
| accessIPv6                          |                                                          |
| addresses                           | ext-net=192.168.200.78                                   |
| config_drive                        |                                                          |
| created                             | 2019-12-15T03:57:49Z                                     |
| flavor                              | myflavor (0086f4af-ea70-4a57-8920-eb90f2829dcc)          |
| hostId                              | 0c1a7786589bff645646e345f1bb56c4ef255dd6d06d2fc825094339 |
| id                                  | 3585fd55-8803-4dbf-b8f3-d3b7409631cc                     |
| image                               | cirros-d190515 (da18f4c8-3131-4fb7-8b0a-015ec1605010)    |
| key_name                            | None                                                     |
| name                                | test                                                     |
| progress                            | 0                                                        |
| project_id                          | dab3559d05534ff9815746bb3bf20864                         |
| properties                          |                                                          |
| security_groups                     | name='default'                                           |
| status                              | ACTIVE                                                   |
| updated                             | 2019-12-15T03:57:57Z                                     |
| user_id                             | 0165a8c49c4647118053a06ce2fbce68                         |
| volumes_attached                    |                                                          |
+-------------------------------------+----------------------------------------------------------+
```

图 13.3.34　查看 test 虚拟机实例的详细信息

13.3 实训项目 13 Nova 基本运维命令及其应用

图 13.3.35 在 Dashboard 界面上查看 test 虚拟机实例详细信息

图 13.3.36 从不同源启动云主机

从图 13.3.36 中可以看到，除了可以从镜像启动实例外，还可以从快照启动，从一个卷或一个卷快照。当然最直接的方式是从镜像启动。下面通过操作示范并讲解通过命令和在 Dashboard 的界面实现另外两种方式创建实例，即从快照和卷创建启动实例。首先是如何实现从快照创建并启动实例。所谓下快照功能，用过 VMware 的读者应该都对 VMware 虚拟机做过快照，在原生 OpenStack 云平台中的快照与 VMware 的快照并不一样，VMware 中的快照是对 VMDK（VMware 虚拟磁盘文件）在某个时间点的"拷贝"，这个"拷贝"并不是对 VMDK 文件的复制，而是保持磁盘文件和系统内存在该时间点的状态，以便在出现故障后虚拟机能够恢复到该时间点。如果对某个虚拟机创建了多个快照，那么就可以有多个可恢复的时间点。

在 OpenStack 云平台中，实例的快照即是镜像，做快照是单纯地对实例镜像做转换和拷贝，生成一个与实例无关联的镜像，做完后以镜像形式上传并存储在 Glance 服务存储镜像的位置。这种快照只保留磁盘信息，无法回滚至多个快照点，只能采用该快照镜像创建一个新的实例。同时，目前的快照方式都是 cold snapshot（冷快照）的方式，即首先先关机，其次执行如下命令生成一个镜像文件，再次开机，最后再调用 glance api 将镜像上传。当然与 cold snapshot 方式相对应的是 live snapshot 方式，即创建快照时，无须暂停虚拟机。所以，目前的 OpenStack 快照并不是真正意义的快照，其实和关闭虚拟机，拷贝一份，再上传没有本质区别。

接下来，下面如何使用命令制作快照，通过 help 命令查询如下。

```
[root@controller ~]# nova help image-create
usage: nova image-create [--show] [--poll] <server> <name>
Create a new image by taking a snapshot of a running server.
Positional arguments:
<server>   Name or ID of server.
<name>     Name of snapshot.
Optional arguments:
--show     Print image info.
--poll     Blocks while server snapshots so progress can be reported.
```

从反馈信息中可以看到有两个必选参数和两个可选参数，"--show"可以显示制作快照之后的详细信息，快照就是一个以实例为基础的镜像，帮助上使用的是"image"单词。"--poll"可以显示进度。

这里为当前 test 虚拟机实例创建一个镜像，命名为 testsnap1，命令如下，其反馈结果如图 13.3.37 所示。

```
[root@controller ~]# nova image-create --show --poll test testsnap1
```

```
Server snapshotting... 100% complete
Finished
+---------------------+------------------------------------------------------------------+
| Property            | Value                                                            |
+---------------------+------------------------------------------------------------------+
| base_image_ref      | da18f4c8-3131-4fb7-8b0a-015ec1605010                             |
| boot_roles          | admin,member,reader                                              |
| checksum            | c686b1b75a1f67e6a4cd6757162fa718                                 |
| container_format    | bare                                                             |
| created_at          | 2019-12-15T05:44:55Z                                             |
| disk_format         | qcow2                                                            |
| file                | /v2/images/f0c4f3ff-cf59-4156-9c9a-b6ee5860807c/file             |
| id                  | f0c4f3ff-cf59-4156-9c9a-b6ee5860807c                             |
| image_location      | snapshot                                                         |
| image_state         | available                                                        |
| image_type          | snapshot                                                         |
| instance_uuid       | 3585fd55-8803-4dbf-b8f3-d3b7409631cc                             |
| min_disk            | 15                                                               |
| min_ram             | 0                                                                |
| name                | testsnap1                                                        |
| os_hash_algo        | sha512                                                           |
| os_hash_value       | 6a73c2c88716be6da521d9aa2f6d4920bb2cf0beed5367b171fc3aacb57f3946181090a66cebc9c9fb9b8ac470d33605f |
| os_hidden           | False                                                            |
| owner               | dab3559d05534ff9815746bb3bf20864                                 |
| owner_id            | dab3559d05534ff9815746bb3bf20864                                 |
| owner_project_name  | admin                                                            |
| owner_user_name     | admin                                                            |
| protected           | False                                                            |
| schema              | /v2/schemas/image                                                |
| self                | /v2/images/f0c4f3ff-cf59-4156-9c9a-b6ee5860807c                  |
| size                | 162398208                                                        |
| status              | active                                                           |
| tags                | []                                                               |
| updated_at          | 2019-12-15T05:45:54Z                                             |
| user_id             | 0165a8c49c4647118053a06ce2fbce68                                 |
| virtual_size        | -                                                                |
| visibility          | private                                                          |
+---------------------+------------------------------------------------------------------+
```

图 13.3.37　建立快照 testsnap1 的反馈结果

从反馈信息中可以看到，创建的快照状态"image_state"为"available"，表示为可用的。

当然用命令还不能很直观地看到创建实例快照时实例和新建快照的变化情况。如图 13.3.38 所示，下面在 Dashboard 界面上创建并观察它们的变化。

首先单击 test 实例最右边的"创建快照"按钮，进入如图 13.3.38 所示的页面，这里需给新的快照命名为 testsnap2，然后单击右下角"创建快照"按钮。

之后界面后自动调转到镜像页面，如图 13.3.39 所示，此时快照"testsnap2"的状态为"已排队"，即为排队状态，此时系统正在为制作快照做准备，继续单击实例页查看它的状态。

图 13.3.38　在 Dashboard 界面上创建快照

图 13.3.39　testsnap2 快照当前的状态

如图 13.3.40 所示，此时实例的状态依然是"运行"，但任务列中则为"快照中"，该状态持续的时间很短，它与快照 testsnap2 的"已排队"的状态一致。紧接着，虚拟机的任务就会进入如图 13.3.41 所示状态。

	实例名称	镜像名称	IP 地址	实例类型	密钥对	状态	可用域	任务	电源状态	时间	动作
☐	test	cirros-d190515	192.168.200.78	myflavor	-	运行	nova	快照中	运行中	10 hours, 32 minutes	创建快照 ▼
☐	cirros2	cirros-d190515	10.0.0.187	myflavor	-	运行	nova	无	运行中	23 hours, 18 minutes	创建快照 ▼

图 13.3.40　testsnap2 快照制作时镜像 test 的状态变更（1）

正在显示 4 项

	实例名称	镜像名称	IP 地址	实例类型	密钥对	状态	可用域	任务	电源状态	时间	动作
☐	test	cirros-d190515	192.168.200.78	myflavor	-	运行	nova	镜像等待上传	运行中	10 hours, 25 minutes	创建快照 ▼
☐	cirros2	cirros-d190515	10.0.0.187	myflavor	-	运行	nova	无	运行中	23 hours, 11 minutes	创建快照 ▼
☐	test-ext	cirros-d190515	192.168.200.83	myflavor	-	运行	nova	无	运行中	1 day, 5 hours	创建快照 ▼

图 13.3.41　testsnap2 快照制作时镜像 test 的状态变更（2）

虚拟机的任务栏状态为"镜像等待上传"，这时镜像等待上传，此时用终端软件尝试登录 192.168.200.78，发现已经登录不上，说明此时实例不在运行，并处于一个暂停状态。

下一个阶段，如图 13.3.42 所示，系统调用 glance api 将镜像上传，这时任务栏状态为"镜像上传"，此时可以单击镜像页观察快照 testsnap2 的状态。

	实例名称	镜像名称	IP 地址	实例类型	密钥对	状态	可用域	任务	电源状态	时间	动作
☐	test	cirros-d190515	192.168.200.78	myflavor	-	运行	nova	镜像上传	运行中	10 hours, 30 minutes	创建快照 ▼
☐	cirros2	cirros-d190515	10.0.0.187	myflavor	-	运行	nova	无	运行中	23 hours, 16 minutes	创建快照 ▼
☐	test-ext	cirros-d190515	192.168.200.83	myflavor	-	运行	nova	无	运行中	1 day, 5 hours	创建快照 ▼

图 13.3.42　testsnap2 快照制作时镜像 test 的状态变更（3）

13.3 实训项目 13　Nova 基本运维命令及其应用

如图 13.3.43 所示，当系统调用 glance api 上传镜像时，testsnap2 的状态变为"保存中"即正在存储。

图 13.3.43　testsnap2 快照正在存储

如图 13.3.44 所示，最终 testsnap2 变为"运行中"状态，同时，可以看到默认情况下，快照为非公有的即私有状态的，也是不受保护的即可以被任何用户删除。

图 13.3.44　testsnap2 快照制作完毕

使用 help 参数查看下命令的用法如下。

```
[root@controller ~]# usage: openstack image delete [-h] <image> [<image> ...]
Delete image(s)

positional arguments:
  <image>       Image(s) to delete (name or ID)
```

从反馈信息中可以看到，只需要在 image delete 后面加上需要删除的快照即可，可以跟快照的名称，也可以跟快照的 ID。

例如，使用命令删除刚刚创建的实例快照 testsnap2，并进行查看，命令如下，查看结果如图 13.3.45 所示。

```
[root@controller ~]# openstack image delete testsnap2
[root@controller ~]# nova image-list

+--------------------------------------+---------------+--------+
| ID                                   | Name          | Status |
+--------------------------------------+---------------+--------+
| 2d698a77-ffb8-4a42-8890-1b839f67f55e | cirros        | active |
| da18f4c8-3131-4fb7-8b0a-015ec1605010 | cirros-d190515| active |
| 3347a41a-85d9-4998-8633-1f79d9a545d9 | ownerdemo2    | active |
| f0c4f3ff-cf59-4156-9c9a-b6ee5860807c | testsnap1     | active |
+--------------------------------------+---------------+--------+
```

图 13.3.45　删除快照后查看云平台镜像

需要注意，image delete 命令是没有返回结果的，需要通过 openstack image list 命令查看镜像是否还存在，从反馈信息可以看到，已经成功删除 testsnap2 快照（镜像）。

接下来是如何在命令中使用快照创建启动实例。快照创建完成后，会被当作镜像上传，通过上面的反馈信息可以看到，所以只需用最简单创建启动实例的命令即可。首先查看下当前的实例类型模板所对应的 ID，命令如下，查询结果如图 13.3.46 所示。

```
[root@controller ~]# openstack flavor list
```

```
+--------------------------------------+----------+------+------+-----------+-------+-----------+
| ID                                   | Name     | RAM  | Disk | Ephemeral | VCPUs | Is Public |
+--------------------------------------+----------+------+------+-----------+-------+-----------+
| 0086f4af-ea70-4a57-8920-eb90f2829dcc | myflavor | 1024 | 15   | 0         | 1     | True      |
| 98ce29c5-8963-4db3-a19e-6a8aa9ece574 | small    | 1024 | 20   | 0         | 1     | True      |
+--------------------------------------+----------+------+------+-----------+-------+-----------+
```

图 13.3.46 查看类型模板所对应的 ID

接下来使用已经上传的快照镜像创建新的实例，命名为 snapinstance，使用 myflavor 实例类型模板，命令如下。

```
[root@controller ~]# openstack server create snapinstance --flavor myflavor --image testsnap1 --network ext-net
```

从反馈信息中可以看到，实例正在创建中，使用 openstack server list 命令查看是否创建成功，命令如下，其结果如图 13.3.47 所示。

```
[root@controller ~]# openstack server list
```

```
+--------------------------------------+--------------+---------+--------------------------+--------------+----------+
| ID                                   | Name         | Status  | Networks                 | Image        | Flavor   |
+--------------------------------------+--------------+---------+--------------------------+--------------+----------+
| 5b948d0f-8557-4c54-b734-5bb5a34eefc4 | snapinstance | ACTIVE  | ext-net=192.168.200.124  | testsnap1    | myflavor |
| 3585fd55-8803-4dbf-b8f3-d3b7409631cc | test         | ACTIVE  | ext-net=192.168.200.78   | cirros-d190515 | myflavor |
| 9a8887c0-1a63-48a8-99df-764a122dc0ee | cirros2      | ACTIVE  | int-net=10.0.0.187       | cirros-d190515 | myflavor |
| 0a65606c-89c1-43d9-b875-436a03a11d83 | test-ext     | ACTIVE  | ext-net=192.168.200.83   | cirros-d190515 | myflavor |
| 01ebb576-88ec-46e8-b28e-314d873a573c | cirros       | SHUTOFF | int-net=10.0.0.71        | cirros       | small    |
+--------------------------------------+--------------+---------+--------------------------+--------------+----------+
```

图 13.3.47 查看 snapinstance 虚拟机实例是否创建成功

从图 13.3.48 的反馈信息中可以看到，此时 snapinstance 实例已经处于"Active"状态，表明已经成功用快照创建实例。

```
+-------------------------------------+------------------------------------------------------------+
| Field                               | Value                                                      |
+-------------------------------------+------------------------------------------------------------+
| OS-DCF:diskConfig                   | MANUAL                                                     |
| OS-EXT-AZ:availability_zone         | nova                                                       |
| OS-EXT-SRV-ATTR:host                | compute                                                    |
| OS-EXT-SRV-ATTR:hypervisor_hostname | compute                                                    |
| OS-EXT-SRV-ATTR:instance_name       | instance-00000019                                          |
| OS-EXT-STS:power_state              | Running                                                    |
| OS-EXT-STS:task_state               | None                                                       |
| OS-EXT-STS:vm_state                 | active                                                     |
| OS-SRV-USG:launched_at              | 2019-12-15T14:41:24.000000                                 |
| OS-SRV-USG:terminated_at            | None                                                       |
| accessIPv4                          |                                                            |
| accessIPv6                          |                                                            |
| addresses                           | ext-net=192.168.200.124                                    |
| config_drive                        |                                                            |
| created                             | 2019-12-15T14:40:58Z                                       |
| flavor                              | myflavor (0086f4af-ea70-4a57-8920-eb90f2829dcc)            |
| hostId                              | 0c1a7786589bff645646e345f1bb56c4ef255dd6d06d2fc825094339   |
| id                                  | 5b948d0f-8557-4c54-b734-5bb5a34eefc4                       |
| image                               | testsnap1 (f0c4f3ff-cf59-4156-9c9a-b6ee5860807c)           |
| key_name                            | None                                                       |
| name                                | snapinstance                                               |
| progress                            | 0                                                          |
| project_id                          | dab3559d05534ff9815746bb3bf20864                           |
| properties                          |                                                            |
| security_groups                     | name='default'                                             |
| status                              | ACTIVE                                                     |
| updated                             | 2019-12-15T14:41:24Z                                       |
| user_id                             | 0165a8c49c4647118053a06ce2fbce68                           |
| volumes_attached                    |                                                            |
+-------------------------------------+------------------------------------------------------------+
```

图 13.3.48 从 snapoftest 快照启动虚拟机实例的详细信息

13.3 实训项目 13 Nova 基本运维命令及其应用

接下来，观察从快照启动的实例与从镜像启动的实例的关联与区别。登录 Dashboard，分别单击创建快照的虚拟机 test 和从 test 快照启动的实例 snapinstance 的"控制台"按钮（在"更多"按钮里），登录实例进行查看，Linux 系统有一个特点，即对原有系统进行复制并重新启动后，其网卡号会发生变化以及网卡所对应的 MAC 地址都发生变化。分别在两台虚拟机中输入 ip a 命令，查看其网卡信息，如图 13.3.49 和图 13.3.50 所示。

图 13.3.49 test 虚拟机实例的网卡信息

图 13.3.50 snapinstance 虚拟机实例的网卡信息

可以再从其他方面观察它们的关联与区别，如磁盘信息，如图 13.3.51 和图 13.3.52 所示。从图中可以看到，它们在磁盘方面的信息是相同的。由此可以再次证明，由快照启动的实例是基于拍摄快照的实例的。

图 13.3.51 test 虚拟机实例的磁盘分区信息

图 13.3.52 snapinstance 虚拟机实例的磁盘分区信息

接下来是如何从卷创建启动实例。若要实现从"卷"创建启动实例，首先需要部署 OpenStack 的 Cinder 服务（部署 Cinder 服务在本书第 11 章 Cinder 的安装与其配置中有具体讲解和示范)。为理解从"卷"启动实例的全过程，首先在 Dashboard 界面上演示如何通过"卷"创建启动实例。

首先，需要通过"卷"页面创建一个"卷"。在"卷"页面，单击"创建卷"按钮打开如图 13.3.53 所示界面，然后在"卷来源"栏下拉列表中选择"镜像"选项，接着选择相应的镜像，和可用域(如果只有一个，系统会默认选择的)，完成后单击"创建卷"按钮。

图 13.3.53　创建云硬盘

创建一个 1GB，镜像为 cirros，命名为 myInstanceVolume 的云启动盘，创建后的卷在 Dashboard 界面中如图 13.3.54 所示。

图 13.3.54　创建好的名为 myInstanceVolume 的云硬盘

然后选择"计算"选项，在其中实例页面中单击"创建实例"按钮，给虚拟机命名为 myVolumeInstance。然后是选择源，在"选择源"的下拉列表中，选择"卷"选项，接着在出现的"可用"选项中选择刚刚创建的云启动盘 myInstanceVolume，如图 13.3.55 所示。

图 13.3.55　从 myInstanceVolume 卷启动虚拟机实例

选择实例类型、网络和合适的安全组（如之前创建的 testsec，可 ping 可 ssh 登录）之后，单击"创建实例"按钮，即可创建新的实例，实例成功启动后将出现在如图 13.3.56 所示的列表中。

图 13.3.56　成功启动的 myVolumeInstance 虚拟机实例

图 13.3.56 中，虚拟机状态为"运行"，电源状态为"运行中"，已经获取到 IP 地址，表明已经成功运行起来了。由于是从"卷"启动，会发现"镜像名称"列为"-"。如果安全组设置没有问题，可以通过 SSH 方式登录虚拟机。

通过 Dashboard 界面的演示，了解了从"卷"启动实例的过程。通过命令的方式也可以实现从"卷"启动实例。

首先，需要使用镜像创建一个"卷"，通过命令查看可用的镜像，命令如下，其结果如图 13.3.57 所示。

```
[root@controller ~]# nova image-list
```

```
+--------------------------------------+----------------+--------+
| ID                                   | Name           | Status |
+--------------------------------------+----------------+--------+
| 2d698a77-ffb8-4a42-8890-1b839f67f55e | cirros         | active |
| da18f4c8-3131-4fb7-8b0a-015ec1605010 | cirros-d190515 | active |
| 3347a41a-85d9-4998-8633-1f79d9a545d9 | ownerdemo2     | active |
| f0c4f3ff-cf59-4156-9c9a-b6ee5860807c | testsnap1      | active |
+--------------------------------------+----------------+--------+
```

图 13.3.57　查询当前镜像的反馈结果

图 13.3.60 中列出了 cirros-d190515 镜像，接着使用 cirros-d190515 镜像创建一个名为

testvolume，大小为 1GB 的"卷"，命令如下，反馈结果如图 13.3.58 所示。

```
[root@controller ~]# openstack volume create testvolume --image cirros-d190515 --size 1
+---------------------+--------------------------------------+
| Field               | Value                                |
+---------------------+--------------------------------------+
| attachments         | []                                   |
| availability_zone   | nova                                 |
| bootable            | false                                |
| consistencygroup_id | None                                 |
| created_at          | 2019-12-18T15:20:47.000000           |
| description         | None                                 |
| encrypted           | False                                |
| id                  | adde1695-8e13-4b51-9f36-cd29d69cfba4 |
| migration_status    | None                                 |
| multiattach         | False                                |
| name                | testvolume                           |
| properties          |                                      |
| replication_status  | None                                 |
| size                | 1                                    |
| snapshot_id         | None                                 |
| source_volid        | None                                 |
| status              | creating                             |
| type                | None                                 |
| updated_at          | None                                 |
| user_id             | 0165a8c49c4647118053a06ce2fbce68     |
+---------------------+--------------------------------------+
```

图 13.3.58　命令创建云启动盘的反馈结果

使用命令进行查看是否创建成功，命令如下，其反馈结果如图 13.3.59 所示。

```
[root@controller ~]# openstack volume list
+--------------------------------------+------------------+-----------+------+----------+
| ID                                   | Name             | Status    | Size | Attached |
+--------------------------------------+------------------+-----------+------+----------+
| adde1695-8e13-4b51-9f36-cd29d69cfba4 | testvolume       | available | 1    |          |
| 0262faf9-6fa4-4e29-b835-ab84e8768027 | myInstanceVolume | in-use    | 1    | Attached |
+--------------------------------------+------------------+-----------+------+----------+
```

图 13.3.59　确认"卷"创建成功

从图 13.3.59 中可以看到创建的卷 testvolume，接着使用创建好的卷来启动一个新的实例。参考前面在 Dashboard 界面创建的过程，这里需要选择一个合适的类型模板，先查看可用的类型模板，命令如下，其结果如图 13.3.60 所示。

```
[root@controller ~]# openstack flavor list
+--------------------------------------+----------+------+------+-----------+-------+-----------+
| ID                                   | Name     | RAM  | Disk | Ephemeral | VCPUs | Is Public |
+--------------------------------------+----------+------+------+-----------+-------+-----------+
| 0086f4af-ea70-4a57-8920-eb90f2829dcc | myflavor | 1024 | 15   | 0         | 1     | True      |
| 98ce29c5-8963-4db3-a19e-6a8aa9ece574 | small    | 1024 | 20   | 0         | 1     | True      |
+--------------------------------------+----------+------+------+-----------+-------+-----------+
```

图 13.3.60　查看当前类型模板

从图 13.3.60 中可以看到两条实例类型模板 myflavor 和 small，现在创建一个名为 testinstance，使用 small 类型模板和 testvolume 卷的虚拟机实例，命令如下，其反馈结果如图 13.3.61 所示。

```
[root@controller ~]# openstack server create testinstance --flavor small --volume testvolume --network ext-net
```

```
Field                                   | Value
----------------------------------------+------------------------------------------------
OS-DCF:diskConfig                       | MANUAL
OS-EXT-AZ:availability_zone             |
OS-EXT-SRV-ATTR:host                    | None
OS-EXT-SRV-ATTR:hypervisor_hostname     | None
OS-EXT-SRV-ATTR:instance_name           |
OS-EXT-STS:power_state                  | NOSTATE
OS-EXT-STS:task_state                   | scheduling
OS-EXT-STS:vm_state                     | building
OS-SRV-USG:launched_at                  | None
OS-SRV-USG:terminated_at                | None
accessIPv4                              |
accessIPv6                              |
addresses                               |
adminPass                               | jrBGobn87won
config_drive                            |
created                                 | 2019-12-18T15:30:46Z
flavor                                  | small (98ce29c5-8963-4db3-a19e-6a8aa9ece574)
hostId                                  |
id                                      | c10cba66-a229-4bfe-bebc-f8c3880dc559
image                                   |
key_name                                | None
name                                    | testinstance
progress                                | 0
project_id                              | dab3559d05534ff9815746bb3bf20864
properties                              |
security_groups                         | name='default'
status                                  | BUILD
updated                                 | 2019-12-18T15:30:46Z
user_id                                 | 0165a8c49c4647118053a06ce2fbce68
volumes_attached                        |
```

图 13.3.61　创建 testinstance 虚拟机实例的反馈结果

从反馈信息中可以看到，成功创建了实例，通过命令方式创建的实例，可以登录 Dashboard 界面进行查看，如图 13.3.62 所示。

实例名称	镜像名称	IP 地址	实例类型	密钥对	状态	可用域	任务	电源状态	时间	动作
testinstance	cirros-d190515	192.168.200.251	small	-	运行	nova	无	运行中	2 minutes	创建快照

图 13.3.62　在 Dashboard 界面中查看创建好的 testinstance 虚拟机实例

图 13.3.62 显示 testinstance 的状态为"运行"，电源状态为"运行中"，表明已经成功从卷启动实例。

步骤 4：配额管理。

云平台的资源是基于物理设备的，是有限的，为防止在没有任何通知的情况下系统资源被耗尽，用户需要设置配额来管理有限的资源，配额是一个可以进行控制修改的范围。目前在 OpenStack 云平台中配额是基于项目的，不同的项目可以有不同的配额方案，项目中所属的用户使用相同的配额方案。同时，OpenStack 存在一个默认公有配额方案即 default，它是全局配额方案，默认情况下，项目没有配置自己独有的配额管理方案时，就是用系统提供的 default 方案，可以通过命令查看 default 方案的具体细节，命令如下。

```
[root@controller ~]# openstack quota show --default
```

```
+---------------------+----------------------------------------------------------------------+
| Field               | Value                                                                |
+---------------------+----------------------------------------------------------------------+
| backup-gigabytes    | 1000                                                                 |
| backups             | 10                                                                   |
| cores               | 20                                                                   |
| fixed-ips           | -1                                                                   |
| floating-ips        | 50                                                                   |
| gigabytes           | 1000                                                                 |
| groups              | 10                                                                   |
| health_monitors     | None                                                                 |
| injected-file-size  | 10240                                                                |
| injected-files      | 5                                                                    |
| injected-path-size  | 255                                                                  |
| instances           | 10                                                                   |
| key-pairs           | 100                                                                  |
| l7_policies         | None                                                                 |
| listeners           | None                                                                 |
| load_balancers      | None                                                                 |
| location            | Munch({'project': Munch({'domain_name': 'Default', 'domain_id': None, 'name': 'adm |
| name                | None                                                                 |
| networks            | 100                                                                  |
| per-volume-gigabytes| -1                                                                   |
| pools               | None                                                                 |
| ports               | 500                                                                  |
| project             | None                                                                 |
| project_name        | admin                                                                |
| properties          | 128                                                                  |
| ram                 | 51200                                                                |
| rbac_policies       | 10                                                                   |
| routers             | 10                                                                   |
| secgroup-rules      | 100                                                                  |
| secgroups           | 10                                                                   |
| server-group-members| 10                                                                   |
| server-groups       | 10                                                                   |
| snapshots           | 10                                                                   |
| subnet_pools        | -1                                                                   |
| subnets             | 100                                                                  |
| volumes             | 10                                                                   |
+---------------------+----------------------------------------------------------------------+
```

图 13.3.63　查看系统默认配额方案的反馈结果

从图 13.3.63 的反馈信息中可以看到一些系统的默认方案，以下是反馈信息中的常用参数，instance 参数表明了默认情况下单项目能够启动的虚拟机实例的个数，当前值为 10 个，cores 表明单项目所有虚拟机实例所使用的虚拟内核的上限值，这里值为 20。假设虚拟机实例为 10 个，平均每个虚拟机两个内核，需要指出的是这里的内核上限和实际物理服务器的 CPU 内核数并没有必然联系，在 OpenStack 云平台中如果物理服务器是 12 核的，这里的值仍然可以超过 12，系统可以通过虚拟化技术对 CPU 进行分时复用，这里的内核数仅仅是逻辑内核，但是其性能确实受到物理服务器 CPU 性能的影响；ram 参数表明了单个项目下所启用的全部虚拟机内存总和的上限值，单位是 MB；floating_ips 参数标明了单个项目可以使用浮动 IP 的最大值，浮动 IP 在采用 GRE 网络类型的时候用于将实例映射到外网的 IP 地址；security_group 参数标明单个项目可以创建安全组的上限，这里值为 10。通过 admin 用户登录云平台，进入"计算"选项的"概况"页面即可通过 Dashboard 界面查看当前用户对应项目的配额及其使用情况，如图 13.3.64 所示。

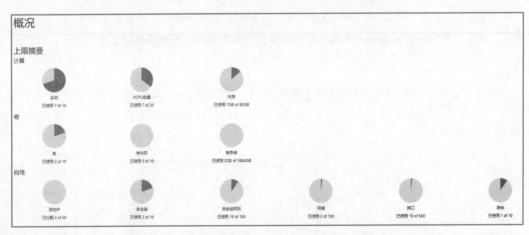

图 13.3.64　通过 admin 用户登录后查看配额及其使用情况

图 13.3.64 显示的是 admin 用户的配额概况。下面使用没有管理权限，属于 project2 项目的用户 user4 登录查看，如图 13.3.65 所示。

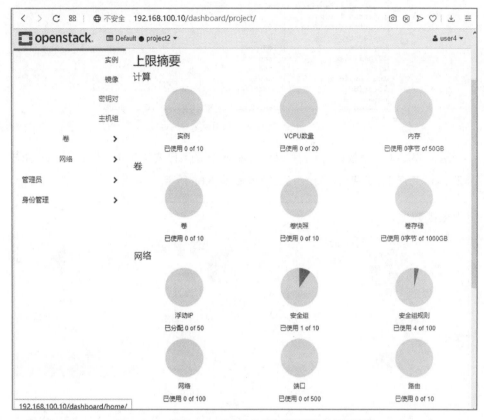

图 13.3.65　通过 user4 用户登录后查看配额及其使用情况

从图 13.3.65 中可以看到 user4 虽然没有正在已经创建的实例，所示与实例相关的配额都是和 admin 用户一样的，在 OpenStack 中同一配额管理方案的具体使用是基于项目的。

作为管理员会发现，默认的配额 default 方案有时不一定是合适的，此时需要对现有 default 方案的具体内容进行相关调整，使用查看 OpenStack 有关配额管理的相关命令，命令如下：

```
[root@controller ~]# openstack quota
openstack: 'quota' is not an openstack command. See 'openstack --help'.
Did you mean one of these?
  quota list
  quota set
  quota show
```

从反馈信息中可以清晰地看到每条命令所对应的作用，在已经知道如何查看 default 方案的内容后，下面修改一下实例的配额。以 project2 项目为例，将其 instances 改为 18，命令如下：

```
[root@controller ~]# openstack quota set --instances 18 project2
[root@controller ~]# openstack quota show project2
```

```
floating-ips         | 50
gigabytes            | 1000
groups               | 10
health_monitors      | None
injected-file-size   | 10240
injected-files       | 5
injected-path-size   | 255
instances            | 18
key-pairs            | 100
l7_policies          | None
listeners            | None
load_balancers       | None
location             | Munch({'project': Munch({'domain_name'
name                 | None
```

图 13.3.66　修改 project2 项目配额后查询结果

从图 13.3.66 的反馈信息中，可以看到 instances 的值已被改为 18，分别用 admin 用户以及 user4 用户登录云平台，如图 13.3.67 和图 13.3.68 所示。

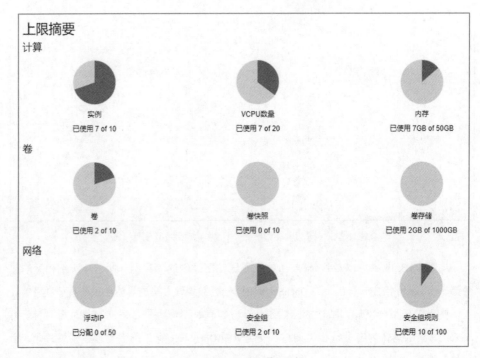

图 13.3.67　admin 项目配额

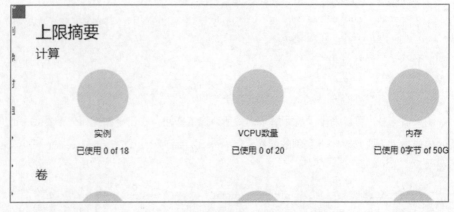

图 13.3.68　project2 项目配额

图 13.3.67 是属于 admin 项目的 admin 用户的配额管理,图 13.3.68 是属于 project2 项目的 user4 用户的配额管理,可以表明配额管理是基于项目的,当只更改指定项目时,非项目所属用户的配额是不变的。目前 OpenStack 云平台暂不支持在 Dashboard 界面更改配额方案。

到此,关于 Nova 的基本运维命令及其应用基本讲解完毕,其余未讲到的部分可以自行参考帮助文档进行试验。

13.4 实训项目 14 Neutron 基本运维命令及其应用

1. 实训前提环境

安装成功的 OpenStack 云平台环境,可以通过 Dashboard 登录云平台,也可以通过终端连接 Controller,进行命令的操作。

电子教案 Neutron 基本运维命令及其应用

2. 实训涉及节点

Controller 节点

3. 实训目标

① 通过命令行对子网进行变更。
② 通过 Dashboard 界面对子网进行变更。
③ Neutron Vlan 网络的配置。

PPT Neutron 基本运维命令及其应用

4. 实训内容

(1)通过命令行对子网进行变更

当用户需要变更自己的当前子网时,需要对当前子网进行删除操作,子网删除可以通过命令和 Dashboard 界面进行操作。

例如,在某个应用场景中,运维人员误操作将 192.168.200.0/26 网络写成了 192.168.200.0/24 网络分配给了虚拟机实例,为保证地址划分正确,必须删除并重新创建正确的子网。下面就以命令操作和 Dashboard 界面操作分别介绍如何删除一个已经存在的子网并创建新的子网。

微课 17 Neutron 基本运维命令及其应用

通过 openstack subnet list 命令查看现有子网,命令如下,如图 13.4.1 所示。

```
[root@controller ~]# openstack subnet list
```

ID	Name	Network	Subnet
cc2b2bba-ef2e-4acf-a1ea-3450ccee5c1a	int-subnet	9067dda9-deeb-4e4c-b135-afbd40c28d84	10.0.0.0/24
f9156ba4-3db7-42bf-90c9-20f0fc015674	external-subnet	6a7e71a0-2d6c-4359-b2ca-3f7559e2ba33	192.168.200.0/24

图 13.4.1 查看现有子网的反馈结果

通过 openstack subnet delete 命令对现有子网进行删除,命令如下。

```
[root@controller jiaoben]# openstack subnet delete external-subnet
```

发现系统提示删除失败:

```
Failed to delete subnet with name or ID 'external-subnet': ConflictException:
```

```
409: Client Error for url: http://controller:9696/v2.0/subnets/f9156ba4-3db7-42bf-
90c9-20f0fc015674, Unable to complete operation on subnet f9156ba4-3db7-42bf-90c9-
20f0fc015674: One or more ports have an IP allocation from this subnet.
1 of 1 subnets failed to delete.
```

根据系统提示，可以判断子网已经被占用，初步判断有虚拟机实例占用该子网，需删除正在使用该子网的虚拟机实例，通过 openstack server list 命令来查看虚拟机实例的运行状态，命令如下，结果如图 13.4.2 所示。

```
[root@controller jiaoben]# openstack server list
```

ID	Name	Status	Networks	Image	Flavor
c10cba66-a229-4bfe-bebc-f8c3880dc559	testinstance	ACTIVE	ext-net=192.168.200.251		small
8072b9fa-75c1-4634-8cb0-33f11f832ab2	myVolumeInstance	ACTIVE	ext-net=192.168.200.184		myflavor
5b948d0f-8557-4c54-b734-5bb5a34eefc4	snapinstance	SHUTOFF	ext-net=192.168.200.124	testsnap1	myflavor
3585fd55-8803-4dbf-b8f3-d3b7409631cc	test	SHUTOFF	ext-net=192.168.200.78	cirros-d190515	myflavor
9a8887c0-1a63-48a8-99df-764a122dc0ee	cirros2	SHUTOFF	int-net=10.0.0.187	cirros-d190515	myflavor
0a65606c-89c1-43d9-b875-436a03a11d83	test-ext	SHUTOFF	ext-net=192.168.200.83	cirros-d190515	myflavor
01ebb576-88ec-46e8-b28e-314d873a573c	cirros	SHUTOFF	int-net=10.0.0.71	cirros	small

图 13.4.2　查看虚拟机实例信息的反馈结果

从图 13.4.2 中可以看到有一台虚拟机实例在使用网络 ext-net，通过 openstack network list 命令可以查看网络所有的网络，命令如下，结果如图 13.4.3 所示。

```
[root@controller jiaoben]# neutron network list
```

ID	Name	Subnets
6a7e71a0-2d6c-4359-b2ca-3f7559e2ba33	ext-net	f9156ba4-3db7-42bf-90c9-20f0fc015674
9067dda9-deeb-4e4c-b135-afbd40c28d84	int-net	cc2b2bba-ef2e-4acf-a1ea-3450ccee5c1a

图 13.4.3　查看网络信息的反馈结果

通过网络信息和子网信息，可以看出有 ext-net 中有一个子网 f9156ba4-3db7-42bf-90c9-20f0fc015674，然后通过这个子网号，在图 13.4.1 中，可以得知子网名称是 external-subnet。下面查看该子网的详细信息，命令如下，结果如图 13.4.4 所示。

```
[root@controller jiaoben]# openstack subnet show external-subnet
```

Field	Value
allocation_pools	192.168.200.2-192.168.200.254
cidr	192.168.200.0/24
created_at	2019-12-14T09:08:41Z
description	
dns_nameservers	
enable_dhcp	True
gateway_ip	192.168.200.1
host_routes	
id	f9156ba4-3db7-42bf-90c9-20f0fc015674
ip_version	4
ipv6_address_mode	None
ipv6_ra_mode	None
location	Munch({'project': Munch({'domain_name': 'Default', 'domain_id': None, 'name': 'admin', 'i
name	external-subnet
network_id	6a7e71a0-2d6c-4359-b2ca-3f7559e2ba33
prefix_length	None
project_id	dab3559d05534ff9815746bb3bf20864
revision_number	0
segment_id	None
service_types	
subnetpool_id	None
tags	
updated_at	2019-12-14T09:08:41Z

图 13.4.4　查看子网 external-subnet 详细信息的反馈结果

由于很多虚拟机实例使用的子网是 external-subnet，所以需要对虚拟机实例 testinstance 进行分离网络操作。可以通过 openstack server remove network 命令对其进行删除，再通过 openstack server list 命令查看实例的网络已经消失，命令如下，结果如图 13.4.5 所示。

```
[root@controller ~]# openstack server remove network testinstance ext-net
```

ID	Name	Status	Networks	Image	Flavor
c10cba66-a229-4bfe-bebc-f8c3880dc559	testinstance	ACTIVE			small
8072b9fa-75c1-4634-8cb0-33f11f832ab2	myVolumeInstance	ACTIVE	ext-net=192.168.200.184		myflavor
5b948d0f-8557-4c54-b734-5bb5a34eefc4	snapinstance	SHUTOFF	ext-net=192.168.200.124	testsnap1	myflavor
3585fd55-8803-4dbf-b8f3-d3b7409631cc	test	SHUTOFF	ext-net=192.168.200.78	cirros-d190515	myflavor
9a8887c0-1a63-48a8-99df-764a122dc0ee	cirros2	SHUTOFF	int-net=10.0.0.187	cirros-d190515	myflavor
0a65606c-89c1-43d9-b875-436a03a11d83	test-ext	SHUTOFF	ext-net=192.168.200.83	cirros-d190515	myflavor
01ebb576-88ec-46e8-b28e-314d873a573c	cirros	SHUTOFF	int-net=10.0.0.71	cirros	small

图 13.4.5　查看实例消失后的反馈结果

当所有虚拟机实例分离网络 ext-net 后就可以对子网进行操作了，首先通过 openstack subnet delete 命令来删除已经存在的子网 external-subnet，命令如下，结果如图 13.4.6 所示。

```
[root@controller ~]# openstack subnet delete external-subnet
[root@controller ~]# openstack subnet list
```

ID	Name	Network	Subnet
cc2b2bba-ef2e-4acf-a1ea-3450ccee5c1a	int-subnet	9067dda9-deeb-4e4c-b135-afbd40c28d84	10.0.0.0/24

图 13.4.6　查看删除子网后的反馈结果

接下来创建新子网，可以通过 openstack subnet create 命令来创建新的子网，在 Controller 节点输入，命令如下。

```
[root@controller ~]# openstack subnet create external-subnet --gateway 192.168.200.1 --network ext-net --subnet-range 192.168.200.0/26
```

需要注意，external-subnet 表示新创建的子网所对应的项目网络的名称，ext-net 是该子网的名称。另外，由于子网掩码向右移动了两位，故 IP 地址要重新划分，可以根据实际情况进行变动。

创建完成后，可以通过 openstack subnet list 命令来查看子网信息，命令如下，结果如图 13.4.7 所示。

```
[root@controller ~]# openstack subnet list
```

ID	Name	Network	Subnet
14580b6e-225a-41d5-bd6a-eb668c926ee0	external-subnet	6a7e71a0-2d6c-4359-b2ca-3f7559e2ba33	192.168.200.0/26
cc2b2bba-ef2e-4acf-a1ea-3450ccee5c1a	int-subnet	9067dda9-deeb-4e4c-b135-afbd40c28d84	10.0.0.0/24

图 13.4.7　查看子网信息反馈结果

在该列表中，可以看到新创建的子网 external-subnet，然后通过 openstack server add network 命令把新网络附加到实例上，为使之生效，需重启虚拟机。

（2）通过 Dashboard 页面对子网进行变更

步骤 1：分离正在使用该子网的虚拟机实例的网络。

首先登录 Dashboard 页面，在"实例"中"项目"栏中可以查看到正在运行的虚拟机实例，如图 13.4.8 所示。

图 13.4.8　查看正在运行的虚拟机实例

可以看到该虚拟机实例使用的网络是用户需要更改的子网，故需先终止该虚拟机实例，单击如图 13.4.8 所示右侧的"更多"下拉按钮▼，然后在弹出的下拉列表中选择"分离接口"选项。

分离接口操作会删除虚拟机实例的网卡，片刻后，虚拟机 IP 地址即被删除，如图 13.4.9 所示。

步骤 2：删除已存在的子网。

此时虚拟机实例与需要更改的子网关联的 IP 地址已被删除，接下来就可以对网络进行操作，在左侧选择"管理员"选项卡，在选项卡中选择"网络"子选项卡，如图 13.4.10 所示。

选择"ext-net"网络，"ext-net"网络选项自动蓝色标亮，自动跳转到网络详情：ext-net 页面，如图 13.4.11 所示。

图 13.4.9　虚拟机删除后实例栏没有任何条目

13.4 实训项目 14 Neutron 基本运维命令及其应用

图 13.4.10 查看 dashboard 界面网络栏目

图 13.4.11 查看 sharednet1 网络条目

在该页面中，可以看到所创建的 192.168.200.0/24 的子网 "ext-subnet"。以下对该子网进行删除操作，单击右侧的"更多"下拉按钮▼，在弹出的下拉列表中选择"删除子网"选项，如图 13.4.12 所示。

图 13.4.12 通过 dashboard 界面删除子网

此时，会提示"成功: 已删除的子网: ext-subnet"，则表示删除成功，如图 13.4.13 所示。

图 13.4.13 提示删除子网成功

步骤 3：创建新子网。

单击左侧的"创建子网"按钮，在弹出的"创建子网"标签页中的"子网"子标签输入子网信息，如图 13.4.14 所示。

图 13.4.14　在标签页中输入子网信息

输入完成后，单击"下一步"按钮，在"子网详情"子标签中输入信息，如图 13.4.15 所示。

上述操作完成后，单击右下方的"创建"按钮，弹出"成功: 已新增子网'ext-subnet2'"提示后，表示新的子网已经成功创建，如图 13.4.16 所示。

步骤 4：重新连接虚拟机接口。

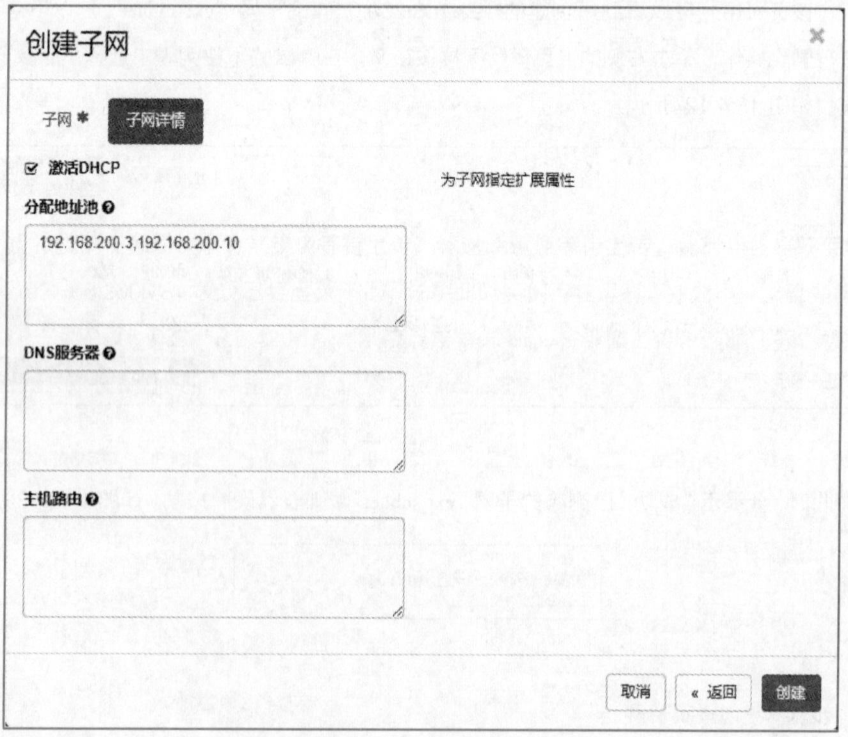

图 13.4.15　输入子网详情

13.4 实训项目 14 Neutron 基本运维命令及其应用

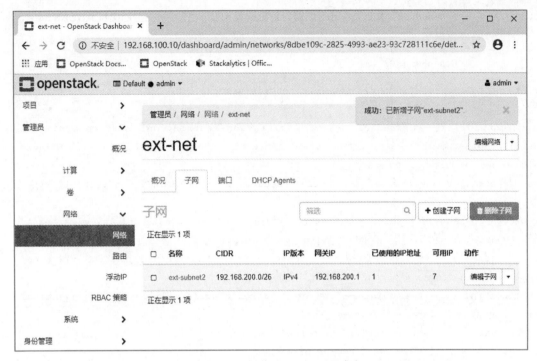

图 13.4.16 新增子网 "subnet2" 成功

选择"项目"→"实例"标签中，单击实例右方下拉菜单的连接接口，再次为实例选择刚刚重新创建子网的网络，可以看到 Neutron 已经为虚拟机实例分配了新创建的子网的 IP 地址，如图 13.4.17 所示，为使之生效，需重启虚拟机。

图 13.4.17 在实例栏中查看启动的虚拟机获取到子网 IP 地址

至此，有关 Neutron 变更子网的操作已经全部结束。

（3）Neutron Vlan 网络的配置

在实际生产环境中，使用 Neutron 搭建 Flat 网络并不常见，Neutron 可以支持多种网络模式，包括 Flat、Vlan、GRE 等，以下是 Vlan 网络的配置过程，以及创建 Vlan 网络。

步骤 1：配置 ML2 插件。

对 Neutron 的 ML2 插件进行配置，首先配置 ML2 支持网络类型，使其为 VLAN 网络，命令行如下。

Controller 节点：

```
[root@controller ~]# crudini --set /etc/neutron/neutron.conf DEFAULT service_plugins router
[root@controller ~]# crudini --set /etc/neutron/neutron.conf service_providers FIREWALL:Iptables:neutron.agent.linux.iptables_firewall.OVSHybridIptablesFirewallDriver:default
[root@controller ~]# crudini --set /etc/neutron/plugins/ml2/ml2_conf.ini ml2_type_flat flat_networks external
[root@controller ~]# crudini --set /etc/neutron/plugins/ml2/ml2_conf.ini ml2 tenant_network_types vlan
[root@controller ~]# crudini --set /etc/neutron/plugins/ml2/ml2_conf.ini ml2_type_vlan network_vlan_ranges external:100:150
[root@controller ~]# crudini --set /etc/neutron/plugins/ml2/openvswitch_agent.ini securitygroup firewall_driver iptables_hybrid
[root@controller ~]# systemctl restart neutron-server
[root@controller ~]# ovs-vsctl add-br br-ex
[root@controller ~]# ovs-vsctl add-port br-ex ens34
[root@controller ~]# cat > /etc/sysconfig/network-scripts/ifcfg-ens34 <<EOF
DEVICE=ens34
TYPE=Ethernet
BOOTPROTO=none
ONBOOT=yes
EOF
[root@controller ~]# systemctl restart network
[root@controller ~]# crudini --set /etc/neutron/l3_agent.ini DEFAULT external_network_bridge br-ex
[root@controller ~]# crudini --set /etc/neutron/plugins/ml2/openvswitch_agent.ini ovs bridge_mappings external:br-ex
```

Compute 节点：

```
[root@controller ~]# crudini --set /etc/neutron/plugins/ml2/ml2_conf.ini ml2_type_flat flat_networks external
[root@controller ~]# crudini --set /etc/neutron/plugins/ml2/ml2_conf.ini ml2 tenant_network_types vlan
[root@controller ~]# crudini --set /etc/neutron/plugins/ml2/ml2_conf.ini ml2_type_vlan network_vlan_ranges external:100:150
[root@controller ~]# crudini --set /etc/neutron/plugins/ml2/openvswitch_agent.ini securitygroup firewall_driver iptables_hybrid
[root@controller ~]# crudini --set /etc/neutron/l3_agent.ini DEFAULT external_network_bridge br-ex
[root@controller ~]# crudini --set /etc/neutron/plugins/ml2/openvswitch_agent.ini ovs bridge_mappings external:br-ex
```

```
[root@controller ~]# ovs-vsctl add-br br-ex
[root@controller ~]# ovs-vsctl add-port br-ex ens34
[root@controller ~]# cat > /etc/sysconfig/network-scripts/ifcfg-ens34 <<EOF
DEVICE=ens34
TYPE=Ethernet
BOOTPROTO=none
ONBOOT=yes
EOF
[root@controller ~]# systemctl restart network
```

同样，也可通过修改 /etc/neutron/plugins/ml2/ml2_conf.ini 文件来实现同样的效果。

步骤 2：重启 Neutron 相关服务。

重启 Neutron 相关服务，需要注意，在重启 Compute 节点的 Neutron 服务前，需要重启 openstack-nova-compute 服务，具体启动命令如下。

Controller 节点：

```
[root@controller ~]# systemctl restart neutron-openvswitch-agent
[root@controller ~]# systemctl restart neutron-metadata-agent
[root@controller ~]# systemctl restart neutron-l3-agent
[root@controller ~]# systemctl restart neutron-server
```

Compute 节点：

```
[root@compute ~]# systemctl restart openstack-nova-compute
[root@compute ~]# systemctl restart neutron-openvswitch-agent
[root@compute ~]# systemctl restart neutron-dhcp-agent
[root@compute ~]# systemctl restart neutron-metadata-agent
```

重启完成后，可以通过在 Controller 节点执行以下命令查看 Neutron 相关服务的运行状态和位置信息，命令如下，结果如图 13.4.18 所示。

```
[root@controller ~]# neutron agent-list
```

```
+--------------------------------------+--------------------+------------+-------------------+-------+
| ID                                   | Agent Type         | Host       | Availability Zone | Alive |
+--------------------------------------+--------------------+------------+-------------------+-------+
| 149a8874-f0e1-4066-a4ba-3039abe2d284 | Metadata agent     | compute    | None              | :-)   |
| 1b670ce4-4eda-415b-a729-9e6b2c13ebb1 | Open vSwitch agent | compute    | None              | :-)   |
| 5bdb8e9d-892c-4c72-a2fd-8d764b36a7b4 | Open vSwitch agent | controller | None              | :-)   |
| 6147b7e3-91fb-4a75-a760-95bfb6d9baee | DHCP agent         | controller | nova              | :-)   |
| 70756160-b688-44b1-bcbc-f6f71170e217 | L3 agent           | controller | nova              | :-)   |
| 8b488e9d-ffc9-4d3e-8dfc-e064877b7adf | Metadata agent     | controller | None              | :-)   |
+--------------------------------------+--------------------+------------+-------------------+-------+
```

图 13.4.18　查看 neutron 相关服务的反馈信息

步骤 3：创建 Vlan 网络。

通过 Neutron 创建一个小型 Vlan 网络。

首先，执行 Neutron 的 net-create 命令来创建虚拟外部网络，命令行如下，结果如图 13.4.19 所示。

```
[root@controller ~]# neutron net-create ext-net --router:external True --provider:physical_network external --provider:network_type flat
```

```
+-------------------------+-------------------------------------------+
| Field                   | Value                                     |
+-------------------------+-------------------------------------------+
| admin_state_up          | True                                      |
| availability_zone_hints |                                           |
| availability_zones      |                                           |
| created_at              | 2019-12-20T09:29:18Z                      |
| description             |                                           |
| id                      | 8067390d-f607-4bf1-92a4-c54a0960e0e6      |
| ipv4_address_scope      |                                           |
| ipv6_address_scope      |                                           |
| is_default              | False                                     |
| mtu                     | 1500                                      |
| name                    | ext-net                                   |
| port_security_enabled   | True                                      |
| project_id              | c492ca8bace64bd3b1f7ba67ea03d23a          |
| provider:network_type   | flat                                      |
| provider:physical_network | external                                |
| provider:segmentation_id |                                          |
| revision_number         | 1                                         |
| router:external         | True                                      |
| shared                  | False                                     |
| status                  | ACTIVE                                    |
| subnets                 |                                           |
| tags                    |                                           |
| tenant_id               | c492ca8bace64bd3b1f7ba67ea03d23a          |
| updated_at              | 2019-12-20T09:29:18Z                      |
+-------------------------+-------------------------------------------+
```

图 13.4.19　查看创建成功的反馈信息

最后，创建外部子网，外部地址在 192.168.200.100~192.168.200.200 区间，且关闭外部网络的 DHCP 功能，创建命令如下，结果如图 13.4.20 所示。

```
[root@controller ~]# neutron subnet-create ext-net --name ext-subnet --allocation-pool start=192.168.200.100,end=192.168.200.200 --disable-dhcp --gateway 192.168.200.1 192.168.200.0/24
```

```
Created a new subnet:
+-------------------+------------------------------------------------------+
| Field             | Value                                                |
+-------------------+------------------------------------------------------+
| allocation_pools  | {"start": "192.168.200.100", "end": "192.168.200.200"} |
| cidr              | 192.168.200.0/24                                     |
| created_at        | 2019-12-20T09:30:50Z                                 |
| description       |                                                      |
| dns_nameservers   |                                                      |
| enable_dhcp       | False                                                |
| gateway_ip        | 192.168.200.1                                        |
| host_routes       |                                                      |
| id                | 168d3018-d3c8-4ff2-9281-e9ee0dc0edee                 |
| ip_version        | 4                                                    |
| ipv6_address_mode |                                                      |
| ipv6_ra_mode      |                                                      |
| name              | ext-subnet                                           |
| network_id        | 8067390d-f607-4bf1-92a4-c54a0960e0e6                 |
| project_id        | c492ca8bace64bd3b1f7ba67ea03d23a                     |
| revision_number   | 0                                                    |
| service_types     |                                                      |
| subnetpool_id     |                                                      |
| tags              |                                                      |
| tenant_id         | c492ca8bace64bd3b1f7ba67ea03d23a                     |
| updated_at        | 2019-12-20T09:30:50Z                                 |
+-------------------+------------------------------------------------------+
```

图 13.4.20　查看子网创建的反馈信息

紧接着，创建虚拟机内部私有网络，命令行如下，结果如图 13.4.21 所示。

```
[root@controller ~]# neutron net-create internalnet
```

13.4 实训项目 14　Neutron 基本运维命令及其应用

```
Created a new network:
+---------------------------+--------------------------------------+
| Field                     | Value                                |
+---------------------------+--------------------------------------+
| admin_state_up            | True                                 |
| availability_zone_hints   |                                      |
| availability_zones        |                                      |
| created_at                | 2019-12-20T09:32:53Z                 |
| description               |                                      |
| id                        | 471e8212-dc14-46a0-9833-236268ec6ba1 |
| ipv4_address_scope        |                                      |
| ipv6_address_scope        |                                      |
| is_default                | False                                |
| mtu                       | 1458                                 |
| name                      | internalnet                          |
| port_security_enabled     | True                                 |
| project_id                | c492ca8bace64bd3b1f7ba67ea03d23a     |
| provider:network_type     | vlan                                 |
| provider:physical_network |                                      |
| provider:segmentation_id  | 85                                   |
| revision_number           | 1                                    |
| router:external           | False                                |
| shared                    | False                                |
| status                    | ACTIVE                               |
| subnets                   |                                      |
| tags                      |                                      |
| tenant_id                 | c492ca8bace64bd3b1f7ba67ea03d23a     |
| updated_at                | 2019-12-20T09:32:53Z                 |
+---------------------------+--------------------------------------+
```

图 13.4.21　创建虚拟机内部私有网络的反馈信息

接着，创建内部子网，虚拟机内部网络在 192.168.200.0/24 网段，命令如下，结果如图 13.4.22 所示。

```
[root@controller ~]# neutron subnet-create internalnet --name demo-subnet --gateway 10.0.0.1 10.0.0.0/24
```

```
Created a new subnet:
+-------------------+--------------------------------------------+
| Field             | Value                                      |
+-------------------+--------------------------------------------+
| allocation_pools  | {"start": "10.0.0.2", "end": "10.0.0.254"} |
| cidr              | 10.0.0.0/24                                |
| created_at        | 2019-12-20T09:36:50Z                       |
| description       |                                            |
| dns_nameservers   |                                            |
| enable_dhcp       | True                                       |
| gateway_ip        | 10.0.0.1                                   |
| host_routes       |                                            |
| id                | 6edccc71-8c19-449c-bb6e-1d2c123e5396       |
| ip_version        | 4                                          |
| ipv6_address_mode |                                            |
| ipv6_ra_mode      |                                            |
| name              | demo-subnet                                |
| network_id        | 471e8212-dc14-46a0-9833-236268ec6ba1       |
| project_id        | c492ca8bace64bd3b1f7ba67ea03d23a           |
| revision_number   | 0                                          |
| service_types     |                                            |
| subnetpool_id     |                                            |
| tags              |                                            |
| tenant_id         | c492ca8bace64bd3b1f7ba67ea03d23a           |
| updated_at        | 2019-12-20T09:36:50Z                       |
+-------------------+--------------------------------------------+
```

图 13.4.22　查看内部子网创建的反馈信息

接下来，创建一个名为 demo-router 的虚拟路由器，连接内外网，命令如下，结果如图 13.4.23 所示。

```
[root@controller ~]# neutron router-create demo-router
```

```
Created a new router:
+-------------------------+--------------------------------------+
| Field                   | Value                                |
+-------------------------+--------------------------------------+
| admin_state_up          | True                                 |
| availability_zone_hints |                                      |
| availability_zones      |                                      |
| created_at              | 2019-12-20T09:37:22Z                 |
| description             |                                      |
| distributed             | False                                |
| external_gateway_info   |                                      |
| flavor_id               |                                      |
| ha                      | False                                |
| id                      | b99a6724-f168-468a-b63b-2c608ae2ebb4 |
| name                    | demo-router                          |
| project_id              | c492ca8bace64bd3b1f7ba67ea03d23a     |
| revision_number         | 1                                    |
| routes                  |                                      |
| status                  | ACTIVE                               |
| tags                    |                                      |
| tenant_id               | c492ca8bace64bd3b1f7ba67ea03d23a     |
| updated_at              | 2019-12-20T09:37:22Z                 |
+-------------------------+--------------------------------------+
```

图 13.4.23　创建 demo-router 虚拟路由器的反馈信息

最后，在虚拟路由器上添加内部接口以及外部网关，添加一个内部接口，命令如下。

```
[root@controller ~]# neutron router-interface-add demo-router demo-subnet
```

添加成功后，可以看到成功提示：

```
Added interface 18978f79-7267-4e20-90d3-de041e15958f to router demo-router.
```

添加一个外部网关，命令如下。

```
[root@controller ~]# neutron router-gateway-set demo-router ext-net
```

添加成功后，同样也可以看到成功提示：

```
Set gateway for router demo-router
```

至此，云平台的 Vlan 类型网络已经全部创建完成。

13.5　实训项目 15　Cinder 基本运维命令及其应用

1. 实训前提环境

安装成功的原生 OpenStack 云平台环境，可以通过 Dashboard 登录云平台，也可以通过终端连接 Controller，进行命令的操作。

2. 实训涉及节点

Controller 节点

3. 实训目标

① 掌握通过命令创建删除云硬盘等基本运维操作。
② 掌握通过 Dashboard 界面创建删除云硬盘等基本运维操作。

4. 实训内容

步骤 1：通过命令行完成云硬盘运维。

可以使用 create 命令创建一个云硬盘，假设云硬盘名称为 volume1，大小为 1GB，命

13.5 实训项目15 Cinder基本运维命令及其应用

令如下,结果如图13.5.1所示。

```
[root@controller ~]# cinder create --display-name volume1 1
```

```
+---------------------------------+--------------------------------------+
| Property                        | Value                                |
+---------------------------------+--------------------------------------+
| attachments                     | []                                   |
| availability_zone               | nova                                 |
| bootable                        | false                                |
| consistencygroup_id             | None                                 |
| created_at                      | 2019-12-18T08:48:29.000000           |
| description                     | None                                 |
| encrypted                       | False                                |
| id                              | 32058478-c745-4b04-b0b5-6888927d9e6a |
| metadata                        | {}                                   |
| migration_status                | None                                 |
| multiattach                     | False                                |
| name                            | volume1                              |
| os-vol-host-attr:host           | None                                 |
| os-vol-mig-status-attr:migstat  | None                                 |
| os-vol-mig-status-attr:name_id  | None                                 |
| os-vol-tenant-attr:tenant_id    | c492ca8bace64bd3b1f7ba67ea03d23a     |
| replication_status              | None                                 |
| size                            | 1                                    |
| snapshot_id                     | None                                 |
| source_volid                    | None                                 |
| status                          | creating                             |
| updated_at                      | None                                 |
| user_id                         | ca4d4da261f5402787b9c0c87f174e58     |
| volume_type                     | None                                 |
+---------------------------------+--------------------------------------+
```

图13.5.1 查看创建云硬盘的返回信息

图13.5.1中云硬盘的名称display_name为volume1,大小size为1GB。可以使用rename命令修改云硬盘的名称,通过list命令查看云硬盘列表,命令如下,结果如图13.5.2所示。

```
[root@controller ~]# cinder list
```

```
+--------------------------------------+-----------+---------+------+-------------+----------+-------------+
| ID                                   | Status    | Name    | Size | Volume Type | Bootable | Attached to |
+--------------------------------------+-----------+---------+------+-------------+----------+-------------+
| 32058478-c745-4b04-b0b5-6888927d9e6a | available | volume1 | 1    | -           | false    |             |
+--------------------------------------+-----------+---------+------+-------------+----------+-------------+
```

图13.5.2 查看云硬盘反馈信息

从反馈的信息可以看出第1列的ID表示云硬盘的ID;第2列Status表示云硬盘当前的状态为available;第3列Name为云硬盘的名称为volume1;第4列Size表示云硬盘的大小为1GB;第5列Volume Type表示云硬盘的类型;第6列Bootable为false表示该云硬盘不可启动;第7列Attached to表示云硬盘挂在的云主机,这里云硬盘还未挂在到云主机上,所以为空。

通过rename命令修改云硬盘名称,再通过list命令查看云硬盘名称变化,命令如下,结果如图13.5.3所示。

```
[root@controller ~]# cinder rename volume1 volume2
[root@controller ~]# cinder list
```

```
+--------------------------------------+-----------+---------+------+-------------+----------+-------------+
| ID                                   | Status    | Name    | Size | Volume Type | Bootable | Attached to |
+--------------------------------------+-----------+---------+------+-------------+----------+-------------+
| 32058478-c745-4b04-b0b5-6888927d9e6a | available | volume2 | 1    | -           | false    |             |
+--------------------------------------+-----------+---------+------+-------------+----------+-------------+
```

图13.5.3 查看修改后云硬盘反馈信息

可以看到第3列设备名称Name已经被修改为volume2,还可以通过show命令查看云硬盘详细信息,命令如下,结果如图 13.5.4 所示。

```
[root@controller ~]# cinder show volume2
```

```
+-------------------------------+--------------------------------------+
| Property                      | Value                                |
+-------------------------------+--------------------------------------+
| attached_servers              | []                                   |
| attachment_ids                | []                                   |
| availability_zone             | nova                                 |
| bootable                      | false                                |
| consistencygroup_id           | None                                 |
| created_at                    | 2019-12-18T08:48:29.000000           |
| description                   | None                                 |
| encrypted                     | False                                |
| id                            | 32058478-c745-4b04-b0b5-6888927d9e6a |
| metadata                      |                                      |
| migration_status              | None                                 |
| multiattach                   | False                                |
| name                          | volume2                              |
| os-vol-host-attr:host         | compute@lvm#LVM                      |
| os-vol-mig-status-attr:migstat| None                                 |
| os-vol-mig-status-attr:name_id| None                                 |
| os-vol-tenant-attr:tenant_id  | c492ca8bace64bd3b1f7ba67ea03d23a     |
| replication_status            | None                                 |
| size                          | 1                                    |
| snapshot_id                   | None                                 |
| source_volid                  | None                                 |
| status                        | available                            |
| updated_at                    | 2019-12-18T08:51:00.000000           |
| user_id                       | ca4d4da261f5402787b9c0c87f174e58     |
| volume_type                   | None                                 |
+-------------------------------+--------------------------------------+
```

图 13.5.4 查看修改后云硬盘详细的反馈信息

从反馈的信息中可以看出第 2 行 attachment_ids 表示云硬盘挂载信息,这里为空;第 3 行 availability_zone 为云硬盘的作用域;第 4 行 bootable 为 false 表示该云硬盘不能作为启动盘;第 6 行 created_at 表示云硬盘创建时间;第 7 行 description 表示云硬盘描述信息;第 8 行 encrypted 为 False 表示云硬盘未加密;第 9 行 id 表示云硬盘的 ID 号;第 10 行 metadata 表示云硬盘的元数据;第 13 行 name 表示云硬盘的名称;第 14 行 os-vol-host-attr:host 表示云硬盘所在主机为 compute 节点;第 15 行 os-vol-mig-status-attr:migstat 表示云主机的迁移状态为没有在迁移;第 16 行 os-vol-mig-status-attr:name_id 表示云硬盘迁移名称;第 17 行表示云硬盘所属项目的 ID;第 19 行表示云硬盘大小为 1GB 等信息。

最后,可以通过 delete 命令删除该云硬盘,通过 cinder list 命令看到云硬盘的状态 status 为 deleting,表示正在删除云硬盘,命令如下,结果如图 13.5.5 所示。

```
[root@controller ~]# cinder delete volume2
[root@controller ~]# cinder list
```

```
+--------------------------------------+----------+--------------+------+
|                  ID                  |  Status  | Display Name | Size |
+--------------------------------------+----------+--------------+------+
| 98a1670f-123a-415f-adfc-53c9ee031a99 | deleting |   volume2    |  1   |
+--------------------------------------+----------+--------------+------+
```

图 13.5.5 查看正在删除的云硬盘的反馈信息

一段时间后可以再次使用 cinder list 命令查看该云硬盘已经被删除。

当需要增加云硬盘的磁盘大小时,可以通过 cinder extend 命令扩展云硬盘的磁盘大小(注意:只能对云硬盘状态为 available 的云硬盘进行扩展),下面将 volume1 这块云硬盘大小修改为 10GB,命令如下,结果如图 13.5.6 所示。

```
[root@controller ~]# cinder extend volume1 10
```

13.5 实训项目 15 Cinder 基本运维命令及其应用

```
+--------------------------------------+-----------+---------+------+-------------+----------+-------------+
| ID                                   | Status    | Name    | Size | Volume Type | Bootable | Attached to |
+--------------------------------------+-----------+---------+------+-------------+----------+-------------+
| d6efe643-cc0f-471b-ab34-9a926bef5bbc | available | volume1 | 10   | -           | false    |             |
+--------------------------------------+-----------+---------+------+-------------+----------+-------------+
```

图 13.5.6 查看扩展云硬盘大小的反馈信息

可以看到 volume1 这块云硬盘的磁盘大小 Size 已经被修改为 10GB。

步骤 2：通过 Dashboard 界面来进行以上的运维操作。

首先登录 Dashboard 用户管理界面，使用 admin 用户登录，密码为 000000，选择"项目"→"卷"标签，如图 13.5.7 所示。

图 13.5.7 登录 Dashboard 界面的云硬盘栏

单击右侧"创建云硬盘"按钮，填写云硬盘名称、云硬盘大小以及可用域。完成后单击"创建卷"按钮，如图 13.5.8 所示。

图 13.5.8 创建云硬盘 Dashboard 界面

299

完成云硬盘创建之后界面返回以下结果，创建云硬盘名称为 volume1，大小为 1GB，状态为 Available，可用域为 nova，如图 13.5.9 所示。

图 13.5.9　创建好的云硬盘信息

完成创建后，单击右侧"编辑卷"按钮，修改云硬盘的名称。将云硬盘的名称修改为 volume2，单击"提交"按钮完成修改，如图 13.5.10 所示。

图 13.5.10　修改云硬盘名称

完成之后，可以看到界面显示的云硬盘名称已经被修改成 volume2，如图 13.5.11 所示。

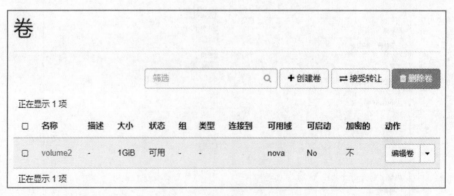

图 13.5.11　修改名称后的云硬盘信息

同时，还可以单击右侧"更多"按钮，查看更多云硬盘的操作。

例如，在"更多"下拉列表中选择"扩展卷"选项，在打开的"扩展卷"页面中输入需要修改的云硬盘大小，这里修改云硬盘的磁盘大小为 10GB，完成之后单击"扩展卷"按钮，如图 13.5.12 所示。

图 13.5.12　扩展卷

完成扩展之后，可以看到云硬盘的配置已经被修改为 10GB，如图 13.5.13 所示。

图 13.5.13　扩展后的云硬盘信息

同样，可以在"更多"下拉列表中选择"管理连接"选项，将云硬盘连接到虚拟机。在打开的"管理已连接卷"页面中，选择将要连接的云主机，这里选择"cirros"云主机。完成后单击"连接卷"按钮完成云硬盘的连接，如图 13.5.14 所示。

图 13.5.14　正在连接云硬盘到云主机

单击"连接卷"后可以看出云硬盘的状态为正在连接状态，当状态为 In-Use 状态时，表示云硬盘已经连接到云主机，如图 13.5.15 所示。

图 13.5.15　连接到云主机的云硬盘信息

当要把云硬盘从云主机断开连接时，可以在云主机右侧"更多"下拉列表中选择"分离卷"选项，进行断开云硬盘连接操作。

单击"分离卷"按钮，如图 13.5.16 所示。

图 13.5.16　断开云硬盘与云主机连接的界面

单击"分离卷"按钮即可在界面右上角看到卷已成功分离实例，如图 13.5.17 所示。

图 13.5.17　确认从云主机断开云硬盘的连接

云硬盘已断开连接，如图 13.5.18 所示。

图 13.5.18　正在断开云主机与云硬盘的连接

待云硬盘完成断开连接之后，云硬盘状态重新回到可用状态，如图 13.5.19 所示。

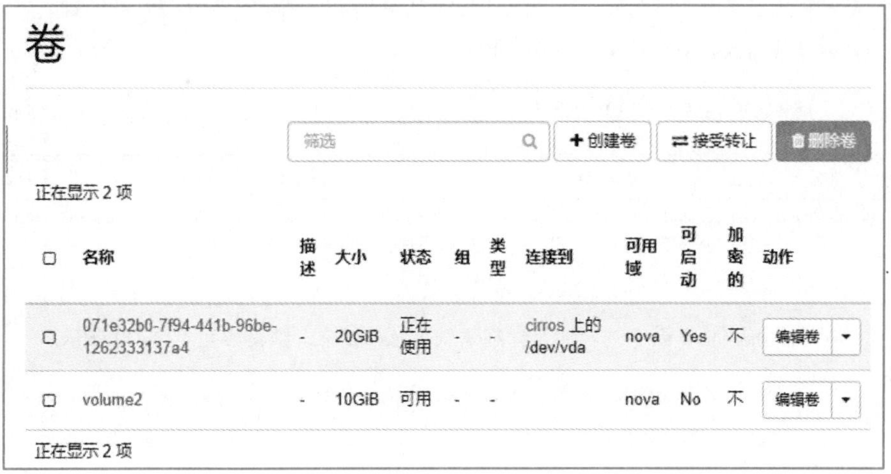

图 13.5.19　断开连接后的云硬盘信息

步骤 3：通过命令行管理快照。

在云硬盘使用时，可以为云硬盘创建快照来保存某一时候的数据状态，使用 snapshot-create 命令为 volume2 云硬盘创建一个名为 volume2-snapshot1 的快照。命令如下，结果如图 13.5.20 所示。

```
[root@controller ~]# cinder snapshot-create --display-name volume2-snapshot1 volume2
```

```
+-------------+--------------------------------------+
| Property    | Value                                |
+-------------+--------------------------------------+
| created_at  | 2019-12-18T09:22:30.554333           |
| description | None                                 |
| id          | db34c043-ca5c-4f3b-825c-178ccdf9e9a8 |
| metadata    | {}                                   |
| name        | volume2-snapshot1                    |
| size        | 10                                   |
| status      | creating                             |
| updated_at  | None                                 |
| volume_id   | 26cab02d-b369-4f0b-9f1b-77f22e86c95f |
+-------------+--------------------------------------+
```

图 13.5.20　创建快照的反馈信息

可以看到图 13.5.20 中快照的 name 为 volume2-snapshot1，还可以使用 snap-rename 命令修改快照的名称。

通过 snapshot-list 命令查看快照列表，命令如下，结果如图 13.5.21 所示。

```
[root@controller ~]# cinder snapshot-list
+--------------------------------------+--------------------------------------+-----------+------------------+------+
| ID                                   | Volume ID                            | Status    | Name             | Size |
+--------------------------------------+--------------------------------------+-----------+------------------+------+
| db34c043-ca5c-4f3b-825c-178ccdf9e9a8 | 26cab02d-b369-4f0b-9f1b-77f22e86c95f | available | volume2-snapshot1 | 10  |
+--------------------------------------+--------------------------------------+-----------+------------------+------+
```

图 13.5.21　查看快照信息

从图 13.5.21 反馈的信息可以看到第 1 列 ID 为快照的 ID 号；第 2 列 volume ID 为云硬盘的 ID 号；第 3 列 status 快照的状态为 available；第 4 列 Name 快照的名称为 volume2-snapshot1；第 5 列 Size 为快照大小 10GB。

通过 snap-rename 命令修改快照名称，命令如下，结果如图 13.5.22 所示。

```
[root@controller ~]# cinder snapshot-rename volume2-snapshot1 volume2-snapshot2
```

再使用 snapshot-list 命令查看快照名称。

```
[root@controller ~]# cinder snapshot-list
+--------------------------------------+--------------------------------------+-----------+------------------+------+
| ID                                   | Volume ID                            | Status    | Name             | Size |
+--------------------------------------+--------------------------------------+-----------+------------------+------+
| db34c043-ca5c-4f3b-825c-178ccdf9e9a8 | 26cab02d-b369-4f0b-9f1b-77f22e86c95f | available | volume2-snapshot2 | 10  |
+--------------------------------------+--------------------------------------+-----------+------------------+------+
```

图 13.5.22　查看快照信息

通过 snapshot-show 命令查看快照详细信息，命令如下，结果如图 13.5.23 所示。

```
[root@controller ~]# cinder snapshot-show volume2-snapshot2
+--------------------------------------------+--------------------------------------+
| Property                                   | Value                                |
+--------------------------------------------+--------------------------------------+
| created_at                                 | 2019-12-18T09:22:30.000000           |
| description                                | None                                 |
| id                                         | db34c043-ca5c-4f3b-825c-178ccdf9e9a8 |
| metadata                                   | {}                                   |
| name                                       | volume2-snapshot2                    |
| os-extended-snapshot-attributes:progress   | 100%                                 |
| os-extended-snapshot-attributes:project_id | c492ca8bace64bd3b1f7ba67ea03d23a     |
| size                                       | 10                                   |
| status                                     | available                            |
| updated_at                                 | 2019-12-18T09:23:58.000000           |
| volume_id                                  | 26cab02d-b369-4f0b-9f1b-77f22e86c95f |
+--------------------------------------------+--------------------------------------+
```

图 13.5.23　查看快照详细信息

从图 13.5.23 反馈的信息中可以看到第 1 行 created_at 为快照创建时间；第 2 行 description 为快照描述信息；第 3 行 id 为快照的 ID 号；第 4 行 metadata 为快照的元数据；第 5 行 name 为快照的名称；第 6 行 os-extended-snapshot-attributes: progress 为快照创建的进度为 100%；第 7 行 os-extended-snapshot-attributes: project_id 为快照创建的项目 ID 号；第 8 行 size 表示快照大小为 10GB；第 9 行 status 表示快照状态为 available；第 11 行 volume_id 表示创建快照的云硬盘 ID 号。

还可以通过快照来创建新的云硬盘。首先通过 snapshot-list 命令查看快照的 id，命令如下，结果如图 13.5.24 所示。

```
[root@controller ~]# cinder snapshot-list
```

ID	Volume ID	Status	Name	Size
db34c043-ca5c-4f3b-825c-178ccdf9e9a8	26cab02d-b369-4f0b-9f1b-77f22e86c95f	available	volume2-snapshot2	10

图 13.5.24 查看云硬盘快照反馈信息

然后通过快照创建一个新的云硬盘 volume3，命令如下，结果如图 13.5.25 所示。

```
[root@controller ~]# cinder create --snapshot-id 75e006f3-cb9f-4f31-bcbc-27ec138d62ed --display-name volume3 1
```

Property	Value
attachments	[]
availability_zone	nova
bootable	false
consistencygroup_id	None
created_at	2019-12-18T09:29:29.000000
description	None
encrypted	False
id	916b2ea9-6dd2-4bca-bd3f-19f93b8441b8
metadata	{}
migration_status	None
multiattach	False
name	volume3
os-vol-host-attr:host	None
os-vol-mig-status-attr:migstat	None
os-vol-mig-status-attr:name_id	None
os-vol-tenant-attr:tenant_id	c492ca8bace64bd3b1f7ba67ea03d23a
replication_status	None
size	10
snapshot_id	db34c043-ca5c-4f3b-825c-178ccdf9e9a8
source_volid	None
status	creating
updated_at	None
user_id	ca4d4da261f5402787b9c0c87f174e58
volume_type	None

图 13.5.25 通过快照创建云硬盘的反馈信息

可以通过 snapshot-delete 命令删除快照，通过 snapshot-list 命令查看快照已经被删除，命令如下，结果如图 13.5.26 所示。

```
[root@controller ~]# cinder snapshot-delete volume2-snapshot2
```

ID	Volume ID	Status	Name	Size

图 13.5.26 删除快照的反馈信息

步骤 4：通过 Dashboard 界面管理快照。

首先单击右侧"更多"按钮，在弹出的下拉列表中选择"创建快照"选项，创建云硬盘的快照。

输入快照名称为 volume2-snapshot1，单击"创建卷快照"按钮。完成云硬盘快照的创建，如图 13.5.27 所示。

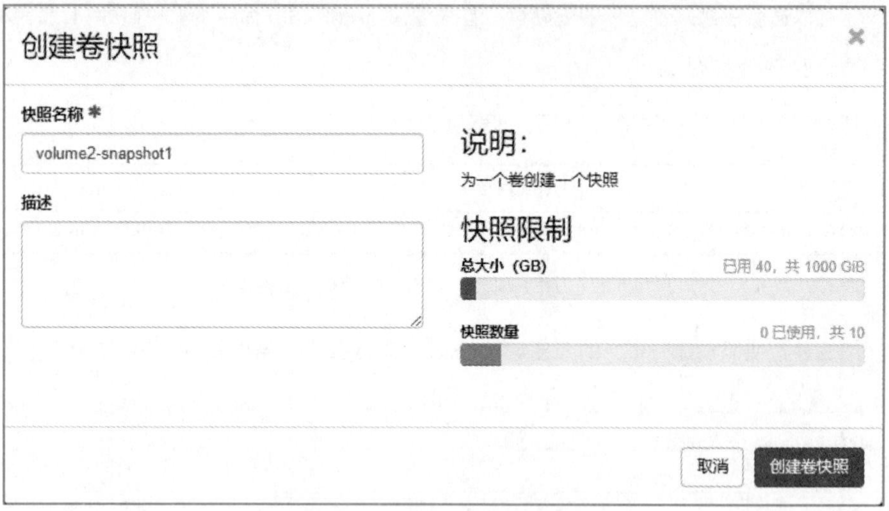

图 13.5.27　创建云硬盘界面

单击界面上方的"快照"标签,进入云硬盘快照页面。可以看到新建的云硬盘快照,如图 13.5.28 所示。

图 13.5.28　查看已经创建好的卷快照状态信息

还可以通过快照来创建一个新的云硬盘,这时快照所能访问到的数据将会装载到新创建的云硬盘内。单击右侧的"创建卷"按钮,进行创建云硬盘的相关参数设置,如图 13.5.29 所示。

图 13.5.29　通过快照创建云硬盘界面

这里输入创建的云硬盘的名称、磁盘大小、使用作为源的快照等信息,这里使用默认的设置信息。单击"创建卷"按钮,完成云硬盘的创建,如图 13.5.30 所示。

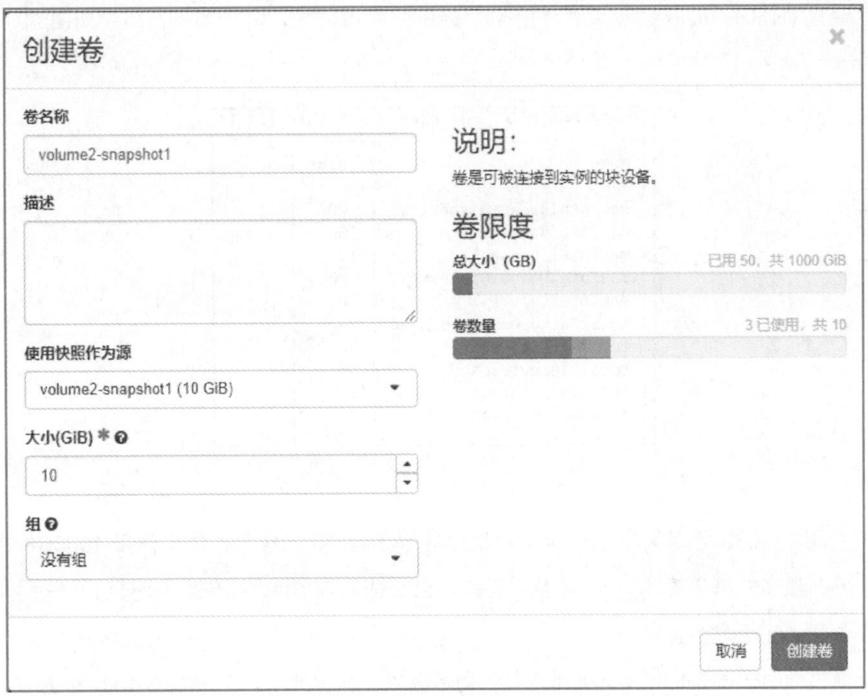

13.5 实训项目 15 Cinder 基本运维命令及其应用

图 13.5.30 通过快照创建云硬盘的界面

单击"卷"标签，可以看到正在创建云硬盘，如图 13.5.31 所示。

图 13.5.31 正在创建的云硬盘信息

待云硬盘创建完成后，可以看到云硬盘的状态为"可用"状态，如图 13.5.32 所示。

图 13.5.32 创建完成的云硬盘信息

步骤 5：cinder 其他服务管理。

通过 absolute-limits 命令查看当前用户的所有使用额度，命令如下，结果如图 13.5.33 所示。

```
[root@controller ~]# cinder absolute-limits
+-------------------------+-------+
| Name                    | Value |
+-------------------------+-------+
| maxTotalBackupGigabytes | 1000  |
| maxTotalBackups         | 10    |
| maxTotalSnapshots       | 10    |
| maxTotalVolumeGigabytes | 1000  |
| maxTotalVolumes         | 10    |
| totalBackupGigabytesUsed| 0     |
| totalBackupsUsed        | 0     |
| totalGigabytesUsed      | 60    |
| totalSnapshotsUsed      | 1     |
| totalVolumesUsed        | 4     |
+-------------------------+-------+
```

图 13.5.33　查看使用额度的反馈信息

由图 13.5.33 可以看出，当前用户最多能够创建 10 个快照，最多使用 1000GB 的磁盘空间，最多能够创建 10 个云硬盘，目前，已经使用 60GB 大小磁盘空间、1 个快照和已经创建了 4 个云硬盘。

通过 cinder availability-zone-list 查看可用域列表。命令如下，结果如图 13.5.34 所示。

```
[root@controller ~]# cinder availability-zone-list
+------+-----------+
| Name | Status    |
+------+-----------+
| nova | available |
+------+-----------+
```

图 13.5.34　查看 cinder 可用域的反馈信息

由图 13.5.34 中可以看出作用域为 nova 域。

通过 nova service-list 命令查看 cinder 子服务状态，命令如下，结果如图 13.5.35 所示。

```
[root@controller ~]# cinder service-list
```

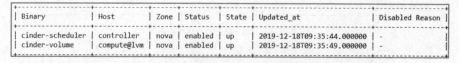

图 13.5.35　查看 cinder 子服务反馈信息

由图 13.5.35 中可以看出 cinder 的两个子服务 cinder-scheduler 服务和 cinder-volume 服务所在的主机 Host 为 controller 节点和 compute 节点，作用域 Zone 都为 nova 域，状态 State 为开启状态 UP 等信息。到此，cinder 服务的运维结束。

习题
原生 OpenStack 云平台各组件运维

第 14 章 虚拟机镜像文件的制作

 本章导读:
　　本章介绍云平台中实例镜像文件的制作,包括准备虚拟机镜像环境以及云平台 qcow2 格式 Windows 镜像制作。

电子资源:
　　电子教案　虚拟机镜像文件的制作
　　PPT　虚拟机镜像文件的制作
　　习题　虚拟机镜像文件的制作

14.1 实训项目 16 准备虚拟机镜像环境

电子教案 准备虚拟机镜像环境

PPT 准备虚拟机镜像环境

微课 19 准备虚拟机镜像环境

1. 实训前提环境

安装成功的原生 OpenStack 云平台环境，可以通过 Dashboard 登录云平台，也可以通过终端连接 controller，进行命令的操作。

2. 实训涉及节点

controller 节点

3. 实训要求

掌握制作虚拟机镜像所需环境。

4. 实训内容

在前面实训过程中，使用的镜像是制作好的。虚拟机镜像是一个文件，该文件中包含了已经安装好可启动操作系统的虚拟磁盘，目前 Stein 版本的 OpenStack 支持以下所列的镜像格式，如 ami、ari、aki、vhd、vmdk、raw、qcow2、vdi 和 iso。在 OpenStack 里，对于 KVM，应用到的镜像格式主要是 raw 和 qcow2 两种，由于 qcow2 独有的特性，人们通常情况下会使用 qcow2 格式的镜像。

手动创建虚拟机，有两种方式进行安装，即使用 virt-manager 或者 virt-install 工具。如果使用 virt-manager，通常需要有一个可以运行 X11 程序的机器。如果在无图形界面的服务器上创建虚拟机镜像，但在本地机器上有 X Server 启动，则可以用 virt-manager，通过使用 ssh X11 转发访问图形界面。同时，因为 virt-manager 和 libvirt 直接交互，通常需要 root 权限来访问。

不想安装相关依赖包到服务器上，且本地机器没有 X Server 或者 X11 转发工作不正常，可以使用 virt-install 工具通过 Libvirt 启动虚拟机，然后使用本地 VNC 客户端连接到虚拟机的图形控制台。本书中首先将使用 virt-install 加 TigerVNC 的方式进行镜像创建的演示，当然也可以使用其他本地 VNC 客户端；之后再使用 virt-manager 的方式进行创建镜像的演示。

以下首先以创建 Windows Server 2012 的 qcow2 镜像为例，演示使用 virt-install 以及 KVM 虚拟化创建 Windows 镜像；其次将使用 virt-manager 图形化工具创建 CentOS 镜像。

在正式开始之前，要确保实验主机或服务器开启虚拟化功能，同时还需要确认是否安装了 KVM 虚拟机，命令如下。

```
[root@controller ~]# lsmod | grep kvm
```

没有任何反馈信息，表明没有安装 KVM 虚拟机，这里还需要安装 KVM 虚拟机以及图形界面，命令如下，结果如图 14.1.1 所示。

```
[root@controller ~]# yum -y groupinstall Virtual*
[root@controller ~]# yum -y groupinstall "GNOME Desktop" "X Window System" "Desktop"
```

14.1 实训项目 16 准备虚拟机镜像环境

```
urw-fonts.noarch 0:2.4-10.el6
vgabios.noarch 0:0.6b-3.7.el6
vte.x86_64 0:0.25.1-8.el6_4
xml-common.noarch 0:0.6.3-32.el6
xorg-x11-font-utils.x86_64 1:7.2-11.el6
yajl.x86_64 0:1.0.7-3.el6
Complete!
```

图 14.1.1 安装完成软件包反馈信息

再次查看，命令如下，结果如图 14.1.2 所示。

```
[root@controller ~]# lsmod | grep kvm
```

```
kvm_intel           54285  0
kvm                333172  1 kvm_intel
```

图 14.1.2 查看 KVM 环境反馈结果

上述命令安装并进行了检验，表明 KVM 已经成功安装。安装完成后，可以重启服务器达到让操作系统内核更新状态的目的。

接着需要检查 libvirtd 服务的运行状态，在下面创建镜像时需要依赖该服务，查看服务运行状态，命令和结果如下。

```
[root@controller ~]# systemctl start libvirtd
[root@controller ~]# systemctl status libvirtd
```

从反馈信息可以看到，libvirtd 服务正在运行。

为确保创建镜像命令的顺利执行，还需要对 /etc/libvirt/qemu.conf 配置文件内容做修改，否则在执行创建镜像命令时会报错，这里以实验过程中创建 Windows 镜像时遇到的报错为例，如图 14.1.3 所示。

```
Starting install...
ERROR    internal error Process exited while reading console log output: char device re
directed to /dev/pts/1
qemu-kvm: -drive file=/root/winserver2012.qcow2,if=none,id=drive-virtio-disk0,format=qc
ow2,cache=none: could not open disk image /root/winserver2012.qcow2: Permission denied

Domain installation does not appear to have been successful.
If it was, you can restart your domain by running:
  virsh --connect qemu:///system start windows2012
otherwise, please restart your installation.
[root@controller ~]#
```

图 14.1.3 创建 Windows 镜像时出现的报错信息

为避免如图 14.1.3 所示错误，需要修改 /etc/libvirt/qemu.conf 文件，内容如下。

```
user = "root"
group = "root"
dynamic_ownership = 0
```

修改时可以直接去掉以上内容前的"#"号，然后把 dynamic_ownership 的值改为 0，目的是禁止 libvirtd 动态修改文件的归属，然后重启 libvirtd 服务。

此时，VNC 默认绑定的是本机 127.0.0.1 地址，如果其他机器想用 VNC 客户端访问这台 KVM 服务器正在安装的镜像，需要把 VNC 绑定到服务器的 IP 地址或者绑定到全局即把地址改为 0.0.0.0，修改 qemu.conf 文件，取消 vnc_listen 行前面的注释，内容如下。

```
vnc_listen = "0.0.0.0"
```

修改完成后，再次重启 libvirtd 服务即可。

完成以上操作后，创建镜像的环境准备工作就完成了。

14.2 实训项目 17 云平台 qcow2 格式 Windows 镜像制作

1. 实训前提环境

完成实训项目 16，即准备好制作虚拟机镜像所需要的环境。

2. 实训涉及节点

controller 节点

3. 实训要求

掌握 virt-install 制作 Windows 镜像的方法。

4. 实训内容

步骤 1：准备所需环境。

在使用 libvirt 启动虚拟机前，检查它的 default 默认网络是否启动。虚拟机要连接到外网，其默认网络必须激活。启动 libvirt 默认网络将创建 linux 网桥 (通常名称是 virbr0)、iptables 规则，以及 dhcp 服务器进程 dnsmasq。

检查 default 默认网络是否激活，命令如下，结果如图 14.2.1 所示。

```
[root@controller ~]# virsh net-list
```

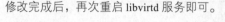

图 14.2.1　检查 libvirt 默认网络状态

首先，需要下载一个 Windows Server 2012 安装版 ISO 镜像，这里使用 cn_windows_server_2012_r2_with_update_x64_dvd_4048415.iso 镜像。接着需要下载一个 VirtIO 驱动 ISO 包，这里下载一个 virtio-win-0.1-52.iso 版本的文件。

步骤 2：制作 Windows 镜像。

首先需要创建一个 15GB 大小的磁盘镜像 (当然也可以创建更大的磁盘)，并命名为 "windows2012.qcow2"，命令如下，结果如图 14.2.2 所示。

```
[root@controller ~]# qemu-img create -f qcow2 winserver2012.qcow2 15G
```

```
Formatting 'winserver2012.qcow2', fmt=qcow2 size=16106127360 encryption=off
cluster_size=65536
```

图 14.2.2　制作磁盘镜像反馈结果

接着，使用 virt-install 命令开始 Windows Server 2012 的安装，命令如下，结果如图 14.2.3 所示。

```
[root@controller ~]# virt-install --connect qemu:///system --name windows2012 --ram 2048 --vcpus 2 --network network=default,model=virtio --disk path=windows2012.qcow2,format=qcow2,device=disk,bus=virtio --cdrom cn_windows_server_2012_r2_with_update_x64_dvd_4048415.iso --disk path=virtio-win-0.1-52.iso,device=cdrom --vnc --os-type windows --os-variant win2k8
```

```
Starting install...
Creating domain...
Cannot open display:
Run 'virt-viewer --help' to see a full list of available command line options
Domain installation still in progress. Waiting for installation to complete.                    | 0 B      00:00
```

图 14.2.3　安装 Windows server 2012 过程中的反馈结果

当出现如上反馈信息时，表明接下来可以通过本地 VNC 连接虚拟机了。

以下是对上述命令中的一些参数的解释，--name windows2012 是虚拟机的名称，这里定义为 window2012，--ram 2048 分配 2GB 内存，--vcpus 2 表示两个 cpu，--network network=default,model=virtio 网络使用的是默认网络，模式为 virtio，网络也可以使用桥接（在下面的镜像创建过程中讲解），--disk path=windows2012.qcow2,format=qcow2,device=disk,bus=virtio 使用之前创建的 qcow2 格式的磁盘，--disk path=virtio-win-0.1-52.iso,device=cdrom 加载 virtio 驱动，--cdrom cn_windows_server_2012_r2_with_update_x64_dvd_4048415.iso 指向 ISO 安装镜像位置，--vnc 使用 VNC 连接，--os-type windows 指定镜像类型为 Windows 类型，--os-variant win2k8 由于 KVM 版本原因，没有 Windows server2012，所以这里用 win2k8 代替（也可以省略）。

使用 TigerVNC 客户端进行连接，单击客户端图标，在"VNC server："文本框中输入启动虚拟机的地址和 VNC 端口，如图 14.2.4 所示。

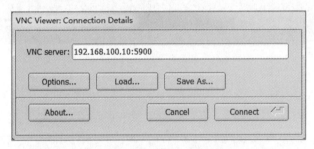

图 14.2.4　TigerVNC 连接界面

单击"Connect"按钮，即可连接到虚拟机的图像控制台，下面操作系统安装的过程与正常安装相同，如图 14.2.5 所示。

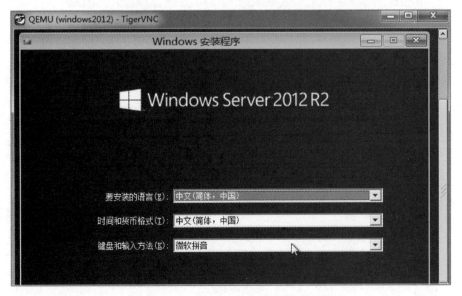

图 14.2.5　通过 TigerVNC 连接镜像安装界面

如图 14.2.5 所示,已经可以进行 Windows Server 2012 的安装了,选择默认即可,单击"下一步"按钮。

单击"现在安装"按钮,如图 14.2.6 所示。

图 14.2.6　开始安装 Windows Server 2012

为方便操作,选择第 2 个选项"Windows Server 2012 R2 Datacenter(带有 GUI 的服务器)",单击"下一步"按钮,如图 14.2.7 所示。

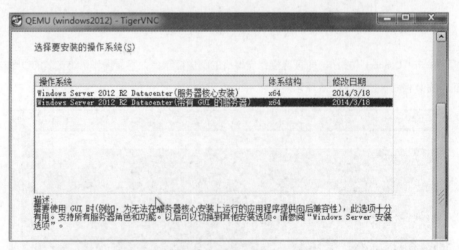

图 14.2.7　选择要安装的系统

选中"我接受许可条款"复选框,单击"下一步"按钮,如图 14.2.8 所示。

至此,会发现并没有正常情况下安装系统所需要的磁盘(驱动器),因为 Windows 安装程序默认不识别硬盘,此时单击"加载驱动程序"按钮,如图 14.2.9 所示。

在浏览文件夹时,选择驱动器(E:)下的 WIN8 文件夹中的 AMD64 文件夹,然后单击"确定"按钮,如图 14.2.10 所示。

14.2 实训项目 17 云平台 qcow2 格式 Windows 镜像制作

图 14.2.8 安装 Windows Server 2012 的同意条款

图 14.2.9 出现找不到驱动器结果

图 14.2.10 选择驱动器所在文件夹

然后会出现驱动列表，其中前两个是"Red Hat VirtIO Ethernet Adapter"网络驱动和"Red Hat VirtIO SCSI controller"，如图 14.2.11 所示。

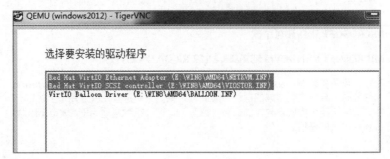

图 14.2.11　选择需要安装的驱动程序

如图 14.2.11 所示，之前创建的 15GB 的硬盘空间就可以看到了，因为没有分配，所以需单击"新建"按钮，选择合适的分区形式，如图 14.2.12 所示。

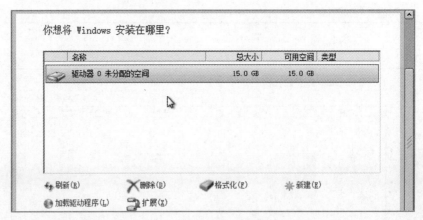

图 14.2.12　找到之前创建的磁盘空间

图 14.2.12 中在实验环境下选择把所有空间一直划分为一个分区，此时系统会自动分出一个保留分区，如图 14.2.13 所示。

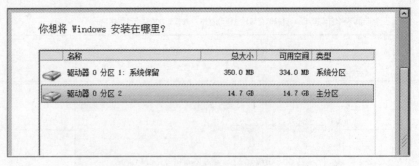

图 14.2.13　划分磁盘空间

下面开始正式安装，需要耐心等待一段时间，完成安装后，虚拟机会自动重启，如图 14.2.14 所示。

重启完后才能后，需要为 administrator 用户设置密码，密码需要符合一定的复杂度，如果密码设置不符合系统要求，会出现提示，如图 14.2.15 所示。

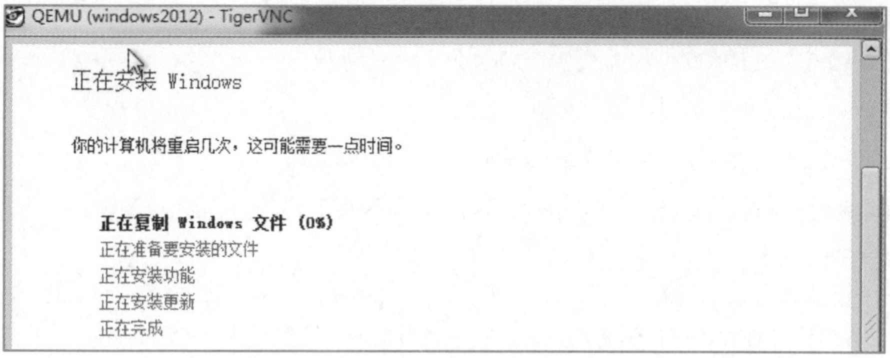

14.2 实训项目 17 云平台 qcow2 格式 Windows 镜像制作

图 14.2.14 等待安装过程

图 14.2.15 安装完成后设置用户信息

设置成功后就可以登录了，由于 Windows Server 2012 登录之前需要按 Ctrl+Alt+Delete 组合键，然而在 VNC 连接的状态，这样不仅会影响本机的状态而且虚拟机并不会起任何作用，所以使用 Windows 自带的屏幕键盘登录，如图 14.2.16 所示。

图 14.2.16 使用屏幕键盘

成功登录后，需要使用命令完成 VirtIO 驱动的安装，如图 14.2.17 所示。

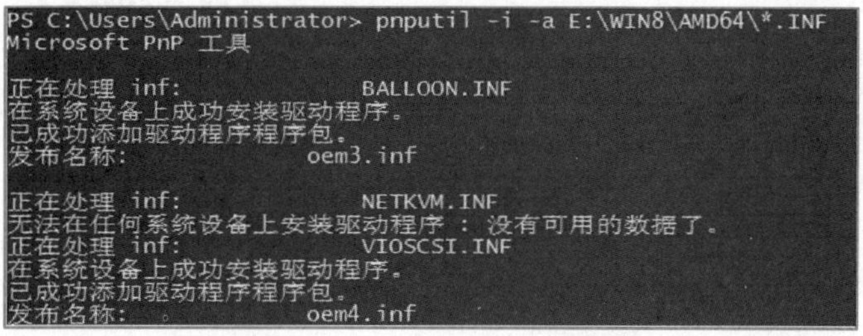

图 14.2.17 安装 VirtIO 驱动

在以上的操作过程中，会出现安装提示，选择同意即可。

然后，需要安装 Cloudbase-Init，在安装之前，为了让 Cloudbase-Init 在系统启动时运行

317

脚本，需要设置 PowerShell 执行策略解除限制，命令如下。

```
C:\powershell
C:\Set-ExecutionPolicy Unrestricted
```

完成上述命令后，下载并安装 Cloudbase-Init，命令如下。

```
C:\Invoke-WebRequest -UseBasicParsing http://www.cloudbase.it/downloads/
CloudbaseInitSetup_Stable_x64.msi -OutFile cloudbaseinit.msi
C:\.\cloudbaseinit.msi
```

在下载过程中由于网络有问题，没有下载成功，在这里使用另一种方法，先通过本机下载，然后放到 Windows server 2012 虚拟机中，可以通过在本地创建共享文件，远程登录或搭建 ftp 等多种方式使虚拟机获取到下载的 Cloudbase-Init，然后点击安装即可，如图 14.2.18 所示。

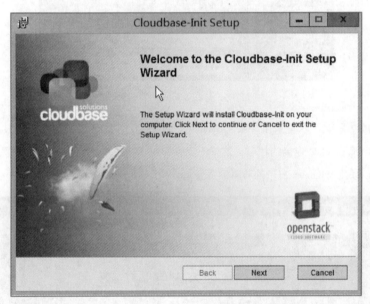

图 14.2.18　安装 Cloudbase-Init 的界面

一步步按提示操作，单击"Next"按钮即可，需要注意，在"Configuration options"窗口，将设置修改为如图 14.2.19 所示。

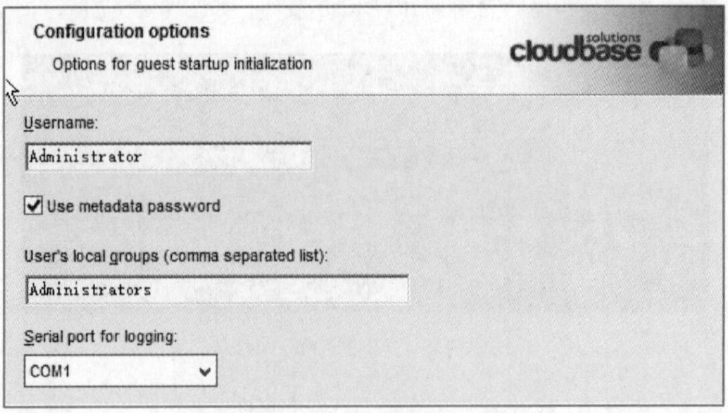

图 14.2.19　配置选项

完成安装后，在"Complete the Cloudbase-Init Setup Wizard"窗口选中"Run Sysprep…"和"Shutdown"复选框，然后单击"Finish"按钮，如图 14.2.20 所示。

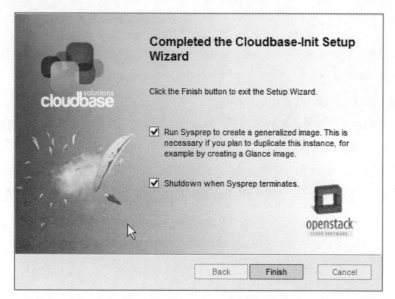

图 14.2.20　Cloudbase-Init 安装完成

然后会自动运行 Sysprep 程序，等待完成（执行 Sysprep 过程中，可能会出错，可以忽略），完成后，关闭虚拟机。这样 Windows 镜像就创建完成了，用户只需通过 glance 命令上传到服务器上即可。上传完成后，即可登录到 Dashboard 界面使用 Windows 镜像创建实例。

习题　虚拟机镜像文件的制作

参 考 文 献

[1] 英特尔开源技术中心. OpenStack 设计与实现 [M].2 版 . 北京: 电子工业出版社 ,2017.

[2] 奥马尔·海德希尔，坚登·杜塔·乔杜里 . 精通 OpenStack[M].2 版 . 山金孝，刘世民，肖力，译 . 北京：机械工业出版社，2019.

[3] 文婷婷，李洪赭，李赛飞 . 基于 OpenStack 虚拟化网络管理平台的设计与实现 [J]. 电子制作，2019(10):47-49.

[4] 黄兴 .OpenStack 私有云在企业信息化建设中的应用 [J]. 中国信息化 ,2019(4):46-47.

[5] 魏迎 .OpenStack 云计算平台的研究与实现 [J]. 电子设计工程 ,2019,27(6):152-155.

郑重声明

高等教育出版社依法对本书享有专有出版权。任何未经许可的复制、销售行为均违反《中华人民共和国著作权法》，其行为人将承担相应的民事责任和行政责任；构成犯罪的，将被依法追究刑事责任。为了维护市场秩序，保护读者的合法权益，避免读者误用盗版书造成不良后果，我社将配合行政执法部门和司法机关对违法犯罪的单位和个人进行严厉打击。社会各界人士如发现上述侵权行为，希望及时举报，我社将奖励举报有功人员。

反盗版举报电话　（010）58581999　58582371

反盗版举报邮箱　dd@hep.com.cn

通信地址　北京市西城区德外大街4号　高等教育出版社法律事务部

邮政编码　100120